Partially Ordered Systems

Partially Ordered Systems
Editorial Board: L. Lam • D. Langevin

Solitons in Liquid Crystals
Lui Lam and Jacques Prost, Editors

Bond-Orientational Order in Condensed Matter Systems
Katherine J. Strandburg, Editor

Diffraction Optics of Complex-Structured Periodic Media
V.A. Belyakov

Fluctuational Effects in the Dynamics of Liquid Crystals
E.I. Kats and V.V. Lebedev

Nuclear Magnetic Resonance of Liquid Crystals
Ronald Y. Dong

Electrooptic Effects in Liquid Crystal Materials
L.M. Blinov and V.G. Chigrinov

Liquid Crystalline and Mesomorphic Polymers
Valery P. Shibaev and Lui Lam, Editors

Preface

Among the various new directions in modern polymer science, the design and investigation of liquid crystal (LC) polymers have been the ones growing most actively and fruitfully. In spite of that, the possible formation of an anisotropic LC phase was only demonstrated theoretically for the first time in the 1950s by Onsager [1] and Flory [2], and then experimentally verified in the studies with polypeptides solutions. In essence, the studies of these LC lyotropic systems did not deviate from the theme of purely academic interest.

It was at the beginning of the 1970s that the experimental "explosion" occurred, when aromatic polyamides were synthesized and their ability to form LC solutions in certain very aggressive solvents was discovered. The search for practical applications of such LC systems was crowned with the successful creation of the new generation of ultrastrong high-modulus thermostable fibers, such as the Kevlar, due to the high degree of order of the macromolecules in the anisotropic LC state.

In fact, these investigations coincided with the swift emergence on the practical "scene" of thermotropic low-molar-mass liquid crystals, with the use of these materials in microelectronics and electrooptics (figures and letters indicators, displays in personal computers, and flat TV, etc.). Polymer scientists also began to develop methods of synthesizing thermotropic LC polymers by incorporating mesogenic fragments in the main (main-chain LC polymers) or side branchings of the macromolecules (side-chain or comb-shaped polymers).

Starting with the first publications on the synthesis of thermotropic LC polymers (1974–1976), an avalanche of publications dedicated to different aspects of the synthesis and study of LC polymers appeared. Since 1977 two to three international conferences and symposiums devoted solely to LC polymers are convened annually. Besides, for a long time, LC polymers have already been included in the programs of various international and national conferences on macromolecules or liquid crystals. In a comparatively short period of time, a number of books and reviews devoted to the synthesis, study, and practical application of lyotropic and thermotropic LC polymers have been published [3–18].

Taking these circumstances into account, we try to include in this book those aspects of LC polymers, which either have not been considered before or need particular attention owing to their real, complex character. It is not by chance that the book is called *Liquid Crystalline and Mesomorphic Polymers*. The point is that two types of polymers are actually considered in this book. The first type consists of truly LC polymers containing fragments of low-molar-mass liquid crystals—the main- and side-chain LC polymers. From numerous studies it was established that such polymers actually form thermodynamically stable phases similar, in many respects, to low-molar-mass liquid crystals.

On the other hand, it was found lately that a number of flexible polymers without any mesogenic groups (e.g., polysiloxanes and polyorganophosphazenes) are able to exist in a state which is neither truly crystalline nor describable in the framework of a LC state. The term mesomorphic is more appropriate for these systems, as well as for other systems such as graphitizable cokes and some cellulose derivatives. All these systems form the second type of polymers considered in this book. We can imagine that in reality the range of such polymers is much wider. The term mesomorphic, wider in meaning than the term liquid crystalline, is probably quite pertinent for the description of any system with some type of structure intermediate between the crystalline and isotropic ones.

It should be mentioned that the terminological questions relating to low-molar-mass liquid crystals and LC polymers are undoubtedly noted by researchers working in these scientific disciplines. We can hope that the work recently started by the Committee on Macromolecular Nomenclature of the International Commission of Pure and Applied Chemistry will be successful, leading to the creation of a special nomenclature for the terms and definitions of LC compounds.

This book consists of ten chapters. Each chapter provides a self-contained state-of-the-art review of one important area of the field. The first part of the book consists of six chapters. In Chapter 1, the theoretical aspects of LC phase formation in polymeric cholesterics (M.A. Osipov) are considered. In the next three chapters, the latest achievements in molecular design and structure of the main-chain polymers (A.H. Windle) and side-chain polymers (V.P. Shibaev, Ya.S. Freidzon, and S.G. Kostromin) as well as the study of mixtures of side-chain polymers with low-molar-mass liquid crystals (F. Hardouin, G. Sigaud, and M.F. Achard) are considered.

In Chapter 5, by G.P. Montgomery, Jr., G.W. Smith, and N.A. Vaz, problems concerning the preparation of a practically very important type of LC polymeric materials, polymer-dispersed LC films, are considered. Chapter 6 was written by L. Stroganov and devoted to the study of the dynamics of main-chain and side-chain polymers by NMR spectroscopy.

Thematically, the second part of the book deals with the description of mesomorphic systems, such as mesophases of graphitizable carbons (H. Marsh and M.A. Diez), polyorganophosphazenes (V.G. Kulichikhin, E.M. Antipov, E.K. Borisenkova, and D.R. Tur), and cellulose derivatives (D.G. Gray and B.R. Harkness). Finally, the book ends with a chapter by L. Lam on the highly exotic bowlics—monomers and polymers consisting of bowl-shaped molecules.

Some words about the character of such a compilation of presentations. The chapters in this book are written by different contributors, which carry styles varying from comprehensive to meticulous. The editors, while preserving the contributors' original styles as much as possible, have tried to improve the uniformity relating to a number of wordings and expressions, although it was not always easy to do so.

We would consider our job as editors very successful if young researchers starting to study LC polymers, would get interested in these new unusual systems and become actively engaged in their investigation. We also hope that specialists and experts in this area would be able to find new stimulating ideas in this book.

Unfortunately, Dr. Stroganov, who worked at the Polymer Department of Moscow State University during the last twenty-five years, was not able to see his work in this book. This very erudite person and nice friend of one of us passed away on July 28, 1992, after a serious and prolonged illness, at the age of fifty-one. He was a person of manifold gifts in all aspects of his life. His devotion to scientific work won him the regard and admiration of all who knew him. His chapter undoubtedly makes a valuable contribution to the field of LC polymers.

Lastly, we would like to take this opportunity to express our gratitude to the contributors for their efforts, and to the editors and staff of Springer-Verlag for efficient cooperation.

Moscow   Valery P. Shibaev
San Jose   Lui Lam

References

1. L. Onsager, Ann. N.Y. Acad. Sci. **5**, 627 (1949).

2. P. Flory, Proc. Roy. Soc. London A **234**, No. 1, 73 (1956).

3. V.P. Shibaev and N.A. Platé, Vysokomolek. Soedin. A **19**, No. 5, 923 (1977) (in Russian); Polymer Sci. USSR A **19**, No. 5, 1065 (1978) (English translation).

4. S.P. Papkov and V.G. Kulichikhin, *Liquid Crystalline State of Polymers* (Chemistry, Moscow, 1977) (in Russian).

5. N.A. Platé and V.P. Shibaev, *Comb-Shaped Polymers and Liquid Crystals* (Chemistry, Moscow, 1980) (in Russian, English translation published by Plenum, New York, 1987).

6. *Mesomorphic Order in Polymers*, edited by A. Blumstein (ACS Symp. Ser., No. 74, Washington, 1978).

7. *Liquid Crystalline Order in Polymers*, edited by A. Blumstein (Academic Press, New York, 1978).

8. Ya.B. Amerik and B.A. Krentzel, *Chemistry of Liquid Crystals and Mesomorphic Polymer Systems* (Chemistry, Moscow, 1981) (in Russian).

9. *Polymer Liquid Crystals*, edited by A. Ciferri, W.R. Krigbaum, and R.B. Meyer (Academic Press, New York, 1982).

10. *Liquid Crystal Polymers*, I, II/III, edited by M. Gordon and N.A. Platé, Advances in Polymer Science (Springer-Verlag, Berlin, 1984).

11. *Polymeric Liquid Crystals*, edited by A. Blumstein (Plenum, New York, 1985).

12. *Liquid Crystalline Polymers*, edited by N.A. Platé (Chemistry, Moscow, 1988) (in Russian, English translation published by Plenum, New York, 1993).

13. *Side Chain Liquid Crystal Polymers*, edited by C. McArdle (Blackie, London, 1989).

14. V.N. Tsvetkov, *Rigid-Chain Polymers* (Consultants Bureau, New York, 1989).

15. *Applied Liquid Crystal Polymers*, edited by M. Takeda, K. Iimura, N. Koide, and N.A. Platé, Mol. Cryst. Liq. Cryst. **169**, 1–192 (1989).

16. V.P. Shibaev and S.V. Belaev, *Prospects of Application of Functional Liquid Crystalline Polymers and Compositions*, Vysokomolek. Soedin. A **32**, 2266 (1990). English translation published by Pergamon Press in Polymer Science USSR **32**, 2361 (1991).

17. *Fundamentals of Polymer Liquid Crystals*, edited by A. Ciferri (VCH, Weinheim, 1991).

18. A.M. Donald and A.H. Windle, *Liquid Crystalline Polymers* (Cambridge University Press, Cambridge, 1992).

Contents

Contributors

Achard, M.F.
   Centre de Recherche Paul Pascal, Université de Bordeaux I,
   Avenue A. Schweitzer, F-33600 Pessac, France

Antipov, E.M.
   Institute of Petrochemical Synthesis, Russian Academy of Sciences,
   Leninsky Prospect 29, Moscow 117912, Russia

Borisenkova, E.K.
   Institute of Petrochemical Synthesis, Russian Academy of Sciences,
   Leninsky Prospect 29, Moscow 117912, Russia

Diez, M.A.
   Consejo Superior de Investigationes Cientificas, Instituto Nacional del
   Carbon, La Corredoria s/n, Apartado 73, 33080 Oviedo, Spain

Freidzon, Ya.S.
   Department of Chemistry, Moscow State University, Moscow 119899,
   Russia

Gray, D.G.
   Paprican and Department of Chemistry, McGill University,
   Pulp and Paper Research Centre, Montreal, P.Q. H3A 2A7, Canada

Hardouin, F.
   Centre de Recherche Paul Pascal, Université de Bordeaux I,
   Avenue A. Schweitzer, F-33600 Pessac, France

Harkness, B.R.
   Dow Corning Japan Ltd., Research and Information Center, 603 Kishi,
   Yamakita-Machi, Ashigarakami-Gun, Kanagawa 258-01, Japan

Kostromin, S.G.
   Department of Chemistry, Moscow State University, Moscow 119899,
   Russia

Kulichikhin, V.G.
Institute of Petrochemical Synthesis, Russian Academy of Sciences,
Leninsky Prospect 29, Moscow 117912, Russia

Lam, L.
Department of Physics, San Jose State University, San Jose,
California 95192-0106, USA

Marsh, H.
Department of Mechanical Engineering and Energy Processes,
Southern Illinois University, Carbondale, Illinois 62901-4613, USA

Montgomery, Jr., G.P.
Department of Physics, General Motors Research and Development
Center, Warren, Michigan 48090-9055, USA

Osipov, M.A.
Institute of Crystallography, Russian Academy of Sciences,
Leninsky Prospect 59, Moscow 117333, Russia

Shibaev, V.P.
Department of Chemistry, Moscow State University, Moscow 119899,
Russia

Sigaud, G.
Centre de Recherche Paul Pascal, Université de Bordeaux I,
Avenue A. Schweitzer, F-33600 Pessac, France

Smith, G.W.
Department of Physics, General Motors Research and Development
Center, Warren, Michigan 48090-9055, USA

Stroganov, L.*
Department of Chemistry, Moscow State University, Moscow 119899,
Russia

Tur, D.R.
Institute of Elementaorganic Compounds, Russian Academy of Sciences,
Vavilov Str. 20, Moscow 117813, Russia

Vaz, N.A.
Department of Physics, General Motors Research and Development
Center, Warren, Michigan 48090-9055, USA

Windle, A.H.
Department of Materials Science, University of Cambridge,
Pembroke Street, Cambridge CB2 3QZ, United Kingdom

---

* Deceased.

1

Molecular Theory of Cholesteric Polymers

M.A. Osipov

1.1 Introduction

The macroscopic cholesteric structure in polymer liquid crystals is formed when the corresponding macromolecules are chiral. Taking into account the origin of chirality of the polymer chain, it is possible to divide the cholesteric polymers into two classes. The polymers of the first class possess the liquid crystalline properties due to the interaction between relatively small rigid mesogenic units in the main chain or in the side chains of the long and flexible macromolecule. The chirality of such a macromolecule is determined by asymmetric carbons in the flexible spaces or tails (this is analogous to the structure of low molecular weight chiral nematics), or by the chiral mesogenic units which are usually represented by familiar cholesterol derivatives. The corresponding polymer cholesterics are mainly thermotropic and their properties are very similar to the properties of low molecular weight liquids crystals. At the same time there is a number of important differences which are related to the influence of the polymer chain [1]. It should be noted that at present there is a growing interest in this kind of cholesteric polymers, and during the past decade hundreds of various main- and side-chain cholesteric polymers have been synthesized and investigated. The achievements in the chemistry and physics of such cholesteric polymers have been summarized in recent reviews [1,2].

The cholesteric polymers of the second class are composed of macromolecules with natural chiral structure. The majority of such macromolecules adopt the helical conformation in appropriate solvents and form lyotropic liquid crystals. The most important examples are the synthetic polypeptides [3,4], which can be called the classical mesogenic polymers (especially the polybenzylglutaminate). The cholesteric properties of synthetic polypeptides were discovered long ago (see, e.g., [5–7]), and systematic experimental investigations have been performed by different authors during the past 15 years (see [3,4] and references therein). The experimental data indicate that the cholesteric properties of polypeptide solutions differ substantially from the properties of low molecular weight cholesterics. For example, a change of temperature induces the sense inversion of the macroscopic helical structure in various PBLG solutions [8,9]. The same inversion can also be induced by a change of solvent parameters. On the contrary, in the case of thermotropic cholesterics, the helix inversion was observed only in certain nematic–cholesteric mixtures. As far as we know, the helix inversion has not also been observed in side- or main-chain thermotropic cholesteric polymers.

In recent years it has also been shown that a cholesteric phase can be formed in the solutions of DNA, RNA, and other double-stranded poly-nucleotides [10,11]. It is important to note that all significant biopolymers are chiral and thus the cholesteric properties of the corresponding liquid crystalline solutions can be very important for the understanding of some general biological problems, including the possibility and function of the liquid crystal state in biological systems.

It should be noted that there are some natural chiral macromolecules which do not possess the helical shape. Typical examples are the cellulose derivatives which form both the lyotropic and thermotropic cholesteric phases. The cholesteric ordering in these polymers has also been extensively investigated during the past decade (see, e.g., [12,13]).

The first molecular theory of cholesteric polymers was developed by Straley [14], who had considered the system of long rigid helices, which interact only via the steric repulsion. This theory has been generalized by Kimura et al. [15] who have also taken into account the intermolecular attraction, using the simple model potential, and described the variation of the pitch caused by a change of temperature and concentration. During the same period there was also progress in the statistical theory of thermotropic cholesterics (for reviews, see [16,17]) based on the ideas of Straley [14] and on the model of chiral dispersion intermolecular interaction, developed by Goossens [18] and Van der Meer et al. [19].

It should be noted that the development of the consistent molecular–statistical theory of thermotropic cholesterics is a more difficult problem than the statistical description of cholesteric ordering in the solution of rod-like macromolecules. Indeed, the behavior of the pitch in the thermotropic choles-teric phase is affected by different additional factors including the smectic fluctuations, temperature variation of the nematic order parameter, and the many-body intermolecular correlations, typical for molecular liquids [16]. For example, the temperature range of the cholesteric phase in thermotropic liquid crystals is usually not very large and hence the smectic fluctuations can be very important throughout the whole phase. This assumption can help us to interpret the complicated temperature variation of the pitch in cholesterol derivatives [17] and it is also used in the case of thermotropic side-chain cholesteric polymers which always demonstrate a strong tendency to form smectic phases [1]. On the contrary, the temperature range of the cholesteric phase in polypeptide solutions is usually much larger (about 100°) and the smectic phases are not formed. Also taking into account the low concentration of the polymer it is reasonable to assume that the smectic fluctuations are very weak. Second, the liquid crystal state is lyotropic and the orientational order parameter weakly depends on temperature. Thus, in the case of polypeptide cholesterics, the temperature variations of the pitch are mainly determined by the packing effects and the chiral interaction between helical macromolecules which are directly taken into account in the statistical theory.

This partially explains why the molecular theory of cholesteric ordering in polymers has been developed mainly for the solutions of chiral macromolecules like synthetic polypeptides. There were no attempts to develop a molecular theory of thermotropic cholesteric polymers, as far as we know.

One of the most complicated problems in the theory of cholesteric ordering in ploymer solutions is the description of the solvent effect. For example, in polypeptide solutions the critical concentrations of the polymer (which corresponds to the formation of the liquid crystal state) are rather insensitive to the solvent, while the macroscopic helical structure of the cholesteric liquid crystal is strongly influenced by the properties of the solvent. For instance, the cholesteric sense of PBLG liquid crystal depends on the nature of the solvent, and the change of component concentrations in a mixed solvent can induce the sense inversion of the macroscopic helical structure [9,20]. The solvent effect was first taken into account by Samulski an Samulski [21] who suggested considering the neutral solvent as a dielectric media, which effects the intermolecular dispersion interaction, and who made the first attempt to show that the sense of the cholesteric structure can be influenced by the relation between the dielectric constant of the solvent and the dielectric parameters of the macromolecule. Starting from the ideas of Samulski and Samulski, the theory of the solvent effect on the cholesteric ordering in polymer solutions has been developed further in [22–24].

In this chapter we present the results of the modern statistical theory of cholesteric ordering in polymer solutions and in thermotropic side-chain polymers, with special emphasis on the effect of solvent on the behavior of the macroscopic helical structure. We also consider the general theory of the chiral dispersion interaction between macromolecules in the dielectric solvent, which is the starting point of the molecular theory for cholesterics. In Section 1.5 we consider the influence of molecular flexibility on the cholesteric ordering in the solution of persistent chains and discuss the cholesteric properties of cellulose derivatives.

1.2 Elementary Models for Chiral Molecules

The physical molecular models should be simple enough to be in an analytical theory and, at the same time, they should represent the principle features of molecular structure which are important for the description of the phenomena under consideration. For example, in the statistical theory of nematic–isotropic phase transition it is sufficient (in the first approximation) to represent the molecule as a uniaxial rod-like rigid particle. However, detailed description of this phase transition is possible only when we take into account molecular biaxiality, the flexibility of the alkyl tail, and permanent molecular dipoles. The molecular theory of cholesteric ordering is based on molecular models which represent the molecular chirality. In the case of synthetic polypeptides, DNA, RNA, and similar helical molecules, the obvi-

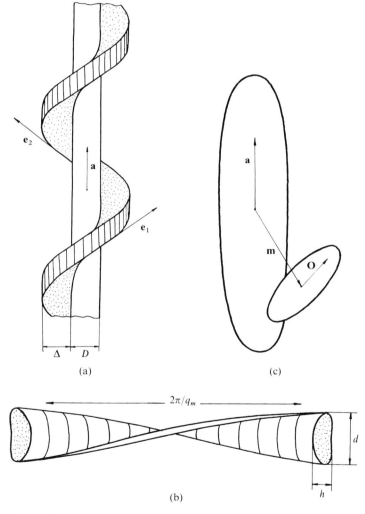

FIGURE 1.1. Simple models for chiral molecules. (a) Schematic representation of a helical macromolecule. (b) "Twisted belt" as a simple model for the cellulose chain. (c) Primitive model for a molecule of the chiral nematic with a substitution group.

ous simple model is the rigid helix, which is presented in Fig. 1.1(a). This well-known model, however, is not universal since some chiral polymer ma-cromolecules do not possess the helical shape. For example, the cellulose derivatives and other polysakharides can be conveniently represented by the model of the "twisted belt" [25] presented in Fig. 1.1(b).

It should be noted that the models, presented in Fig. 1.1(a, b), reflect main-ly the chiral molecular shape. However, the molecular chirality manifests itself also in the distribution of the polarizability, and therefore the dielectric properties of a chiral macromolecule are characterized by both the polariz-ability and optical activity tensors.

In Fig. 1.1(c) we present for comparison the simple model of a low molecu-lar weight chiral nematic proposed by Van der Meer and Vertogen [26]. In the synthesis of chiral nematics the hydrogen atom in the alkyl chain is substituted for a certain group which is presented on Fig. 1.1(c). Note that this molecule is chiral when the three vectors $\mathbf{a}$, $\mathbf{m}$, and $\mathbf{O}$ are not parallel to the same plane and $\mathbf{a} \cdot \mathbf{O} \neq 0$. Here the unit vectors $\mathbf{a}$ and $\mathbf{O}$ are in the direction of the long axis of the mesogenic molecule and the substitution group, respectively, and the vector $\mathbf{m}$ is pointing from the center of mass of the molecule to that of the substitution group. Then it is reasonable to introduce the pseudoscalar quantity $\Delta_i = (\mathbf{a}_i \cdot \mathbf{o}_i)(\mathbf{a}_i \times \mathbf{o}_i \cdot \mathbf{m}_i)$ which can be considered as a measure of the chirality of the molecular structure. Note that the parameter Δ_i changes sign when the handedness of the molecule is changed. The parameter Δ_i can also easily be defined for the models of a rigid helix and the twisted belt. For example, in the case of the helix, the three characteristic vectors are shown in Fig. 1.1(a).

It is convenient to use the parameter Δ also in the case when the molecular chirality is determined by different orientations of the polarizability tensors in different parts of the rigid molecule (i.e., for Kuhn models). Indeed, when it is possible to consider two parts of the molecule with polarizability tensors $\alpha^{(1)}_{\alpha\beta}$ and $\alpha^{(2)}_{\alpha\beta}$, the molecular chirality is determined by the parameter

$$\Delta = \varepsilon_{\alpha\beta\gamma}\alpha^{(1)}_{\beta\nu}\alpha^{(2)}_{\gamma\nu}m_\alpha,$$

where the vector $\mathbf{m}$ points from the first polarizability center to the second one.

1.3   Chiral Anisotropic Interaction Between Macromolecules in the Solvent

1.3.1   Model Potentials

The statistical theory of cholesteric ordering in liquid crystals is based on the model potentials of interaction between chiral molecules. The relative orien-tation of the two uniaxial molecules can be specified by three unit vectors: $\mathbf{a}_1$, $\mathbf{a}_2$, and $\mathbf{u}_{12}$, where the vector $\mathbf{a}_i$ is in the direction of the long axis of the molecule "i" and $\mathbf{u}_{12} = \mathbf{R}_{12}/|\mathbf{R}_{12}|$ where $\mathbf{R}_{12}$ is the intermolecular vector.

Then the total intermolecular interaction energy depends on $\mathbf{a}_1, \mathbf{a}_2, \mathbf{u}_{12}$, and $\mathbf{R}_{12}$ and can be split into two parts

$$V(1, 2) = V(\mathbf{a}_1, \mathbf{R}_{12}, \mathbf{a}_2) = V_{\text{ach}}(1, 2) + V_{\text{ch}}(1, 2), \qquad (1.1)$$

where $V_{\text{ch}}(1, 2) = V_{\text{ch}}(\mathbf{a}_1, \mathbf{R}_{12}, \mathbf{a}_2)$ is the energy of chiral interaction (i.e., the interaction energy which changes sign when the chirality of any interacting molecule is reversed).

The arbitrary interaction between uniaxial molecules can be expanded in the complete set of the orthogonal basis functions $T^{lL\lambda}(\mathbf{a}_1, \mathbf{u}_{12}, \mathbf{a}_2)$ which are called rotational invariants [27]

$$V(\mathbf{a}_1, \mathbf{R}_{12}, \mathbf{a}_2) = \sum_{l, L, \lambda = 0}^{\infty} J^{lL\lambda}(R_{12}) T^{lL\lambda}(\mathbf{a}_1, \mathbf{u}_{12}, \mathbf{a}_2), \qquad (1.2)$$

where $|l - \lambda| \leq L \leq l + \lambda$. The functions $T^{lL\lambda}$ do not depend on the particular Cartesian coordinate system and can be expressed in terms of the spherical harmonics [27]. Explicit expressions for the relevant $T^{lL\lambda}(\mathbf{a}_1, \mathbf{u}_{12}, \mathbf{a}_2)$ are given, for example, in [28]. Note that because of the indistinguishability of the director states $\mathbf{n}$ and $-\mathbf{n}$, the odd terms in $\mathbf{a}_1$ and $\mathbf{a}_2$ (i.e., odd in l and λ [27]) must be discarded. At the same time the functions $T^{lL\lambda}$, with $l + L + \lambda$ odd, are pseudoscalars and can appear only for chiral molecules. Then the chiral interaction potential $V_{\text{ch}}(1, 2)$ can be expanded in basic functions $T^{lL\lambda}(1, 2)$ with l, λ even, and L odd. Taking into account only the first term of this expansion, the chiral interaction potential $V_{\text{ch}}(1, 2)$ is approximately written as

$$V_{\text{ch}}(1, 2) \simeq J^*(R_{12})(\mathbf{a}_1 \cdot \mathbf{a}_2)(\mathbf{a}_1 \times \mathbf{a}_2 \cdot \mathbf{u}_{12}), \qquad (1.3)$$

where the coefficient $J^*(R_{12})$ is also a pseudoscalar and represents the molecular chirality.

The expression (1.3) can be considered as a convenient model potential which can cause the cholesteric twist in a liquid crystal. The model potential (1.3) has been used both in the theory of thermotropic [16] and lyotropic polymer [15] cholesterics. Note that the potential (1.3) is minimal when the long axes of the two neighboring molecules are not parallel.

It can readily be shown that the potential (1.3) can be responsible for the helical twisting of long molecular axes in the cholesteric phase. Let us assume, for simplicity, that the long axes $\mathbf{a}_1$ and $\mathbf{a}_2$ are normal to $\mathbf{u}_{12}$. Then the energy of the chiral interaction can be written as $V_{\text{ch}}(1, 2) = \frac{1}{2}J^* \sin \omega$, where ω is the angle between $\mathbf{a}_1$ and $\mathbf{a}_2$, $\mathbf{a}_1 \cdot \mathbf{a}_2 = \cos \omega$.

In the first approximation the total intermolecular interaction potential can be written in the form

$$V(1, 2) = J_0 \cos^2 \omega + \frac{1}{2}J^* \sin 2\omega, \qquad (1.4)$$

where we have used the simple form of the achiral interaction potential (i.e., the Maier–Saupe model for the anisotropic interaction energy). Now the minimum of the potential (1.4) corresponds to the nonzero $\omega_0 =$

arc $\tan(J^*/J_0) \simeq J_0/J^*$ since the chiral interaction is much weaker than the achiral one [16] and therefore $J^* \ll J_0$. Thus, in a chiral liquid crystal the long axes of neighboring molecules are not parallel, and this "microscopic twist" results finally in the formation of the macroscopic helical structure. It should be noted, however, that the elementary twisting angle ω_0 for neighboring molecules is extremely small and can be estimated as $\omega_0 \sim R_0/P$, where R_0 is the average intermolecular distance and p is the pitch of the macroscopic helix.

It is important that the model potential (1.3) be constructed with the help of simple symmetry arguments and is not related to the particular model of chiral intermolecular interaction. At the same time, the coupling constant J^* can be calculated only with the help of such a model. In many cases, calculations of this kind can help to establish the dependence of the coupling constant J^* (and hence the cholesteric pitch) on some important parameters of the system. For example, it will be shown below that the chiral interaction between macromolecules in polymer solutions strongly depends on the dielectric constant of the solvent.

According to the theory of low molecular weight cholesterics [16], the chiral interaction potential (1.3) is determined mainly by the dipole–dipole, dipole–quadrupole dispersion interaction between chiral molecules. This interpretation corresponds to the model of interaction between fluctuating induced dipoles and quadrupoles. The generalized theory of dispersion forces can also be used to describe the chiral interaction between macromolecules in the dielectric solvent, and is discussed in the next section.

1.3.2 Solvent Effect on the Chiral Intermolecular Interaction

The interaction between macromolecules can be considered within the framework of the general theory of dispersion interaction between macroscopic bodies [53,54]. In this theory, a molecule is represented by a simple body filled by the dielectric with the effective polarizability, and the interaction between such bodies is determined by long-wave fluctuations of the electromagnetic field. As a result the coupling constant of such a dispersion interaction is a function of the frequency-dependent effective polarizability of a macromolecule which can be measured experimentally.

The general theory of dispersion interaction between macroscopic bodies was first established by Dzyaloshinskii et al. [51] using the methods of quantum field theory. On the other hand, McLachlan developed an elementary theory [52] and succeeded in recovering most of the results already obtained in [51]. Using the McLachlan theory, Imura and Okano [53] have obtained the following expressions for the energy of the interaction between two ellipsoidal particles "i" and "j" in a solution:

$$U(i, j) = -\frac{h v_i v_j}{32_\pi^2 R_{ij}^6} J_{\alpha\beta} C_{\alpha\beta}^2; \qquad (1.5)$$

where

$$C_{\alpha\beta} = \mathbf{e}_\alpha^i \cdot \mathbf{e}_\beta^j - 3(\mathbf{e}_\alpha^i \cdot \mathbf{u}_{ij})(\mathbf{e}_\beta^j \cdot \mathbf{u}_{ij}), \qquad J_{\alpha\beta} = \int_0^\infty d\xi \, \sigma_{\alpha\alpha}^*(i\xi)\sigma_{\beta\beta}^*(i\xi). \qquad (1.6)$$

Here $\sigma_{\alpha\alpha}^*(i\xi)$ $(\alpha = x, y, z)$ are the components of the effective molecular polarizability in a solution, $\mathbf{e}_\alpha$ is the set of unit vectors in the direction of the principal axes of the ellipsoid "i," v_i is the particle volume, and $\mathbf{R}_{ij} = R_{ij}\mathbf{u}_{ij}$ is the intermolecular vector. The effective polarizabilities of a spheroidal particle are given by

$$\sigma_{\alpha\alpha}^* = \frac{(\varepsilon_\alpha^i - \varepsilon_\alpha^m)\varepsilon_\alpha^m}{\varepsilon_\alpha^i + q_\alpha(\varepsilon_\alpha^i - \varepsilon_\alpha^m)}, \qquad (1.7)$$

where ε_α is the permittivity and the index i refers to the molecule, while the index m refers to the medium (solvent). The quantities q_α are the depolarization coefficients of the ellipsoid. In the derivation of (1.7) it was assumed that the principal axes of the dielectric permittivity tensor coincide with those of the spheroid. The theory of macroscopic dispersion forces has been used in the description of the interaction between biological macromolecules and colloid particles (see, e.g., [54] and references therein).

Note that the expression for the energy of dispersion interaction, (1.5), and also other similar expressions for the interaction potential, does not represent the molecular chirality, since in this model the macromolecule is characterized only by the permittivity tensor $\hat{\varepsilon}^i(\omega)$. It is well known that molecular chirality weakly affects the total intermolecular interaction potential. However, the relatively weak chiral interaction is extremely important in some cases. Indeed, it is the chiral intermolecular interaction which is responsible for the discrimination of enantiomers and determines the inherent dissymmetry of living systems. The chiral interaction is also responsible for the helical twisting in cholesteric liquid crystals.

It is interesting that the McLachlan theory enables us to take into account the effects of spatial dispersion, i.e., to use the nonlocal susceptibility tensor $\hat{\varepsilon}(\mathbf{r} - \mathbf{r}', \omega)$ which is sensitive to the molecular chirality. In the general case, the nonlocal polarizability $\hat{\alpha}(\mathbf{r} - \mathbf{r}', \omega)$ also determines the optical activity of the medium. The latter can easily be shown with the help of the "multipole" expansion of the polarizability tensor

$$\hat{\alpha}(\mathbf{r} - \mathbf{r}', \omega) = [\hat{\alpha}_i(\mathbf{r}, \omega) + \hat{\beta}_i'(\mathbf{r})\nabla - \beta_i''(\mathbf{r})\nabla']\delta(\mathbf{r} - \mathbf{r}_i)\delta(\mathbf{r} - \mathbf{r}'), \qquad (1.8)$$

where ∇ and ∇' are differential operators with respect to $\mathbf{r}$ and $\mathbf{r}'$, respectively, and $\hat{\beta}(\mathbf{r}, \omega)$ is the optical activity tensor

$$\beta_{\alpha\beta\gamma}' = \int \alpha_{\alpha\beta}(\mathbf{r} - \mathbf{r}', \omega)r_\gamma' \, d\mathbf{r}', \qquad \beta_{\alpha\beta\gamma}' = \beta_{\beta\alpha\gamma}''. \qquad (1.9)$$

Now it is possible to obtain the following expression for the energy of the chiral interaction between macromolecules "i" and "j" in terms of the local polarizability and optical activity of the molecules:

$$U^*(i, j) = -\frac{\hbar}{2\pi} \int d\omega \, dl_i \, dl_j \, \varepsilon_m^{-1}(\omega) \, \mathrm{Tr}[(\hat{\beta}_i^*(l_i)\mathbf{V}_i)\hat{T}(\mathbf{R}_{ij}) \cdot \hat{\alpha}_i^+(l_j) \cdot \hat{T}(\mathbf{R}_{ij})$$

$$+ \hat{\alpha}_i^* \cdot \hat{T}(\mathbf{R}_{ij}) \cdot (\hat{\beta}_j^*(l_j)\mathbf{V}_j) \cdot \hat{T}(\mathbf{R}_{ij})], \qquad (1.10)$$

where

$$T_{\alpha\beta}(\mathbf{R}_{ij}) = -R_{ij}^{-3}(\delta_{\alpha\beta} - 3u_{ij\alpha}u_{ij\beta}), \qquad \hbar\omega \gg R_0 kT. \qquad (1.11)$$

In (1.10) the integration is performed along the contours of the macromolecules i and j. Then $\hat{\alpha}_i^*(l_i, \omega)$ is the effective local polarizability and $\hat{\beta}_i^*(l_i, \omega)$ is the effective local optical activity of the macromolecule i in point l_i along the molecule. Thus, in this model, the macromolecule is represented as a thread of arbitrary configuration, characterized by the effective local polarizability and optical activity which are sensitive to the dielectric properties of the solvent.

The general equation (1.10) can be simplified if we assume that the local rigid fragment of the macromolecule is uniaxial. This approximation is reasonable in the majority of cases considered in this chapter, including the solutions of helical molecules. Then we arrive at the expression

$$U^*(i, j) = -J_{ij}^* \int dl_i \, dl_j \, R_{ij}^{-7}[\mathbf{a}_i \cdot \mathbf{a}_j - 6(\mathbf{a}_i \cdot \mathbf{u}_{ij})(\mathbf{a}_j \cdot \mathbf{u}_{ij})](\mathbf{a}_i \times \mathbf{a}_j \cdot \mathbf{u}_{ij}), \quad (1.12)$$

with

$$J_{ij}^* = (6\hbar/\pi) \int_0^\infty d\omega \, \Delta\alpha^*(\omega)\Delta g^*(\omega)\varepsilon_m^{-2}(\omega). \qquad (1.13)$$

Here $\mathbf{a}_i$ is the unit vector in the direction of the long axis of the rigid fragment and Δg^* is the anisotropy of the effective gyration tensor $\hat{g}^*$, which is related to the optical activity tensor $\hat{\beta}^*$ as $\beta_{\alpha\beta\gamma}^* = \delta_{\alpha\beta\eta}g_{\gamma\eta}^*$, where $\delta_{\alpha\beta\eta}$ is the Levi-Civita tensor.

The effective polarizability $\hat{\alpha}^*$ and the gyration tensor $\hat{g}^*$ are determined both by the dielectric properties of a chiral macromolecule and the solvent. The anisotropic parts of these tensors can be written as

$$\Delta\alpha^* = (D^2/8)[\varepsilon_m^2 + \varepsilon_m(\varepsilon_\| - 3\varepsilon_\perp) + \varepsilon_\|\varepsilon_\perp]/(\varepsilon_\perp + \varepsilon_m), \qquad (1.14)$$

$$\Delta g^* = (D^2/4)\varepsilon_m[(2g_\| - g_\perp)\varepsilon_m - g_\perp]/(\varepsilon_m + \varepsilon_\perp)^2, \qquad (1.15)$$

where $\varepsilon_\|$ and $\varepsilon_\perp$ are the longitudinal and transverse susceptibility of the macromolecule, respectively, $g_\|$ and $g_\perp$ are the longitudinal and transverse component of the gyration tensor, respectively, and ε_m is the dielectric constant of the solvent.

The general expression (1.12) for the chiral interaction energy depends on the particular configuration of a macromolecule. When the intermolecular distance is large enough, the molecular shape is insignificant and the chiral interaction potential for two uniaxial molecules can be written in the simple form

$$U^*(i, j) = -J_{ij}^* R_{ij}^{-7} [\mathbf{a}_i \cdot \mathbf{a}_j - 6(\mathbf{a}_i \cdot \mathbf{u}_{ij})(\mathbf{a}_j \cdot \mathbf{u}_{ij})](\mathbf{a}_i \times \mathbf{a}_j \cdot \mathbf{u}_{ij}), \qquad (1.16)$$

where the coupling constant J_{ij}^* is given by (1.13). Note that the coupling constant J_{ij}^* in (1.12), (1.16) is a pseudoscalar and changes sign under the transition between enantiomers. The constant J_{ij}^* is a function of the anisotropy of the gyration tensor which determines the optical activity of a macromolecule. It should be noted also that the anisotropy of the effective polarizability and the gyration tensor strongly depends on the solvent dielectric constant ε_m. According to (1.12), (1.13) the sign inversion of the polarizability anisotropy and the chiral intermolecular interaction can be induced by a change of the solvent dielectric constant.

It is interesting to note that (1.12)–(1.14) confirm the basic idea of Samulski and Samulski [21], since the coupling constant J^* of the chiral interaction potential really changes sign at a certain value of the solvent dielectric constant $\varepsilon_m = \varepsilon_m^*$

$$\varepsilon_m^* = \tfrac{1}{2}(3\varepsilon_\perp - \varepsilon_\parallel) \pm \tfrac{1}{4}(9\varepsilon_\perp^2 + \varepsilon_\parallel^2 - 2\varepsilon_\parallel \varepsilon_\perp)^{1/2}.$$

Note that the relation between the solvent dielectric constant ε_m and the components of the dielectric susceptibility of the media "inside" the model macromolecule differ substantially from the result obtained in [21] with the help of the crude model.

1.4   Statistical Theory of Cholesteric Ordering

1.4.1   General Theory

From the phenomenological point of view the cholesteric ordering corresponds to the twist orientational deformation of the director field, i.e., to the twisting of the director around the axis of the cholesteric helix. In the case of purely twist deformation the distortion free energy of the cholesteric is written as

$$F_d(r) = \tfrac{1}{2}K_{22}(\mathbf{n} \cdot \nabla \times \mathbf{n})^2 + \lambda(\mathbf{n} \cdot \nabla \times \mathbf{n}), \qquad (1.17)$$

where K_{22} is the twist elastic constant and λ is the pseudoscalar quantity which is determined by the chirality of a liquid crystal. The inverse pitch of the macroscopic helix is proportional to the parameter λ

$$q = 2\pi/p = \lambda/K_{22}.$$

The purpose of the molecular–statistical theory is the calculation of the macroscopic parameters λ and K_{22}. The most general expressions for λ and K_{22} can be derived with the help of the density functional theory, which has been successfully used in the description of liquid crystals during the past decade [29–31,55,56].

The density functional theory is based on the assumption that the free energy of a condensed system can be represented as a functional of the one-particle density $\rho(\mathbf{r}, \boldsymbol{\omega})$, $F = F\{\rho(\mathbf{r}, \boldsymbol{\omega})\}$. In the case of a liquid crystal $\rho(\mathbf{r}, \boldsymbol{\omega}) = \rho_0 f_1(\mathbf{r}, \boldsymbol{\omega})$, where ρ_0 is the number density of molecules and $f_1(\mathbf{r}, \boldsymbol{\omega})$ is the one-particle distribution function which determines the probability of finding a molecule at position $\mathbf{r}$ with orientation $\boldsymbol{\omega}$. Then the free energy of a liquid crystal can be written as a sum of two parts

$$F = F_{id} + H, \tag{1.18}$$

where F_{id} is the free energy of the ideal system without intermolecular interactions

$$F_{id} = kT \int d\mathbf{x} \, \rho(\mathbf{x})\{\ln \rho(\mathbf{x})\Lambda - 1 + U_e(\mathbf{x})/kT\}, \tag{1.19}$$

and $U_e(\mathbf{x})$ is the potential of interaction with the external field $\mathbf{x} = \{\mathbf{r}, \boldsymbol{\omega}\}$.

The potential H is determined by intermolecular interactions and the functional derivatives of H are related to the direct correlation functions of the system. In this chapter we will need only the expressions for the first two derivatives

$$\delta H/\delta\rho(x) = kT[-\ln \rho(\mathbf{x}) - U_e(\mathbf{x})/kT + \mu/kT + \text{const.}], \tag{1.20}$$

$$\delta^2 H/\delta\rho(\mathbf{x}_1)\delta\rho(\mathbf{x}_2) = -kTC_2(\mathbf{x}_1, \mathbf{x}_2), \tag{1.21}$$

where $C_2(\mathbf{x}_1, \mathbf{x}_2)$ is the pair direct correlation function related to the full pair correlation function $g_2(\mathbf{x}_1, \mathbf{x}_2)$ by the Ornshtein–Zernike equation

$$g_2(\mathbf{x}_1, \mathbf{x}_2) = C_2(\mathbf{x}_1, \mathbf{x}_2) + \int C_2(\mathbf{x}_1, \mathbf{x}_3)g_2(\mathbf{x}_2, \mathbf{x}_3)\rho(\mathbf{x}_3) \, d\mathbf{x}_3.$$

Equations (1.20), (1.21) enable us to perform the functional Taylor expansion of the free energy.

Now the elastic free energy of the cholesteric can be calculated in the following way. Let us expand the free energy F of the distorted liquid crystal around the value F_0 in the homogeneously aligned sample. Using (1.18)–(1.21), the distortion free energy can be written in the form [31,32]

$$F - F_0 = -\tfrac{1}{2}kT \int [C_2(\mathbf{x}_1, \mathbf{x}_2) - \rho_0^{-1}\delta(\mathbf{x}_1 - \mathbf{x}_2)]$$

$$\times \delta\rho(\mathbf{x}_1)\delta\rho(\mathbf{x}_2) \, d\mathbf{x}_1 \, d\mathbf{x}_2 + \cdots, \tag{1.22}$$

where $\delta\rho(\mathbf{x}) = \rho(\mathbf{x}) - \rho_0(\mathbf{x})$ is the difference of the one particle densities in the distorted and homogeneous states. Note that the variation $\delta\rho(\mathbf{x})$ is proportional to the gradients of the director $\nabla_\alpha n_\beta(\mathbf{r})$, and hence the higher-order terms in the expansion (1.22) should not be taken into account since they determine the nonlinear corrections to the elastic free energy (1.17).

In the case of weak orientational deformations (this is the general case for real distorted liquid crystals) the single-particle distribution function depends on the position $\mathbf{r}$ only through the director $\mathbf{n}(\mathbf{r})$. Then the distribution function of the distorted liquid crystal can be represented as $f_1 = f_0[1 + h]$, where f_0 is the distribution function of the homogeneous sample and the small correction h is proportional to the gradients of the director. The function $f_0(\omega, \mathbf{n}(\mathbf{r}_2))$ can be further expanded in the gradients of the director

$$f_0(\omega, \mathbf{n}(\mathbf{r}_2)) = f_0(\omega, \mathbf{n}(\mathbf{r}_1)) + (\mathbf{r}_{12} \cdot \mathbf{V})\rho(\omega, \mathbf{n}(\mathbf{r}_1)) + \tfrac{1}{2}(\mathbf{r}_{12} \cdot \mathbf{V})^2\rho(\omega, \mathbf{n}(\mathbf{r}_1)) + \cdots.$$
$$(1.23)$$

In the case of uniaxial molecules $\mathbf{V}\rho(\omega, \mathbf{n}) = \rho_0 f_1'(\mathbf{a} \cdot \mathbf{n})\mathbf{V}(\mathbf{a} \cdot \mathbf{n})$ where $f_1' = \partial f_1/\partial(\mathbf{n} \cdot \mathbf{a})$ and the unit vector $\mathbf{a}$ is in the direction of the long molecular axis.

Substituting (1.23) into the general equation (1.22), it is possible to obtain, after some mathematical manipulations, the following general expressions for the parameters λ and K_{22}:

$$\lambda = -kT\rho_0^2 \int d\mathbf{a}_1\, d\mathbf{a}_2\, d\mathbf{r}_{12}\, C_2(\mathbf{a}_1, \mathbf{a}_2, \mathbf{r}_{12})r_{12}^x f_1(\mathbf{a}_1 \cdot \mathbf{n})f_1'(\mathbf{a}_2 \cdot \mathbf{n})a_{2y}, \quad (1.24)$$

$$K_{22} = -\tfrac{1}{2}kT\rho_0^2 \int d\mathbf{a}_1\, d\mathbf{a}_2\, d\mathbf{r}_{12}\, C_2(\mathbf{a}_1, \mathbf{a}_2, \mathbf{r}_{12})(r_{12}^x)^2$$
$$\times f_1'(\mathbf{a}_1 \cdot \mathbf{n})f_1'(\mathbf{a}_2 \cdot \mathbf{n})a_{1y}a_{2y}, \quad (1.25)$$

where the axis $\mathbf{z}$ is parallel to the local director $\mathbf{n}$. The expression (1.25) for the twist elastic constant has been obtained by different authors [33,34], and the expression for the pseudoscalar parameter λ has been obtained in [31,55].

It is interesting to note that the parameter λ (and hence the helical pitch) is determined by the asymmetrical part of the direct correlation function $C_2(1, 2)$ of a liquid crystal, i.e., the part which changes sign under the inversion transformation $\mathbf{r}_{12} \leftrightarrow -\mathbf{r}_{12}$. Thus, in the general case, the cholesteric ordering is determined by chiral intermolecular correlations in a liquid crystal. On the contrary, the twist elastic constant K_{22} is determined by the achiral correlations according to (1.25).

Equations (1.17), (1.24), and (1.25) present the general formal expression for the pitch of the cholesteric structure which can be used in the theory of both low molecular weight and polymer cholesterics. These expressions, however, can be simplified in the case of polymer solutions composed of rod-like macromolecules, since it is possible to obtain the analytical expression for the direct correlation function of such a system. Indeed, the free energy of the solution of long hard rods is given approximately by the first two terms of the virial expansion [35,36], since the higher-order terms are proportional to φ^n, where φ is the volume fraction of the polymer. In the system of sufficiently long rods the nematic (cholesteric)–isotropic phase transition corresponds to the low values of $\varphi \sim D/L$, where D is the molecular diameter and L is the molecular length [36]. In this case, the direct correlation function is given by the following simple expression:

$$C_2(\mathbf{x}_1, \mathbf{x}_2) \simeq \Omega(\xi_{12} - r_{12})[1 + V(\mathbf{x}_1, \mathbf{x}_2)/kT], \quad (1.26)$$

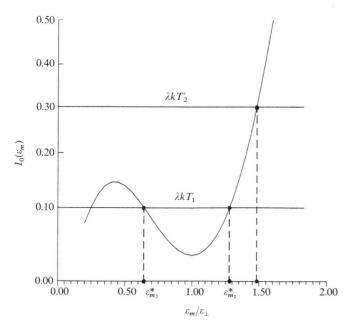

FIGURE 1.2. Critical values of the solvent dielectric constant which correspond to the sense inversion in cholesteric polymer solutions.

where $V(\mathbf{x}_1, \mathbf{x}_2)$ is the energy of attraction between rigid macromolecules and $\Omega(\xi_{12} - r_{12})$ is a step function, $\Omega(\xi - r) = 0$ when $\xi < r$, and $\Omega(\xi - r) = -1$ when $\xi > r$. The function $\xi_{12}(\boldsymbol{\omega}_1, \boldsymbol{\omega}_2, \mathbf{u}_{12})$ is the closest distance of approach between the centers of mass of the molecules "1" and "2" at the fixed orientation and is completely determined by the molecular shape. The first term in (1.26) is determined by the steric repulsion and makes a contribution to the helical pitch if the molecule possesses the chiral shape.

Let us first consider the contribution from the steric interaction of chiral macromolecules to the pseudoscalar parameter. This contribution is given as follows:

$$\lambda_s = -\rho_0^2 kT \int d\mathbf{a}_1 \, d\mathbf{a}_2 \, a_{2y} f_1(\mathbf{a}_1 \cdot \mathbf{n}) f_1'(\mathbf{a}_2 \cdot \mathbf{n}) \int d\mathbf{u}_{12} \, u_{12}^x \xi_{12}^3(\mathbf{a}_1, \mathbf{a}_2, \mathbf{u}_{12})/3,$$
(1.27)

and depends on the chirality of the molecular shape. The calculation of the parameter λ_s is simplified if we assume the ideal orientational order in the polymer liquid crystal. This assumption is reasonable in the case of long rods since in such system the nematic order parameter is close to unity [36]. Then the calculation of λ_s is a purely geometrical problem which is rather difficult, however, even in the case of simple chiral particles. For example, in the case of rigid helices the parameter can only be estimated as [14]

$$\lambda_s \simeq L^2 D \Delta / 2,$$
(1.28)

where the parameter Δ is the height of the ridge of the coil and is shown in Fig. 1.1(a). The parameter Δ corresponds to the length of the side chain.

On the contrary, in the model of twist belts it is possible to obtain the asymptotically exact expression [25]

$$\lambda_s = -\tfrac{1}{2}kT\rho_0^2 q_m h^2 (d + 5h/3\pi)L^2, \qquad (1.29)$$

which is valid when $q_m(d + h) \gg 1$. Here h is the thickness of the belt and d is the broadth, $q_m = 2\pi/p_m$ where p_m is the period of the twist.

The helical ordering is determined only by the chiral part of the total dispersion interaction energy $V(1, 2)$ in (1.26). In the case of rigid rod-like macromolecules this chiral interaction potential is given by (1.15). Substituting (1.15), (1.26) into the general equation (1.24), it is possible to obtain the following general expression for the pitch of the cholesteric polymer solution, composed of rod-like macromolecules [22,24]:

$$q = 2\pi/p = (\rho_0^2/6)[I_0(\varepsilon_m) - \lambda_s kT]S(S + 2)K_{22}^{-1}, \qquad (1.30)$$

with

$$I_0(\varepsilon_m) = (11\pi E_0/180\varepsilon_m)\Delta\alpha^*(0)\Delta g^*(0)D^{-3}, \qquad (1.31a)$$

and

$$\Delta\alpha^*\Delta g^* = (D^4/32)[\varepsilon_m^2 + \varepsilon_m(\varepsilon_\| - 3\varepsilon_\perp) + \varepsilon_\|\varepsilon_\perp]$$
$$\times [(2g_\| - g_\perp)\varepsilon_m - \varepsilon_\perp g_\perp]/(\varepsilon_\perp + \varepsilon_m)^3. \qquad (1.31b)$$

Here S is the nematic order parameter and E_0 is the average excitation energy of the molecule.

The first term in (1.30) represents a contribution from the chiral dispersion interaction which is determined by the coupling constant (1.12) and the second term is a contribution from steric repulsion.

1.4.2   Temperature and Solvent Effect on the Cholesteric Pitch in Polypeptide Solutions

The pitch of the cholesteric helix in the solution of helical rigid macromolecules is given by (1.30), (1.31), where the parameter $\lambda_s \simeq L^2D\Delta/2$. It is interesting to note that contributions from the chiral dispersion interaction and steric repulsion to the helical pitch have different signs, i.e., these two kinds of chiral intermolecular interactions lead to opposite macroscopic twists. This property was noted in [15] and has been used to explain the thermally induced sense inversion which has been observed in the solutions of PBLG and other synthetic polypeptides [8,9]. Indeed, according to (1.30), the helix inversion should be observed when

$$T = T^* = I_0(\varepsilon_m)/\lambda_s k_B. \qquad (1.32)$$

and the pitch is a linear function of temperature if the twist elastic constant K_{22} is nearly permanent. The linear temperature variation of the pitch has also been observed in polypeptide solutions [9,3].

It should be noted that the elastic constant K_{22} is also determined both by steric repulsion and dispersion interaction between macromolecules. Substituting (1.26) into (1.25) we again arrive at the conclusion that the constant K_{22} can be represented as a sum of two terms

$$K_{22} = K_{22}^{(d)} + kTK_{22}^{(s)}, \qquad (1.33)$$

where the second term, which is proportional to the absolute temperature, is contributed to from steric repulsion. It is difficult to estimate a priori which contribution to the twist elastic constant is predominant, but the experimentally observed linear temperature variation of the pitch can be naturally described if we assume that the elastic constant is mainly determined by dispersion interaction. On the contrary, is we assume that the steric interaction is predominant, the pitch should be a linear function of the inverse temperature T^{-1}. Then the experimentally observed behavior of the pitch can be described only in the narrow temperature interval $\Delta T \ll T$. On the other hand, in several PBLG solutions, the linear temperature variation of the pitch is observed in the relatively broad interval of about $100°$ [3,9]. The twist elastic constant solution of hard rods has been calculated by many authors, but only taking into account steric repulsion [37,38,15].

It should also be noted that the linear temperature variation of the pitch is typical for lyotropic cholesteric polymers. In thermotropic main-chain polymer cholesterics (with asymmetric centers in flexible chains) the pitch is usually proportional to the inverse temperature (see, e.g., [39]), and in side-chain thermotropic polymer cholesterics the temperature dependence of the pitch is even more complicated [1]. The intermediate case is represented by cellulose derivatives [12,13]. According to [13] the deviations from the linear temperature variation of the pitch in the solutions of some cellulose derivatives are rather large, and the variation is better described in terms of the inverse temperature. From this point of view, the cellulose derivatives are closer to thermotropic liquid crystals, taking into account also that the volume fraction of cellulose in such solutions is large ($\varphi \simeq 0.6-0.8$) compared with typical polypeptide solutions.

It can readily be seen from (1.32), (1.31) that the temperature of the inversion T^* strongly depends on the solvent dielectric constant ε_m. This dependence was observed, for example, in [8,9] using the mixed solvents with two components which have different dielectric susceptibilities. On the other hand, the helix inversion can also be observed at constant temperature if the solvent dielectric constant is changed. Thus, in the general case, the helical pitch of the cholesteric structure, formed in the solution of helical macromolecules, should be a function of both temperature and the solvent dielectric constant, and the sense of the macroscopic helix can be changed by both factors.

The solvent effect on the helical pitch is determined by the function $I_0(\varepsilon_m)$ which is shown in Fig. 1.3 for the typical values of the parameters. Note that the helix inversion should be observed at the critical value of the solvent

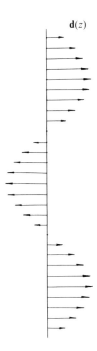

$\mathbf{d}(z)$

FIGURE 1.3. Schematic representation of the helix of dipoles in a macromolecule.

dielectric constant $\varepsilon_m = \varepsilon_m^*$, which corresponds to the point of intersection between the curve $I_0(\varepsilon_m)$ and the horizontal line $\lambda_s kT$. In the general case the inversion parameters are determined by (1.32). At relatively high temperature, (1.32) has the only solution which determines one helix inversion. This is a typical situation for PBLG liquid crystals [3]. Note also that at sufficiently high temperatures there is no sense inversion, since the steric contribution to the helical twisting power is predominant in the whole range of solvent dielectric constants.

At lower temperatures (1.32) has two main solutions (the third solution corresponds to very low values of ε_m) $\varepsilon_m = \varepsilon_{m2}^*$ and $\varepsilon_m = \varepsilon_{m1}^*$ (see Fig. 1.3) and hence the two-fold helix inversion is to be expected. It is interesting to note that such two-fold sense inversion has been observed in the cholesteric phase of PBLG in TCP–m-creasol mixed solvents [9]. Note also that the difference of critical values of the solvent dielectric constant $\varepsilon_{m1}^* - \varepsilon_{m2}^*$ is a growing function of temperature according to the experiment.

It should be noted, however, that the two-fold sense inversion cannot be typical for the cholesteric solutions of helical macromolecules, since it corresponds to extremely low temperatures and can be observed only in particular cases. As far as we know, the two-fold sense inversion has been observed in only one mixed solvent which contains the chemically active m-creasol. Then it is reasonable to assume [22] that the short-range interactions between m-creasol and the side-chains of PBLG can renormalize the effective parameters of the macromolecule (mainly $\varepsilon_\parallel$, $\varepsilon_\perp$ and $g_\parallel$, $g_\perp$) and promote the two-fold sense inversion.

The most consistent test of the existing theory of the solvent effect on the cholesteric ordering in polymer solutions can be achieved for a series of non-interactive chemically similar solvents with different dielectric constants. The corresponding experiment, performed by Toriumi et al. [9] with the series of alkyl chlorides, indicates that the inverse pitch is a decreasing function of the solvent dielectric constant, and becomes negative at sufficiently high values of the dielectric constant (for dichloromethane and dichloroethane).

In real polypeptide solutions the solvent effect is not described completely by the value of the dielectric constant, since the real solvent cannot be considered simply as dielectric media. For example, PBLG forms the left-handed cholesteric structure at 25 °C in dichloromethane ($\varepsilon = 10.5$), and the right-handed structure in m-creasol ($\varepsilon = 12$) and dimethylformamide ($\varepsilon = 16$). This contradiction again can be explained by short-range polymer–solvent interactions which renormalize the effective parameters of the theory. At the same time, the purely dielectric effect, discussed above, is present in the general situation and should be taken into account.

It should be noted that the total chirality of the polypeptide molecule is not permanent in the cholesteric phase. Indeed, while the backbone chirality of the α-helix seems to be unchanged, the side chains of the PBLG molecule possess a high degree of orientational freedom. In particular, deuterium NMR studies on a labeled PBLG sample have demonstrated that the average orientation of the labeled benzyl fragment is effected by temperature and solvent [40,50]. As a result it has been assumed that the chirality of the side chains is reversed and this determines the inversion of the macroscopic helical structure [40]. However, deuterium NMR properties of PBLG–d_7 have been reexamined recently by Toriumi et al. [41]. The results indicate that the sense inversion in PBLG solutions do not necessarily require side-chain conformational transitions, and thus the anisotropic dispersion interaction between chiral macromolecules in the dielectric solvent remains the only general mechanism of sense inversion in cholesteric polymer solutions.

1.4.3   Cholesteric Ordering in Biopolymer Solutions

In recent years the liquid crystalline state has been discovered in solutions of nucleic acids, including the homogeneous water–salt solutions [44], and the dispersed phase which is formed during the intermolecular condensation of nucleic acid molecules of relatively low molecular mass in water–salt solutions containing poly(ethylenglicol) [10,11,42,43,58]. In both systems the cholesteric ordering has also been observed. For concentrated homogeneous solutions of DNA the cholesteric structure has been revealed by studying the textures of thin layers of such solutions [44]. In the dispersed phase the cholesteric twisting is indicated by the appearance of an abnormally intense band in the CD spectrum in the absorption region of chromophores incorporated in the macromolecules [10]. This interpretation is confirmed by specific textures of phases obtained after sedimentation of the DNA microphases. It should be noted that the cholesteric properties of the dispersed phase are very interesting from the biological point of view, since the small droplets of DNA in the polymer matrix can be considered as simple models for chromosomes.

The molecular theory of cholesteric ordering in solutions of helical macromolecules, discussed above, can also be applied to the solutions of nucleic acids. Then we can predict the sense inversion of the cholesteric structure in such solutions caused by a change in temperature and the dielectric properties of the solvent.

Let us consider first the influence of temperature. The thermally induced sense inversion of the helical structure has not yet been observed in the solutions of DNA, but was found in the solutions of complexes of DNA + drug [45] and in the solutions of double-stranded polyribonucleotide poly-(I)poly(C) [11]. In [11,45] the helix inversion has been indicated by the sign inversion of the intense band in the CD spectra. At the same time x ray scattering data indicated that the orientational (nematic) ordering of the macromolecules has not been affected.

The anomalous optical activity of the dispersed phase, formed by DNA macromolecules in the solutions containing a polymer, also depends on the composition of the mixed solvent [42]. In this case, the sense inversion of the cholesteric structure is achieved at a critical concentration of the additive. According to [42] the inversion points in different solvents correspond to approximately the same value of the solvent dielectric constant. This behavior is in accordance with the theory of solvent-induced sense inversion presented in the previous section.

It should be noted however, that the liquid crystalline solutions of nucleic acids are rather complex objects and can be described qualitatively by the existing molecular theory. In such systems some other kinds of chiral intermolecular interactions, which have not been taken into account in the theory, can be very important. For example, a helical macromolecule can also include the helix of permanent dipoles (see Fig. 1.3). Then the electrostatic dipole–dipole interaction will effect the cholesteric ordering. Such a model was considered by Kim [46]. Now the energy of the system of helics is given by the thermodynamic perturbation theory

$$F = F_0 + \langle U_{dd} \rangle - (2kT)^{-1} \langle [U_{dd} - \langle U_{dd} \rangle]^2 \rangle + \cdots, \qquad (1.34)$$

where F_0 is the free energy of the system without dipoles and U_{dd} is the total potential of the dipole–dipole interaction. According to [46] $\langle U_{dd} \rangle = 0$, and the third term in (1.34) yields the following contribution to the distortion free energy of the cholesteric phase:

$$\Delta F_d \approx -(2a^4 d^4 l^3 \rho_0^3/\pi kT) q_m q \int_{Dq}^{\infty} d\xi \, K_1^2(\xi)\xi, \qquad (1.35)$$

where l is the length of the molecule, d is the absolute value of the permanent dipole, a is the length of the elementary segment of the macromolecule which contains one dipole, D is the molecular diameter, and $q_m = 2\pi/p_m$ where p_m is the pitch of the dipolar helix. Note that this contribution is negative and strongly depends on the absolute value of the permanent dipole ($\sim d^4$).

The DNA molecules in a solution possess the effective negative charges, related to the phosphate groups, which form a helix on the "surface" of the macromolecule. These charges determine the corresponding dipoles (with respect to the molecular axis) which take part in the chiral electrostatic interaction considered by Kim. In the general case, the dipole–dipole interaction makes an additional negative contribution to the expression (1.30) for the inverse pitch and promotes the sense inversion of the cholesteric structure. It should be noted, however, that the effect of the shielding of the charges by the counter ion atmosphere in the water–salt solutions has not been taken into account in [46]. Thus (1.35) cannot be used directly in the description of cholesteric ordering in the solutions of nucleic acids.

1.4.4 Influence of Solvent Chirality on the Cholesteric Ordering

Shiau and Labes have shown experimentally that the addition of the chiral solvent (phenethyl alcohol) to the cholesteric lyotropic phases of synthetic polypeptides causes a strong perturbation of the pitch [47]. It has also been found that l-alcohol increases and d-alcohol decreases the pitch relative to the recemic alcohol. These effects can be naturally explained by the general theory of cholesteric ordering in polymer solutions if we consider the influence of the chiral solvent on the effective dispersion intermolecular interaction [23].

Equation (1.7) indicates that the effective polarizability of a macromolecule in the dielectric media depends on the dielectric permittivity of the media. Then it is natural to expect that the achiral molecule should possess the effective optical activity in the chiral solvent. In [23] we have obtained the following expression for the effective gyration tensor of a rod-like macromolecule in the chiral dielectric solvent:

$$g_{\perp}^* = \frac{V}{2\pi}\frac{g_{\perp}\varepsilon_m}{\varepsilon_m + \varepsilon_{\perp}} - \frac{V}{4\pi}\beta_0 \frac{\varepsilon_m(5\varepsilon_{\perp} - \varepsilon_{\parallel}) - \varepsilon_{\perp}(\varepsilon_{\parallel} + 3\varepsilon_{\perp})}{(\varepsilon_m + \varepsilon_{\perp})^2}, \qquad (1.36a)$$

$$\Delta g^* = \frac{V}{4\pi(\varepsilon_m + \varepsilon_{\perp})^2}\{2\varepsilon_m[(2g_{\parallel} - g_{\perp})\varepsilon_m - g_{\perp}\varepsilon_{\perp}]$$

$$- \beta_0[\varepsilon_m(5\varepsilon_{\perp} - \varepsilon_{\parallel}) - \varepsilon_{\perp}(3\varepsilon_{\perp} + \varepsilon_{\parallel})]\}, \qquad (1.36b)$$

where β_0 is the optical activity of the solvent.

The second term in (1.36b) is determined by the solvent chirality and does not depend on the optical activity of the isolated macromolecule (i.e., on the components of the gyration tensor $g_{\parallel}$, $g_{\perp}$). This term makes an additional contribution to the inverse pitch according to the general equation (1.31a)

$$\Delta q = \Delta I_0(\varepsilon_m)/K_{22} = -\kappa_0 \Delta\alpha^*(0)\Delta g_{sl}^*(0), \tag{1.37}$$

with $\kappa_0 = 11\pi E_0/180\varepsilon_m K_{22}$, where $\Delta g_{sl}^*(0)$ is a contribution from solvent chirality

$$\Delta g_{sl}^*(0) = \beta_0 \frac{V[\varepsilon_m(5\varepsilon_\perp - \varepsilon_\|) - \varepsilon_\perp(3\varepsilon_\perp + \varepsilon_\|)]}{4\pi(\varepsilon_m + \varepsilon_\perp)^2}, \tag{1.38}$$

Now the total inverse pitch can be written in the form

$$q = q_0 + \Delta q; \tag{1.39}$$

where q_0 is given by (1.30) for the achiral solvent and Δq is proportional to the optical activity of the solvent according to (1.37), (1.38).

With the help of (1.37)–(1.39) it is possible to explain qualitatively the main experimental results of [47]. Indeed, the inverse pitch q appears to be a function of the optical activity of the chiral solvent β_0, and the parameter β_0 has different signs for left- and right-handed solvents. Then the addition of chiral alcohol to the solvent would yield the contribution to the inverse pitch Δq which also has different signs for l- and d-alcohol. Note that the parameter β_0 is the optical activity of the mixed solvent, and in the first approximation it is proportional to the concentration of the chiral additive. The approximately linear concentration dependence has been observed in experiment [47] in the middle concentration range.

It is interesting to consider the dispersion interaction of two achiral macromolecules in the chiral solvent. In this case the components of the gyration tensor of the isolated molecules $g_\perp$ and $g_\|$ are equal to zero, but nevertheless the molecule possesses the effective optical activity which is proportional to that of the solvent. According to (1.31b), there exists an effective chiral dispersion interaction between such molecules embedded in the chiral media. This interaction results in the cholesteric ordering of the corresponding polymer liquid crystal.

On the other hand, it is reasonable to assume that the chiral solvent, which does not take part in the short-range polymer–solvent interaction, cannot change the effective shape of an achiral macromolecule. Then the contribution from the steric repulsion to the general equation (1.31a) is absent, and the helical pitch is determined only by the effective chiral dispersion interaction

$$q = \Delta I_0(\varepsilon_m, \beta_0)/K_{22} = -\kappa_0 \Delta\alpha^*(0)\Delta g_{ef}^*(0)/K_{22}, \tag{1.40}$$

where the anisotropy of the effective gyration tensor of the molecule in the chiral dielectric solvent is given by (1.38).

Thus it is possible to predict the new class of cholesteric polymers, composed of achiral macromolecules in chiral solvents. Note that, in principle, the cholesteric ordering can be observed not only in polymer liquid crystals, composed of rod-like rigid macromolecules, but also in any nematic polymer in the chiral solvent. At present, we can support this conclusion by the only example of cholesteric ordering of this type which has been observed in poly-1,4-benzamide liquid crystal when a small fraction of the chiral solvent was added [48].

1.5 Influence of Molecular Flexibility on the Cholesteric Ordering in Polymer Solutions

Cholesteric ordering was also observed in the liquid crystalline solutions of cellulose and its derivatives, which are composed of relatively flexible macromolecules [12,13,49]. However, so far, the theory of cholesteric ordering was developed only in the case of rigid rods and, only recently, the theory has been generalized to polymer liquid crystals composed of long persistent chains [25].

Let us consider a dilute solution of long persistent chains with persistent length l and diameter D, $D/l \ll 1$. Then we can consider the segments of length $l_0 \ll l$, which are rigid. In this case, the free energy of the polymer solution can be written in the form

$$F/kT = -(\rho^2/2) \int f_1(\mathbf{r}_1, \boldsymbol{\theta}_2)\Omega(\mathbf{r}_{12}, \boldsymbol{\theta}_1, \boldsymbol{\theta}_2) f_1(\mathbf{r}_2, \boldsymbol{\theta}_2) \, d\mathbf{r}_1 \, d\mathbf{r}_2 \, d\boldsymbol{\theta}_1 \, d\boldsymbol{\theta}_2$$

$$+ c \int_0^N dt \int \psi^+ (l(\mathbf{a} \cdot \boldsymbol{\nabla}_r) - \Delta_q + \partial/\partial t)\psi \, d\boldsymbol{\theta} \, d\mathbf{r}$$

$$+ c \int [f_1(0, \mathbf{r}, \boldsymbol{\theta}) \ln \psi(0, \mathbf{r}, \boldsymbol{\theta})$$

$$+ f_1(N, \mathbf{r}, \boldsymbol{\theta}) \ln \psi^+(N, \mathbf{r}, \boldsymbol{\theta})] \, d\boldsymbol{\theta} \, d\mathbf{r}, \qquad (1.41)$$

where c is the number density of the chains, $\rho = cL/l_0$ is the number density of the segments where L is the total length of the chain, $f_1(\mathbf{r}, \boldsymbol{\theta})$ is the one-particle distribution function, and $\Omega(\mathbf{r}_{12}, \boldsymbol{\theta}_1, \boldsymbol{\theta}_2)$ is the step function defined in Section 1.4.2. Note that the second and third terms in (1.41) describe the packing entropy which is related to the flexibility of the chains. In (1.41) the function $\psi(t, \mathbf{r}, \boldsymbol{\theta})$ is related to the distribution function $f(t, \mathbf{r}, \boldsymbol{\theta})$ [36]

$$f_1 = \psi\psi^+, \qquad \psi^+(t, \mathbf{r}, \boldsymbol{\theta}) = \psi(N - t, \mathbf{r}, \boldsymbol{\theta}), \qquad (1.42)$$

where the variable t is changing along the chain. Now it is possible to expand the one-particle distribution function in powers of the gradients of the director (as was discussed in Section 1.4.1) and to derive an expression for the deformation free energy.

According to [25], the deformation free energy of the solution of flexible chains is given by

$$F_d(r) = (F(\mathbf{r}) - F_0)/kT = -(\rho^2/2) \int f_1(\mathbf{n} \cdot \mathbf{a}_1)$$

$$\times (\mathbf{C}(1, 2)\nabla) f_1(\mathbf{n} \cdot \mathbf{a}_2) \, d\mathbf{a}_1 \, d\mathbf{a}_2 + (\text{elastic terms}), \qquad (1.43)$$

with

$$\mathbf{C}(1, 2) = \int \mathbf{r}_{12} \Omega(\mathbf{r}_{12}, \boldsymbol{\theta}_1, \boldsymbol{\theta}_2)[1 + \beta V(\mathbf{r}_{12}, \boldsymbol{\theta}_1, \boldsymbol{\theta}_2)] \, d\mathbf{r}_{12} \, d\mathbf{b}_1 \, d\mathbf{b}_2,$$

$$(1.44)$$

where $V(\mathbf{r}_{12}, \boldsymbol{\theta}_1, \boldsymbol{\theta}_2)$ is the attraction energy of the interaction between the segments of the chain and $\mathbf{b}_1$, $\mathbf{b}_2$ are the unit vectors in the direction of the short axes of the segments.

It is interesting to note that the configuration entropy does not contribute to the first order term (in the gradients of the director) in the expansion (1.43) which determines the cholesteric twisting, and hence the molecular flexibility does not influence substantially the cholesteric properties of the solution. At the same time, the elastic constants of polymer solutions strongly depend on the chain flexibility [38] and the account of configurational entropy is the most serious problem of the theory.

The function $\mathbf{C}(1, 2)$ can be calculated analytically in the athermal solution of long twisted belts which present the simple model for cellulose derivatives [25]

$$\mathbf{C}(1, 2) \simeq \tfrac{1}{2}\mathbf{e}(\mathbf{a}_1 \cdot \mathbf{a}_2)q_m h^2(d + 5h/3\pi)l_0^2, \qquad (1.45)$$

where $\mathbf{e} = \mathbf{a}_1 \times \mathbf{a}_2 / |\mathbf{a}_1 \times \mathbf{a}_2|$. Equation (1.45) is asymptotically exact in the limit $q_m(h + d) \ll 1$.

With the help of (1.44) it is possible to obtain the expression for the pseudo-scalar parameter λ in the distortion free energy of the cholesteric in (1.17)

$$\lambda = -\tfrac{1}{2}kT\rho^2 q_m h^2(d + 5h/3\pi)l_0^2 \int f_1(\mathbf{a}_1 \cdot \mathbf{n}) f_1(\mathbf{a}_2 \cdot \mathbf{n})$$

$$\times [3(\mathbf{n} \cdot \mathbf{a}_1)(\mathbf{n} \cdot \mathbf{a}_2) - (\mathbf{a}_1 \cdot \mathbf{a}_2)](\mathbf{a}_1 \cdot \mathbf{a}_2)[1 - (\mathbf{a}_1 \cdot \mathbf{a}_2)^2]^{-1/2} \, d\mathbf{a}_1 \, d\mathbf{a}_2. \quad (1.46)$$

In the case of perfect orientational order ($S \to 1$), (1.46) can be simplified to

$$\lambda = -(kT/2)\rho^2 q_m h^2(d + 5h/3\pi)l_0^2, \qquad (1.47)$$

and the inverse pitch of the athermal solution of long flexible twisted belts can be written as [25]

$$q = 2\pi/p = -6\varphi q_m h^2(d + 5h/3\pi)/(\pi d/4 + h) \, dl(1 - S), \qquad (1.48)$$

where $\varphi = cLd(\pi d/4 + h)$ is the volume fraction of the polymer. Note that the nematic order parameter S is also a function of φ and usually $1 - S = \text{const.} \, \varphi^{5/3}$ when $\varphi \ll 1$ and const. $\ll 1$ [36].

In real systems the attraction of chains is also important and hence it is necessary to use the general equation (1.44). Then we arrive at the following expression for the helical pitch, which is similar to (1.30) derived in the case of polypeptide solutions:

$$q = -\tfrac{1}{2}\rho^2(\kappa - \lambda_s kT)K_{22}^{-1}. \tag{1.49}$$

where the parameter λ_s is given by (1.47) and κ is related to the attraction interaction.

Thus the behavior of the pitch in cholesteric solutions of long flexible chains should be qualitatively similar to the behavior in solution of rigid helical macromolecules, including the temperature variation of the pitch and solvent induced sense inversions of the cholesteric structure. However, as far as we know, such inversions have not yet been observed in liquid crystalline solutions of cellulose derivatives.

References

1. Ya.S. Freidzon and V.P. Shibaev, in *Liquid Crystal Polymers*, edited by N.A. Platé (Plenum, New York, 1993).

2. V.P. Shibaev and Ya.S. Freidzon, in *Side-Chain Liquid Crystal Polymers*, edited by C. McArdle (Blackie, Glasgow, 1989).

3. T. Uematsu and Y. Uematsu, Adv. Polymer Sci. **59**, 37 (1984).

4. E.T. Samulski and D.B. Dupre, J. Chim. Physique **80**, 25 (1983).

5. C. Robinson, Tetrahedron **13**, 219 (1961).

6. C. Robinson, J.C. Ward, and R.B. Beevers, Faraday Discuss. Chem. Soc. **25**, 29 (1958).

7. C. Robinson, Mol. Cryst. **1**, 467 (1966).

8. H. Toriumi, Y. Kuzumi, Y. Uematsu, and I. Uematsu, Polymer. J. **11**, 836 (1979).

9. H. Toriumi, K. Yahagi, I. Uematsu, and Y. Uematsu, Mol. Cryst. Liq. Cryst. **94**, 267 (1983).

10. Yu.M. Evdokimov, S.G. Skuridin, and V.I. Salyanov, Liq. Cryst. **3**, 1443 (1988).

11. S.G. Skuridin, N.S. Badaev, A.T. Dembo, G.B. Lordkipanidze, and Yu. M. Evdokimov, Liq. Cryst. **2**, 51 (1988).

12. G.V. Laivins and D.G. Gray, Polymer. J. **26**, 1435 (1985).

13. R.S. Werbowyi and D.G. Gray, Macromolecules **17**, 1532 (1985).

14. J.P. Straley, Phys. Rev. A **14**, 1835 (1976).

15. H. Kimura, M. Hosino, and H. Nakano, J. Phys. Soc. Jpn. **51**, 1584 (1982).

16. B.W. Van der Meer and G. Vertogen, in *Molecular Physics of Liquid Crystals*, edited by G.R. Luckhurst and G.W. Gray (Academic Press, New York, 1979).

17. L.M. Lisetsky and G.S. Chilaya, Uspekhi Fiz. Nauk **134**, 279 (1981).

18. W.J.A. Goossens, Mol. Cryst. Liq. Cryst. **12**, 237 (1971).

19. B.W. Van der Meer, G. Vertogen, A.J. Dekker, and J.G.J. Ypma, J. Chem. Phys. **65**, 3935 (1976).

20. D.B. Dupre, R.W. Duke, W.A. Hines, and E.T. Samulski, Mol. Cryst. Liq. Cryst. **40**, 247 (1977).

21. T.V. Samulski and E.T. Samulski, J. Chem. Phys. **67**, 824 (1977).

22. M.A. Osipov, Chem. Phys. **96**, 259 (1985).

23. M.A. Osipov, Vysokomolek. Soedin. **19a**, 1603 (1987).

24. M.A. Osipov, Nuovo Cimento D **10**, 1249 (1988).

25. M.A. Osipov, A.N. Semenov, and A.R. Khokhlov, Khim. Fiz. **6**, 1312 (1987).

26. B.W. Van der Meer and G. Vertogen, Z. Naturforsch. **34a**, 1359 (1979).

27. L. Blum and A.J. Torruella, J. Chem. Phys. **56**, 303 (1972).

28. B.W. Van der Meer, Thesis, Gröningen, 1979.

29. T.J. Sluckin and P. Shukla, J. Phys. A, Math. Gen. **16**, 1539 (1983).

30. Y. Singh, Phys. Rev. A **30**, 583 (1984).

31. M.A. Osipov, Kristallografiya **33**, 817 (1988).

32. S.A. Pikin, M.A. Osipov, and E.M. Terentjev, Soviet Scientific Reviews, **A11**, 193 (1989).

33. M.D. Lipkin, S.A. Rice, and U. Mohanty, J. Chem. Phys. **84** 1796 (1986).

34. A. Poniewierski and J. Stecki, Mol. Phys. **38**, 1931 (1979).

35. J.P. Straley, Mol. Cryst. Liq. Cryst. **24**, 7 (1973).

36. A.Yu. Grossberg and A.R. Khokhlov, Adv. Polymer. Sci. **41**, 53 (1981).

37. J.P. Straley, Phys. Rev. A **8**, 2181 (1973).

38. A.Yu. Grossberg and A.V. Zestkov, Vysokomolek. Soedin. **28**, 86 (1986).

39. Ho-Jin Park, Jung-Il Jun, and R.W. Lenz, Polymer. J. **26**, 1301 (1985).

40. K. Czarniecka and E.T. Samulski, Mol. Cryst. Liq. Cryst. **69**, 205 (1981).

41. H. Toriumi, T. Yamazaki, A. Abe, and E.T. Samulski, Liq. Cryst. **1**, 87 (1986).

42. Yu.M. Evdokimov, V.I. Salyanov, A.T. Dembo, M.I. Schrago, and L.A. Khanina, Kristallografiya **31**, 736 (1986).

43. S.G. Skuridin, A.T. Dembo, M.A. Osipov, H. Damaschun, G. Damacshum, and Yu.M. Evdokimov, Dokl. Akad. Nauk SSSR **285**, 713 (1985).

44. F. Livolant, European J. Cell. Biol. **33**, 300 (1984).

45. V.I. Salyanov and Yu.M. Evdokimov, Vysokomolek. Soedin. A **28**, 971 (1986).

46. Y.H. Kim, J. Phys. (Paris) **43**, 559 (1982).

47. C.C. Shiau and M.M. Labes, Mol. Cryst. Liq. Cryst. **124**, 125 (1985).

48. M. Panar and L.F. Boster, Macromolecules **10**, 1401 (1977).

49. F. Fried, J.M. Gilli, and P. Sixon, Mol. Cryst. Liq. Cryst. **98**, 209 (1983).

50. E.T. Samulski, J. Phys. (Paris) Coll. **40**, C3–471 (1979).

51. I.E. Dzyaloshinsky, L.P. Pitaevsky, and E.M. Lifshitz, Adv. Phys. **10**, 165 (1961).

52. A.D. McLachlan, Faraday Discuss. Chem. Soc. **40**, 239 (1965).

53. H. Imura and K. Okano, J. Chem. Phys. **58**, 2763 (1973).

54. Yu.S. Barash and V.L. Ginzburg, Uspekhi Fiz. Nauk **143**, 345 (1984).

2

Structure of Thermotropic Main-Chain Polymers

A.H. Windle

2.1 Introduction

2.1.1 Scope

This chapter is partly a review, and partly a report of some new work. Where it is in the review mode, it focuses on areas in which liquid crystalline polymers are uniquely different from their small molecule counterparts, and is written particularly to assist those familiar with liquid crystal science in appreciating the polymeric aspects of the subject, although limitations of space preclude a treatment which is as comprehensive or as deep as we might wish. While attention is paid to the application of theories first developed for small molecule systems to the polymeric case, aspects of molecular design aimed at reducing the crystalline melting point of mesogenic polymers are not discussed in any great depth, while lyotropic or chiral systems do not feature at all. Section 2.2 deals with the architecture of nematogenic chain molecules, both in respect to the critical role of persistence length in determining liquid crystalline stability and to the influence of the nematic–isotropic transition on chain length. Section 2.3 introduces a new classification of levels of positional order in main-chain liquid crystalline polymers, exploring the smectic domain and discussing predicted and experimentally observed phases alike.

2.1.2 Main-Chain Mesogenic Systems

The principle of creating a liquid crystalline polymer by taking a series of small rod-like molecules and joining them end-to-end is relatively straightforward, although the synthetic chemistry itself might not be. Short rods can be made into long ones, and it is clear that the stability of the mesophase, expressed, for example, as the nematic to isotropic transition temperature, will be increased.

There is, however, a very significant problem in that the crystal melting point is also increased. This trend is clearly illustrated by the data [1, 2] of

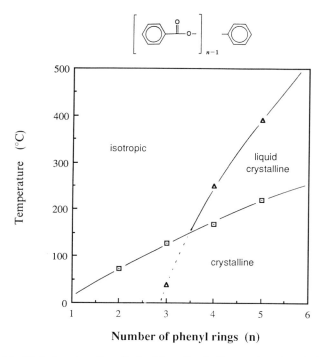

FIGURE 2.1. Plot showing the effect of length of oligomeric ester chains on their melting points and on the nematic–isotropic transition temperatures. Data from [1] and [2] ($\Box$ T_m; $\triangle$ T_{lc-i}).

Fig. 2.1, in which both the nematic–isotropic transition temperature and the melting point are plotted against the number of units in oligomeric ester molecules. In general, the liquid crystalline transition temperature increases more quickly than the melting point so the range of mesophase stability does increase. However, the temperatures themselves become steadily higher, reaching 391 °C and 220 °C, respectively, for the five unit chain, so that as we move toward yet longer chains, processing will become increasingly inconvenient, while the molecule itself will tend to decompose before it melts.

The scheme shown in Fig. 2.2 summarizes the molecular design routes which have been developed for materials based on main-chain mesogenic units. The first step in obtaining a high molecular weight molecule, which will have a crystal melting point in what is normally regarded as the "accessible" range (typically less than 300 °C), is to make the chain somewhat more flexible. The simple thermodynamic relationship

$$T_m = \Delta H / \Delta S$$

shows why this is so. A stiff chain will have comparatively little additional freedom to move in the melt than in the crystal, so ΔS is small and T_m

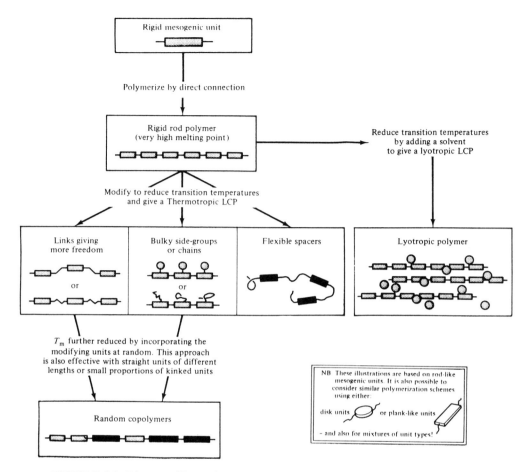

FIGURE 2.2. Diagram illustrating the principles of the molecular design of main-chain liquid crystalline polymers (adapted from [4]).

correspondingly high. On the other hand, a more flexible molecule, while being similarly constrained in the crystal, will have a much greater ability to squirm in the melt, hence a higher ΔS and a lower melting point. The linking of phenyl rings through ester groups, to form polyhydroxybenzoic acid (1), does bring the melting point down compared with a chain formed of rings directly connected together. However, it is still high, certainly above 400 °C, and the polymer can only be processed using solid state powder routes. The introduction of still greater flexibility, typically by inserting short sequences of alkane units into the backbone, e.g., (2), brings both transitions down further. However, it is easy to lose the liquid crystalline phase field altogether as we are inevitably reducing the nematic–isotropic transition faster than the melting point. Note, for example, that PET (3) is not liquid crystalline, as there are too many flexible CH_2 groups per phenyl ring. The presence of large side groups tends to militate against crystallinity more than liquid crystallinity, and the molecule of polychlorophenylene terephthalate (4) has a melting point of 370 °C while still being liquid crystalline. However, some of the contribution to the reduction of the melting point of this molecule comes from the fact that the chlorine atom may be 2 substituted on some rings and 3 substituted on others.

The lack of perfect periodicity along the chain can begin to have a major influence in frustrating crystal perfection, and thus reducing the melting point without having any significant effect on the stability of the mesophase. The destruction of chain periodicity through random copolymerization of para linked units is the most popular route to melting point control in main-chain thermotropic polymers (Fig. 2.2). It is exploited in the creation of the commercially significant random copolyesters, and molecule (5), a random copolymer of hydroxybenzoic and hydroxynaphthoic acids with $x \approx 0.75$, is the basis of an important product range marketed as *Vectra* [3]. As a further stage, units which disrupt crystallinity by virtue of additional flexibility, or by being kinked or having large sidegroups, will, if added in at random positions, have a far stronger influence in reducing the melting point than if positioned regularly along the chain. The molecule (6), which for $x = 0.6$ has a crystal melting point of 190 °C, is an example of this effect. The molecular background to the subject is described in greater detail in Chapter 3 of reference [4]

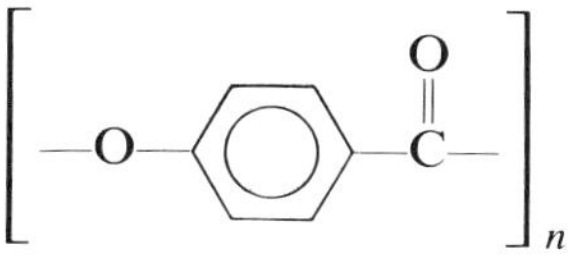

PHBA
poly(hydroxybenzoic acid)   (1)

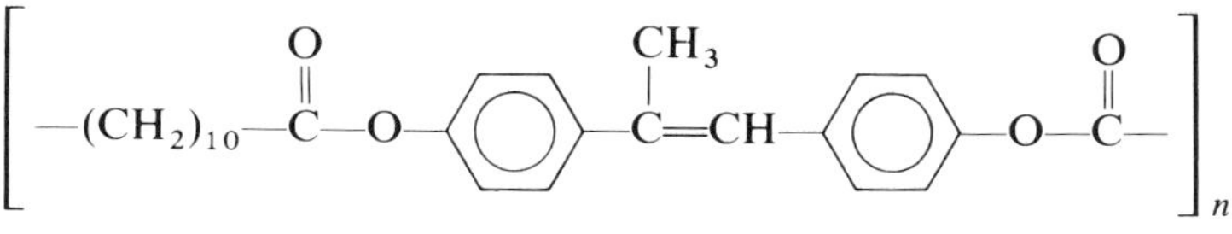

(2)

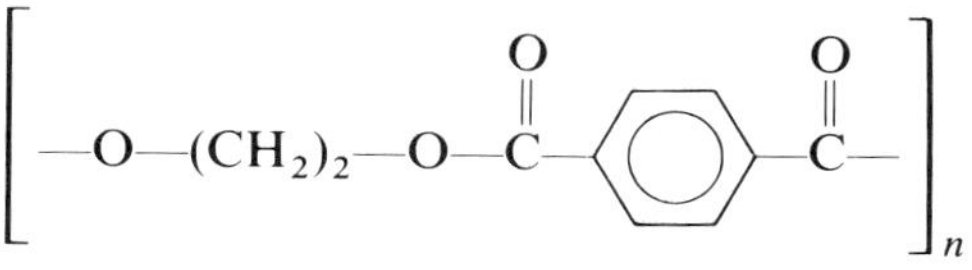

PET   (3)
poly(ethyleneterephthalate)

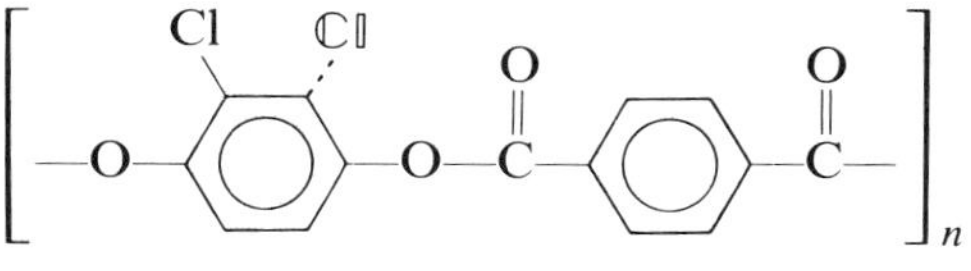

(4)

$$
\left[\overset{\overset{\text{O}}{\|}}{-\text{C}}-\!\!\bigcirc\!\!-\text{O}- \right]_{x}
\quad
\left[\overset{\overset{\text{O}}{\|}}{-\text{C}}-\!\!\bigcirc\!\!\bigcirc\!\!-\text{O}- \right]_{1-x}
\tag{5}
$$

$$
\left[\overset{\overset{\text{O}}{\|}}{-\text{C}}-\!\!\bigcirc\!\!-\text{O}- \right]_{x}
\quad
\left[\overset{\overset{\text{O}}{\|}}{-\text{C}}-\!\!\bigcirc\!\!-\overset{\overset{\text{O}}{\|}}{\text{C}}- \right]_{(1-x)/2}
$$

$$
\left[-\text{O}-(\text{CH}_2)_2-\text{O}- \right]_{(1-x)/2}
\tag{6}
$$

2.2   Molecular Architecture

2.2.1   Theoretical Background

As with conventional, small molecule liquid crystals, liquid crystallinity in polymers is the result of long-range cooperation in molecular orientation. In the small molecule case, this order depends on a sufficient level of anisotropy in the short-range intermolecular interactions. The order may be seen to stem from either packing considerations consequent on the rod-like shape of the molecule, or from longer-range anisotropic attractive forces, or from some measure of both. Models to predict thermotropic behavior are well known.

The earliest due to Onsager [5] is appropriate for long, rigid rods, in that the transition from isotropic to liquid crystalline phases occurs at very low concentrations. For thermotropic systems without solvent, in which the rods are either short or not straight, the model is less suitable as it assumes a dilute approximation involving only the second virial coefficient. Its extension to more concentrated solutions, and hence to the thermotropic melt, depends on the introduction of a decoupling approximation [6]. The lattice model of Flory and coworkers, extends readily to thermotropic melts based on rods of different aspect ratios. In the treatment of Flory and Ronca [7] it is predicted that the critical aspect ratio of the rod molecules for liquid crystallinity is 6.42. It should be emphasized, however, that in their simplest forms both theories are *athermal*, in that while their predictions depend on the concentration and aspect ratio of the rod molecules they do not account for the influence of temperature. In more recent developments of the Flory lattice model, in particular, temperature is brought in the form of an orientationally dependent energy function. On the other hand, the mean field theory of Maier and Saupe [8] is based primarily on a long-range orientationally dependent energy function and is not so readily extendible to polymeric systems.

Thus the essential challenge is how to apply theories which were originally formulated to explain the behavior of rigid rods to nematogenic polymer chains which are neither completely rigid nor perfectly straight.

2.2.2   Architecture of Isolated Polymer Molecules

It is possible to synthesize chain molecules which are remarkably stiff, an example of which is polybenzo-*bis*-oxazole (7). What little flexibility there is will come from the bending and stretching of bonds rather than rotation. However, while such polymers would be expected to show a very high nematic–isotropic transition temperature their melting points will also be high, and certainly in the range in which the molecule will thermally decompose.

PBO
poly(*p*-phenylene
benzobisoxazole)

(7)

The answer, in this particular case, is to follow the lyotropic route and process the liquid crystalline material in solution. If we wish to avoid solvents and design a thermotropic polymer, the crystal melting point has to be kept clearly below the degradation temperature. A combination of a degree of chain flexibility, and disruption of the periodicity of the chain by random copolymerization is the route most often taken, e.g., molecule (5). However, the introduction of limited flexibility into the chain raises the significant question as to how flexible the chain can be while still retaining mesogenic behavior, and thus a nematic–isotropic transition which is above the reduced melting point.

It should be noted that the important criterion for liquid crystallinity is chain *straightness* (also described as "rectilinearity" [9]), rather than rigidity per se. It is possible to imagine a rigid chain which is anything but straight; on the other hand a flexible chain will, with time, take up many different conformations and thus trajectories. Figure 2.3(a) shows a simulation of an aromatic copolyester (molecule (5)) at an artificially high temperature (1300 K). The various trajectories can be viewed as snapshots of the molecule taken at random times. It may well be that, by chance, one such snapshot would show a straight molecule, but the fact that such a conformation happens to occur would not, in itself, be a criterion for liquid crystallinity. On the other hand, if the straight conformation has the lowest enthalpy, then reducing the temperature will mean that the width of the distribution of trajectories about the straight one will be reduced and the molecule will become, on average, straighter. Figure 2.3(b) shows trajectories of the same molecule as in Fig. 2.3(a) but simulated for a lower temperature (250 K). Most of the conformations are now fairly straight. The increasing straightness of a chain

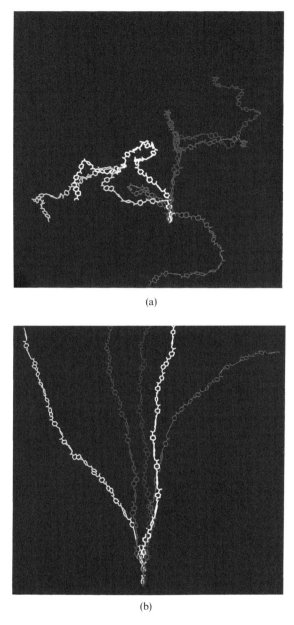

(a)

(b)

FIGURE 2.3. Simulations of random examples of the trajectory of the mesogenic molecule (5). The first bond of each molecule is set vertical on the page: (a) at the artificially high temperature of 1300 K which would be above the nematic–isotropic transition temperature; and (b) at the lower temperature of 250 K.

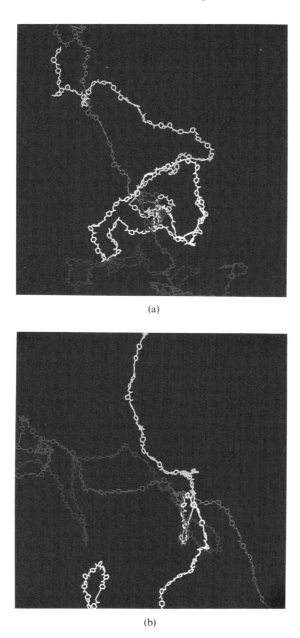

(a)

(b)

FIGURE 2.4. Set of simulations for molecule (8) which contains isophthalic acid units which effectively puts kinks into the chain at random positions. When these units are present in sufficiently low concentration, the polymer exhibits a range of liquid crystallinity above its melting point: (a) at 1300 K, and (b) at 250 K.

with decreasing temperature provides a basis for the isotropic–nematic transition in polymers.

Alternatively, if the lowest enthalpy conformation is not a straight one, then reducing the temperature to make the chain more rigid, will not make it into a straight rod. Figure 2.4(a, b) shows a series of simulations at 1300 K and 250 K, respectively, for molecule (8) in which some of the aromatic rings are meso linked so that they effectively kink the chain at quite low temperatures. Note that, although the straight sequences between the kinks are somewhat more apparent at the lower temperature, the straightness of the chain overall is determined more by the presence of kinked units than the temperature.

Figure 2.5 shows three views of the same molecule (5) at 500 K but at different resolutions. The molecular weight is typical of melt polymerized material. Looking at the whole chain, its trajectory is to all intents and purposes part of a random coil, a closer view emphasizes the tendency the chain has to run straight for sequences of many units, while a closer view still shows the chemical make-up of the individual units. The fact that real mesogenic chains, if long enough, will assume a random coil trajectory is significant as it implies the ability to form entanglements. It is important to decide the extent to which entanglements will be preserved across the isotropic–nematic transition, as their presence in the nematic phase will not only play an important part in determining the viscoelastic behavior, but they may conceivably influence the microstructure also.

The next question which must be asked is, How straight does a chain have to be before it will form a nematic phase? Is it possible, for example, to predict the combination of chemical architecture and temperature which will render the molecule just straight enough for liquid crystallinity?

2.2.3   Persistence Length as a Critical Parameter

It is worthwhile to see whether there is a single parameter which can effectively describe the straightness and thus mesogenicity of an isolated chain as a function of its chemical structure and temperature. A prime candidate is the

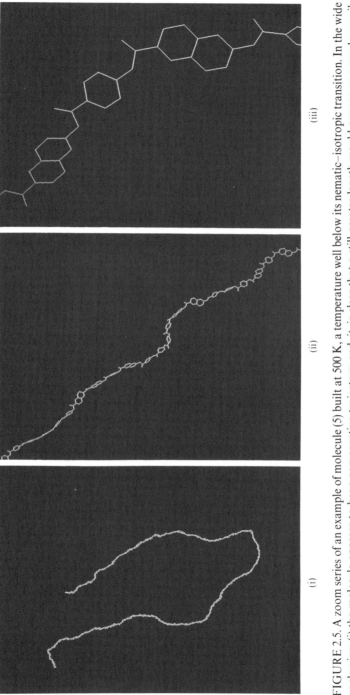

FIGURE 2.5. A zoom series of an example of molecule (5) built at 500 K, a temperature well below its nematic–isotropic transition. In the wide angle view (i) the molecule appears to have comparative trajectory and it is clear that a still greater length would appear as a random coil. The intermediate view (ii) shows a comparatively straight sequence, such as we would expect in a mesogenic molecule. The close up (iii) emphasises the details of the chemical structure and that many of the backbone bonds are far from parallel to the local chain axis.

persistence length, q, which is a defining parameter of the semiflexible "worm-like" chain first discussed by Kratky and Porod [10]. In general terms, it is defined as the mean translation of a chain in the direction of the first unit. A useful aspect of this parameter is that it can be estimated as a function of temperature by using modern molecular modeling techniques which treat the chain in full chemical detail.

A chain built using the Monte Carlo route in which the probability of a given bond setting is determined by the relative energy of that setting and the temperature will be just one possible conformation sequence amongst many. However, if many chains are built, it is possible to determine average parameters such as the mean square end-to-end length and the persistence length. In the latter case, there can be a difficulty in defining the direction of the first chemical unit, especially if that unit is at all complex. Coulter and Windle [11], in building chains from a common starting point with the first unit in a fixed orientation, defined the reference vector as the mean orientation of the chain, and the persistence length as the mean of the projected lengths of all the chains onto this vector. A further modification was incorporated to allow for the fact that inhomogeneous chains, for example, those consisting of flexible and rigid sequences, would give different apparent persistence lengths depending on whether the first unit of the chain, as built, was at the start of a flexible sequence or of a rigid sequence. The method was to repeat the model measurements starting at various different units of the chain at random, with the probability skewed in proportion to the length of the unit. The values were recorded for the chain as it was built, giving rise to plots such as

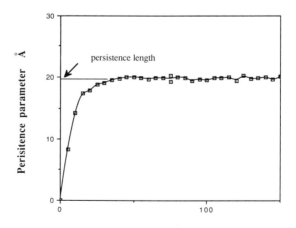

Number of rotatable units in chain

FIGURE 2.6. Plot of the persistence parameter, calculated as the average of many examples of molecule (5) at 300 K. The plot shows the parameter determined as a function of the length of the chain built. The plateau region of the curve occurs at the persistence length, q, of the long-chain molecule.

that shown in Fig. 2.6. The rising portion of the curve represents chains shorter than the persistence length, while the plateau gives the true persistence length. Modeling of a range of aromatic copolyester systems for which the temperature of the nematic–isotropic transition has been measured, showed that liquid crystallinity was first observed when the ratio of persistence length to diameter (the persistence ratio, q/d) approach 5. Examples of this comparison are shown in Fig. 2.7.

The predictive potential of this critical ratio was tested against the experi-

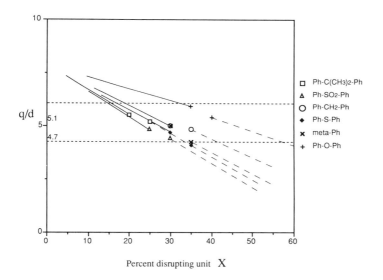

FIGURE 2.7. A comparison of the persistence lengths calculated for a range of random copolyesters as a function of content of non-mesogenic, disrupting units [11]. The full curves correspond to the range of compositions for which the polymers are observed experimentally to be mesogenic, while the dashed lines represent compositions for which the polymers are observed to be isotropic. Note that the transition corresponds to a q/d ratio of between 4.5 and 5. Experimental data from [27].

mentally determined mesogenic–isotropic transition temperatures of molecules (6) and (8). At the transition temperature of 400 °C of molecule (6) with $x = 0.6$, the persistence ratio was 4.82, while for (8) with $x = 0.2$ at 330 °C, it was 4.6 [12]. On the basis of a critical ratio of 5, the liquid crystalline–isotropic transition temperature of polyhydroxybenzoic acid, (1), was predicted to be 730 °C [12], which, of course, cannot be reached in practice on account of thermal degradation.

2.2.4 Kuhn Chains

For every set of polymer chains, each with total length (or contour length, L_c), there is an equivalent set of Kuhn chains [13] consisting of n_k freely jointed rods each of length l_k, which have not only the same contour length (as the real molecule), but also the same mean square end-to-end length, $\langle L^2 \rangle$ (Fig. 2.8). Considering a molecule in terms of its equivalent Kuhn chain, consisting of linked rigid rods, raises the question as to whether there can be a simple aspect ratio (length/diameter) criterion for liquid crystallinity similar to that implied by the Flory lattice model for unconnected rods. Flory suggested that his model could be applicable to the Kuhn chain [14], in that the well-established criterion of a critical aspect ratio of a short rod molecule should also be relevant to the rigid links of an otherwise flexible chain. The implication that the critical persistence ratio of a worm-like chain will correspond to the condition where its equivalent Kuhn chain will have links of the critical ratio of 6.42, or thereabouts, immediately runs into difficulty, for the

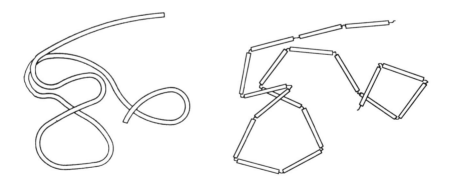

Worm-like chain   Equivalent Kuhn chain

FIGURE 2.8. Examples of a worm-like chain and its equivalent Kuhn chain. The chains each have the same contour length, the length of the rods in the Kuhn chain being chosen so that the mean of the squares of the end-to-end lengths are equal for each type of chain.

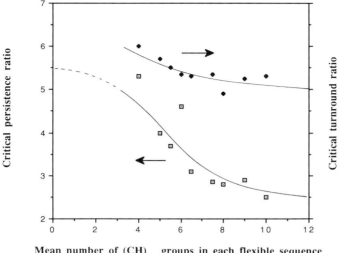

FIGURE 2.9. Plots of the critical persistence ratio, q/d, and the turn-round ratio, l_{t-r}/d, measured from models representing the experimentally observed liquid crystalline–isotropic transition temperature, against the amount of flexible sequence chain inserted between the rigid units expressed as $(x + y)/2$ (9). Data from [16].

persistence length of a Kuhn chain, and thus of a real chain to which it is equivalent, is $l_k/2$.

Hence, if modeling such as that above, suggests a critical persistence ratio for liquid crystallinity of 5, it implies that an equivalent Kuhn chain would have links of critical aspect ratio of about 10, not 6.4. Matheson and Flory [15], have suggested that a flexible sequence of sufficient length between rigid segments is equivalent to a Kuhn chain in which the flexible part behaves as a freely rotatable joint of zero volume. Recent modeling work [12], has shown that the critical persistence ratio decreases as the chain is made more Kuhn-like through the introduction of flexible sequences of increasing length between the rigid units. Figure 2.9 shows this decrease as a function of the mean number of CH_2 units in the flexible sequences of molecule (9). The authors have shown that a different parameter, itself also a measure of straightness, maintains a much more constant critical value irrespective of the degree to which the chain is Kuhn-like. The parameter, which is called the turn-round length l_{t-r}, is the mean distance along the chain in the direction of the first link before it first reverses direction. For a worm-like chain it is equal to the persistence length, but for the Kuhn-like chain it is equal to twice the persistence length, i.e., the Kuhn length l_k. The ratio l_{t-r}/d is plotted on Figure 2.9 also. It can be seen to have a value of between 5 and 6, over the range of flexible sequence concentrations investigated.

$$\left[-(CH_2)_x-O-\underset{}{\bigcirc}-\overset{O}{\underset{\|}{C}}-O-\underset{}{\bigcirc}-O-(CH_2)_y-O-\underset{}{\bigcirc}-\overset{O}{\underset{\|}{C}}-O- \right]_n$$

(9)

2.2.5   Change in Chain Parameters in the Nematic Phase

It should be emphasized that the critical values of the persistence ratio, q/d, derived from calculations made on isolated chains will, by virtue of the Flory theorem, be quite relevant to the isotropic phase. However, once the polymer undergoes a transition to the nematic phase, the chain experiencing the mutual field of its neighboring molecules will be extended along the director axis. It is important to appreciate that, for a nematic, the orienting field is quadrupolar, which means that orientations of chain units both parallel and antiparallel with the director are equally favored, while those normal to it will occur with reduced probability. The orienting field thus affects the energy of a unit as $-\cos^2 \phi$ where ϕ is the angle between the director and the chain axis of the unit. Figure 2.10(a) shows a two-dimensional representation of the influence of a quadrupolar field on a Kuhn chain of freely jointed rods. Note how the chain is not fully extended although, as far as the rods are concerned, it is aligned. The extensional strain associated with this type of orientation corresponds, in the case of rods of high axial ratio, to only 100%. However, a worm-like molecule has no flexible joints at which to fold sharply, and to do so would be at considerable energy cost. Thus, in respect to a response to a quadrupolar field, the Kuhn chain cannot be considered equivalent to its worm-like counterpart. Furthermore, the smoothly curved folds, as depicted in Fig. 2.10(b), will also have a high energy as they result in a significant length of chain being unfavorably oriented. There is thus a driving force for the elimination of the smooth folds, or at least a reduction to a level where their probability corresponds to a balance between their excess energy (in the quadrupolar field) and kT. The end-to-end length should thus increase markedly with the field as the fold density reduces and the molecule extends. Vroege and Odijk [17] have shown that the increase of mean square end-to-end length is exponential with the increasing internal nematic field, which in a lyotropic system would correspond to the increasing polymer concentration.

Flexible sequences inserted into semirigid molecules will provide opportunities for sharp folds in the nematic phase at low energy cost. Such molecules will thus have the capability to respond to the nematic field rather as Kuhn chains, with only a modest change in overall dimensions. It is also possible that kinked units in otherwise linear molecules will provide favorable sites for sharp folds. The significance of sharp folds extends beyond their

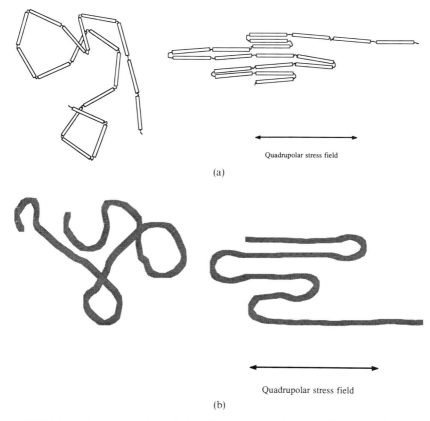

FIGURE 2.10. Representation of the effect of a quadrupolar field on: (a) a Kuhn chain; and (b) a worm-like chain.

influence on the change in molecular dimensions at an isotropic–nematic transition, and they influence the energy of splay distortions in polymer melts as they effectively act as chain ends in compensating for the density gradient associated with this particular distortion. It has also been suggested by Warner [18] that movement of sharp folds along a chain could provide a mechanism by which a dipolar (dc) field could subsequently align the molecules in a polar sense.

2.2.6   Microstructural Considerations Involving Nematic Polymers

Moving up in scale, the next important level of structure for nematic polymers involves what is often known as the optical microstructure. For small

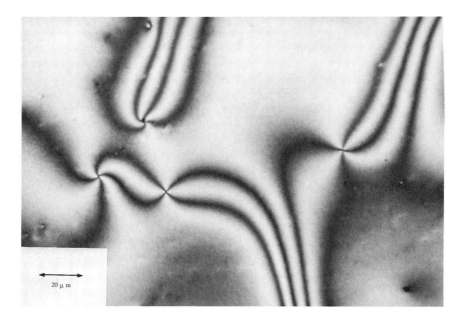

FIGURE 2.11. Example of a Schlieren texture in a thin sample of liquid crystalline polymer which contains flexible links in the backbone (6). The texture is revealed through crossed polars.

molecule nematics the microstructures are quite well characterized. Probably the best known is the Schlieren texture, such as that shown in Fig. 2.11. The contrast, as seen between crossed polars in the transmitted light microscope, reveals the gradual (on a molecular scale) variation in the chain director with position, as with any general series of meandering streamlines, that there are singularities where the change in orientation with position is very rapid. In three dimensions these disclinations can be either line or point defects. In the absence of crossed polars they appear as dark lines or points, as the rapidly changing refractive index at their cores leads to phase contrast. The lines, or *nemata*, seen under these conditions give the nematic phase its name.

The characteristic microstructures of the nematic phase are seen in polymeric as well as small-molecule materials. However, thermotropic polymers, based on comparatively stiff chains without flexible spacers, often show a microstructure on a much finer scale than the typical Schlieren texture. Such textures (Fig. 2.12) often tax the resolution of the light microscope and reveal few specific features. Observations on random copolyesters, e.g., [19], have indicated that although the textures appear mobile in the melt they do not coarsen, at least over a period of several hours. On the other hand, similar textures formed in nematics based on molecules with flexible spacers do

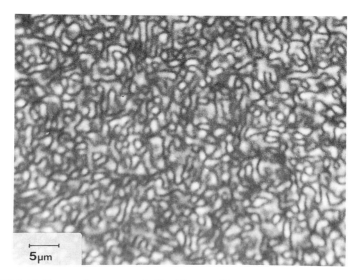

FIGURE 2.12. Fine-scale texture of a liquid crystalline polymer (5) in which the chain does not have any particularly flexible units. The sample is of the order of 3 μm thick and is seen through crossed polars.

appear to relax with time to give a more conventional threaded texture, either when induced by shear [20] or by sample preparation [21].

The energetic basis for the development of textures are the three elastic moduli, K_1, K_2, and K_3, associated, respectively, with splay, twist, and bend distortions. The existence of profoundly different types of textures in some polymer systems prompts the question as to whether there is any significant difference in the values of these constants compared with the small molecule systems.

Odijk [22] has used a scaling approach to derive expressions for the dimensionless elastic constants of systems of rigid rods for each of the three distortions. After Taratuta et al. [23], Odijk's equations can be recast as

$$K_1 \approx (\tfrac{7}{8}\pi)\phi(L/d)(kT/d),$$

$$K_2 \approx \tfrac{1}{3}K_1 \quad \text{(a result first shown by Straley [24]),}$$

$$K_3 \approx (\tfrac{4}{3}\pi^2)\phi^3(L/d)^3(kT/d).$$

The term (kT/d) redimensionalizes the constant to give the units of force, while $\phi = 1$ for thermotropic polymers.

The dependence of the bend constant, K_3, is more steep, increasing as L^3. The reason for this rapid increase is that the rods are deemed to be stiff, so that bending can only be accommodated between their ends and the longer the length the lower the frequency of ends.

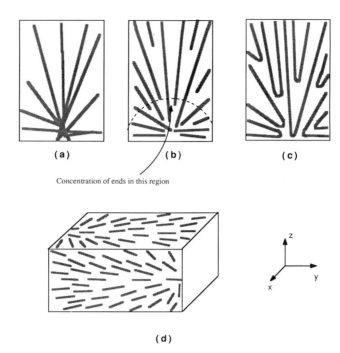

FIGURE 2.13. Splay distortion showing; (a) the difficulty with long rods in avoiding overlap (bottom) and filling space (top) and the need for special positioning of the ends; (b) to obtain a more even distribution; (c) shows that hairpin folds can play an equivalent role to chain ends in compensating for splay; while (d) illustrates the organization of splay–splay compensation.

There is, however, another important contribution to splay energy in polymeric systems, as long rods set at an angle to each other in a plane will gradually move further and further apart, breaking the intermolecular bonds and tending to create a void which would have very high energy (Fig. 2.13(a)). The only way in which the tendency to void can be compensated is for another rod to "start" when a void is able to accommodate it, and so on (Fig. 2.13(b)). This rather special positioning of rods has an entropy price which contributes to the splay energy. Although the magnitude of the additional splay energy, often referred to as ΔK, was first estimated by de Gennes [25], a calculation more relevant to the model depicted is that of Meher [26] which indicates that [23]

$$\Delta K_1 \approx (4/\pi)\phi(L/d)(kT/d),$$

and is thus, like K_1, linearly dependent on L.

The issue which now arises is the value of the elastic moduli in the case of polymer chains which are not completely rigid. Odijk [22] extended his scaling argument to propose that

$$K_2 \approx \phi^{1/3}(q/d)^{1/3}(kT/d)$$

and

$$K_3 \approx \phi(q/d)(kT/d).$$

Note that in each case dependence on the length of the rigid rod molecule is replaced by dependence on the cube root of the length parameter with the length replaced by the persistence length. Hence, as long as the chain length, L, is significantly larger than the persistence length, q, it will not affect either K_2 or K_3. In this case, the persistence length corresponds to that of an isolated molecule at temperature T [23], although the actual value of q in the nematic phase will of course be higher.

As before, we might invoke $K_1 \approx 3K_2$ to give the splay constant, however, the ΔK_1 parameter, dependent on the low entropy associated with the special position of the chain ends to compensate for the density gradient, will remain proportional to the chain length. Thus as the chain length exceeds the persistence length, the values of K_2 and K_3 will tend to constant values while the splay energy, K_1, will increase monotonically with L. For semiflexible chains, with a high molecular weight, the splay constant is thus likely to exceed the other two moduli by a significant margin. The relative magnitude of the elastic constants is represented in Fig. 2.14. There is also the possibility that a structure which consists of extensive splay distortion will require the segre-

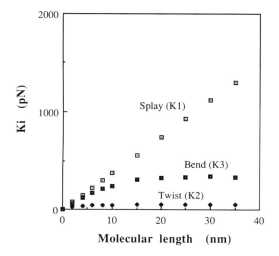

FIGURE 2.14. Diagram illustrating the predicted effect of chain length on the elastic constants: splay K_1; twist K_2; and bend K_3. The values are calculated, using the relations given above, for polymer (5) at 673 K. When the length of the chain exceeds the persistence length, q, the values of K_2 and K_3 become independent of length while K_1 increases monotonically.

gation of a greater number of chain ends than is available. In such circumstances the overall level of splay distortion would be limited.

There are, however, two other mechanisms by which the density gradient associated with splay of long chains can be compensated. If the mesogenic chain is sufficiently flexible, then the energy associated with sharp folds will not be excessive (see discussion in Section 2.2.5 above), and these will be able to act effectively as chain ends and provide the required compensation (Fig. 2.13(c)). The higher conformational energy of the folds will provide yet a further contribution to the splay energy. It follows that the occurrence of sharp folds will be particularly favorable in molecules containing short flexible sequences in an otherwise semirigid backbone. Alternatively, there is the possibility of what is known as splay–splay compensation. In this case, splay in one plane, about, what for a limited field of view could be described as a mean director, can be compensated by splay of the opposite sense in a plane rotated 90° about the vector. Such compensation is depicted in Fig. 2.13(d). In the case of an infinitely long chain, these two mechanisms are the only ones available to permit splay.

2.3   Levels of Order in Mesophases of Main-Chain Liquid Polymers

2.3.1   Classification for Polymers

The purpose of this section is to explore the influence of various levels of positional order on the architecture of liquid crystalline polymer phases. In the case of "conventional" low-molar-mass liquid crystals, the different types of positional order form the various subsets of the smectic class. These are numerous, and have been treated in detail by Gray and Goodby [28] and reviewed more recently by Leadbetter [29]. For these materials, smecticity stems from the segregation of the ends of the rod molecules into layers, which are either normal to the director (S_A) or at particular angle to it (S_C). Other smectic subclasses involve crystalline order between the molecules within the layers, where lateral register *between* the layers is either nonexistent, as in S_B or S_E, or present in some degree as in the so-called hexatic variants.

In main-chain thermotropic *polymers*, the segregation of molecule ends would involve much larger distances, and in any case the molecular weight distribution, which is a normal characteristic of the polymeric state, would seemingly preclude the formation of the well-ordered layers which typify smectics in the Friedelian sense. Nevertheless, there is an increasing body of literature which attests to the occurrence of limited levels of positional order in main-chain liquid crystalline polymers, although the classification established for low-molar-mass compounds does not lend itself readily to the polymer scene. Here we set out a broad classification which is relevant to polymers, and relate the various possibilities to different observations. Cholesterics are not included at this stage, and no examples are drawn from lyotropic systems. It should also be emphasized that the various levels of positional order are not subdivided according to the same scheme as that used for low-molar-mass smectic subclasses, although, a number of polymeric mesophases have a level of positional order which approaches that of a three-dimensional crystal. Indeed, it is not easy to decide just where the classification as a liquid crystal runs out, and for this reason the full range of states from nematic to three-dimensional crystal, is explored. Correspondingly, the term *mesophase* is preferred to *liquid crystal*.

2.3.2   Aspects of Chain Structure

Periodicity Within the Chains and Longitudinal Register

In order to simplify as far as possible the treatment of molecular organization, a number of assumptions are made about the polymer chains. They are assumed at this stage to be homopolymers, and in the case of phases where the lateral packing of the molecules is crystalline, to adopt a regular conformation and thus have axial periodicity. The possibility of neighboring chains being either *in* or *out* of longitudinal register is a significant factor in the classification and is illustrated in Fig. 2.15(a). In many of the diagrams which follow the chains are shown end-on, being represented as schematic cross sections. In such diagrams the presence of longitudinal register is indicated by short dashed lines drawn between the sections (Fig. 2.15(b)).

In the case where there is only short-range positional order between the molecules, such as in a nematic phase, the question of the degree to which a chain is periodic along its axis becomes more open. Very rigid molecules, such as poly-*para*-phenylenbenzobisoxazole (7), would be expected to hold their axial periodicity in the mesophase over comparatively large distances. However, such materials have very high crystal melting points and their liquid crystalline phases are normally only seen as lyotropic solutions. The molecules of thermotropic polymers, where they are homopolymers, such as polyhydroxybenzoic acid (1), will show less than exact periodicity in the mesophase.

This means that there will be a degree of one-dimensional paracrystalline disorder along the chain which will not be perfectly straight. The distance over which order is maintained, which is related to the persistence length (in the mesophase), is essentially derivative, and depends on the nature of the molecule and its state of packing (crystalline or liquid crystalline). It does not in itself lead to separate classifications although it will be discussed again below.

The situation is different in the case of copolymers. Typically, thermotropic block copolymers consist of sequences of straight, comparatively rigid units, interspaced with flexible sequences such as alkane chains as in (9). In this case, there are two levels of longitudinal register. One of these is the

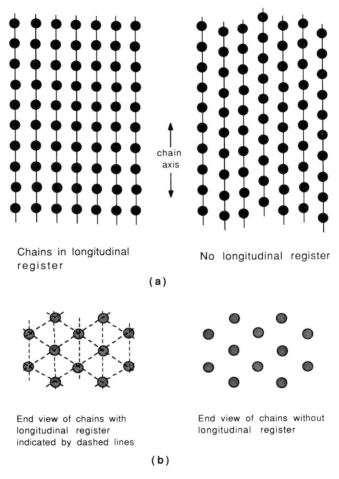

Chains in longitudinal
register

No longitudinal register

(a)

End view of chains with
longitudinal register
indicated by dashed lines

End view of chains without
longitudinal register

(b)

FIGURE 2.15. Illustration of longitudinal register between parallel chain molecules. (a) Two structures, with and without register, looking normal to the chain molecules which are vertical on the page. Note that the register is between the individual units along the chain and is not associated with any correlation in the position of the chain ends. (b) Two cross-sectional views of packed chain molecules showing the use of dashed lines to indicate longitudinal register when this occurs in the "out-of-page" direction.

"course level" in which the different types of sequence are simply segregated so the rigid, usually aromatic parts are neighbors, as are the flexible sequences. Additionally, there is the question as to whether the register is exact on the scale of the length of the individual chemical units making up the sequences. However, for the purpose of the classification below, simple segregation of chemically different sequences which are organized regularly along the chain can be viewed as a case of longitudinal register. Longitudinal register between *random* copolymer chains is only possible for a limited fraction of the polymer, there being a number of regimes which make this organization possible.

Chain Cross Section and Rotational Correlation

The cross sections of chains are highly variable depending on their chemical make-up. Some, such as PTFE, are quite good approximations to cylinders and comparatively smooth, while chains such as those based on planar aromatic groups have quite anisotropic cross sections. In cases where additional side groups prevent free rotation about the chain axis the form becomes ribbon-like or lamellar [30]. Molecules which are rod-like are referred to as *calamitic* (from the Greek for *reed*) when they are approximately cylindrical,

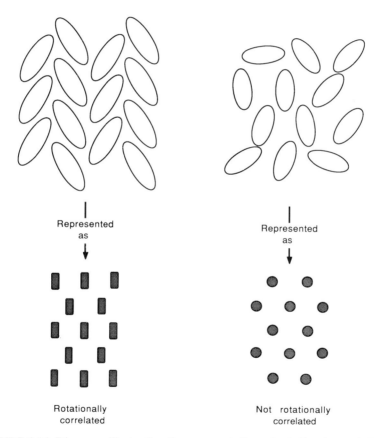

FIGURE 2.16. Diagrams illustrating the representation of rotational correlation between the molecules about the chain axis in cross-sectional view. Where rotational correlation is present, the chain cross sections are drawn as rectangles, all in the same orientation. This does not imply that the correlation of the actual molecules is necessarily as geometrically simple.

and *sanidic* (from the Greek for *plank*) when ribbon-like. Where the cross section of the chain is not symmetrical the question arises of rotational correlation, which may take the form of the mutual alignment of the sections about the chain axis, or of more complicated correlations such as parquet flooring-type arrangements. In the cross-sectional diagrams which follow, the presence of rotational correlation is indicated by rectangular sections which are mutually aligned. Packing arrangements in which there is no rotational correlation are represented by the use of circular cross sections, even though the individual molecules themselves may be far from cylindrical (Fig. 2.16).

2.3.3   Basis for Classification

The two structures which mark the limits of the classification are a full three-dimensional crystal on one hand and a simple nematic on the other. The approach used is to start with the crystal and progressively remove aspects of the order in a logical sequence. An advantage of starting with the crystal is that its unit cell is useful as a basis for orienting the initial disordering elements, although the drawback perhaps is that some of the more important liquid crystalline structures are only reached in the later stages of the process.

Two distinct disordering processes can be identified. The first is the removal of shear correlation between a particular set of adjacent planes, and in a particular direction. This is illustrated in Fig. 2.17(a). Using the nomenclature similar to that applied to (say) a strain tensor, the direction of the disorder is defined by the direction of the disordering shear movement, say $_2$, and the orientation of the shear planes which contain the $_2$ and $_3$ axes, as $_1$. Hence, if the order element is represented by O_{21}, then its loss is represented as $\mathcal{O}_{21}$. A second aspect of disorder has been described by Helfrich [31]. It is the loss of order in a direction normal to a set of planes, so that the periodicity of their spacing is lost, and their long-range positional order degrades to short-range positional order. It is referred to here as paracrystalline disordering. In one dimension it is disorder based on a distribution of lattice spacings which means that correlation is lost after a certain distance [32]. It is equally applicable to the spacing of a single set of planes in three dimensions, although matters become more complicated for the loss of order in more than one dimension. It is illustrated in Fig. 2.17(b). In notational terms the loss of periodicity in the spacing of the $_1$ planes is represented as a disordering motion in direction $_1$, and thus the order element which is lost is O_{11}.

The order elements can be organized as a matrix

$$\begin{bmatrix} O_{11} & O_{12} & O_{13} \\ O_{21} & O_{22} & O_{23} \\ O_{31} & O_{32} & O_{33} \end{bmatrix}$$

The three rows represent the three directions in which the disordering translations occur, while the three columns represent the planes on which the

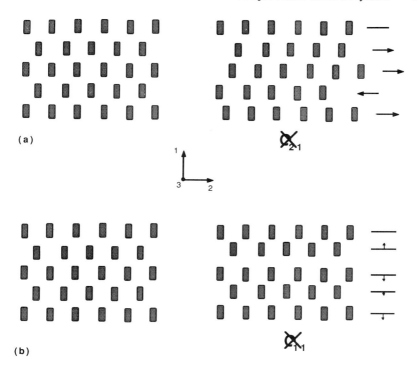

FIGURE 2.17. The two types of disordering operation: (a) the loss of shear correlation between planes; and (b) the loss of periodic spacing (long-range correlation) between planes.

disordering acts. The first of the subscripts of a disordered element thus refers to the direction of the disorder, the second to the orientation of the planes involved. Increasing stages of shear disorder will systematically eliminate the "off-diagonal" terms, while paracrystalline disordering will eliminate the diagonal terms.

The value of formalizing the stages of disorder using the matrix arrangement above is that it enables the process to be followed in a systematic way and above all will ensure that no conditions of partial order are omitted.

2.3.4   Mesophases Formed by Successive Shear Disorder of the Crystal

Order Retained by Virtue of Chain Rigidity

First consider axially periodic chains which, in the notation used, lie along the $_3$ axis. When the chains are packed on a crystal lattice, we assume that their axial periodicity is maintained and paracrystalline disordering along this direction is not possible. Also, the order elements O_{13} and O_{23} are considered to be immune from the various incremental components of shear disorder on planes which do not contain the chains.

Loss of Longitudinal Register in Rotationally Correlated Structures: Crystal, Sheet Crystal, and Sanidic 2 Classes

The top row of structures in Fig. 2.18 represents full crystalline order and two initial stages of shear disorder. Remembering that the dashed lines represent longitudinal register between the rotationally correlated chains, the first disordering procedure is randomizing shear movements on the $_2$ planes (containing the axes $_1$ and $_3$) in the $_3$ (chain) direction which is out of the page. This disordering component represents the loss of the shear order element O_{32} or, equivalently, O_{31}. The resultant structure is a form of sheet crystal, and it is reminiscent of structures seen in polymer crystals such as Nylon 66, where the intact sheets consist of hydrogen bonded chains, although the randomization is almost certainly not complete in the case of Nylon.

In the third example on the top row of Fig. 2.18 all longitudinal register has been removed by the randomization of both O_{3i} shear order elements. The loss of all longitudinal register produces a structure with the chains packed on a two-dimensional lattice and with rotational correlation about the chain axes. It is classified as a *sanidic*. It is the most ordered example of this class of structure and is distinguished here from its less well-ordered counterparts by the designation *sanidic 2*.

Loss of Longitudinal Register in the Absence of Rotational Correlation: Plastic Crystal and Canonic Classes

The second row of Fig. 2.18 shows equivalent structures in which rotational correlation is absent. The loss of longitudinal register on only one set of planes is not consistent with the rotational symmetry of the chains. Such a structure, although depictable, is thus omitted from the classification (bottom-center). The equivalent of the full crystal structure, although without rotational correlation, is an example of a *plastic crystal* in which there is full three-dimensional order but one component of rotational disorder. Moving now to the case of no longitudinal register (bottom-right), we have the rotationally symmetric equivalent of the sanidic 2 which corresponds to the crystalline packing of completely smooth rods. It fits well with the classification of *canonic* (Greek *rod* or *rule*) introduced by Frank [33] in the context of columnar phases of discotic liquid crystals.

Intermediate Degrees of Rotational Correlation

Main-chain aromatic polyesters, such as polyhydroxybenzoic acid, polyhydroxynaphthoic acid, and polyphenylene terephthalate, all show high temperature crystalline forms which at first sight appear to show crystalline hexagonal packing and full longitudinal register. For example, the wide angle

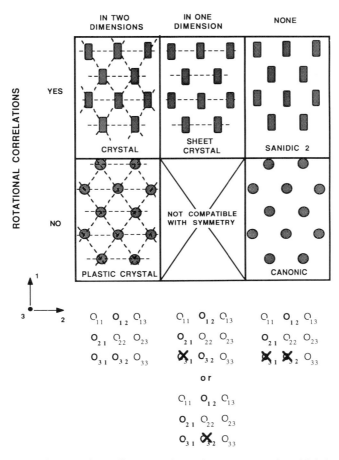

FIGURE 2.18. The top three diagrams show the two stages in which longitudinal register can be lost between chains in a crystal. The two-dimensional lattice describing the packing remains intact after each stage. The bottom two diagrams represent crystals without rotational correlation, before and after the loss of longitudinal register. Note how the loss of longitudinal register in one dimension is not compatible with the increased symmetry.

X-ray diffraction scan of polyhydroxybenzoic acid at 400 °C shown in Fig. 2.19 [34] has a strong single interchain peak indicating the high level of symmetry, with well-defined 002, 004, and 006 meridional peaks which mark the longitudinal register between the chains. However, closer examination shows that there are clear 211 and 213 peaks which should be forbidden under simple hexagonal symmetry, while exact measurement of all the peak positions shows that the hexagonal cell is very slightly distorted toward

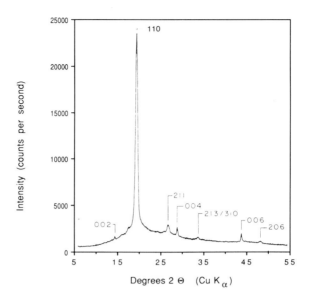

FIGURE 2.19. Wide angle x ray diffraction scan of polyhydroxybenzoic acid at 400 °C. There is one strong hk0 peak characteristic of a hexagonal cell. However, the 211 and 213 peaks are also present which indicates that the contents of the cell do not have hexagonal symmetry There is a significant degree of rotational correlation between the chains which accounts for this apparent incompatibility.

orthorhombic, the structure of the low temperature crystal phase. The evidence is that there is still short-range rotational correlation between adjacent chains which results from the interlocking of neighboring aromatic rings. The rotations are constantly changing due to thermal motion, and the structure has been described as a *molecular gearbox*. The correlations mean that the contents of any unit cell do not have hexagonal symmetry although the cell shape itself does, or nearly so.

The possibility of having to introduce yet another class of order, short-range rotational correlation into the overall classification is highlighted by these studies of plastic crystals. But more significantly, it underlines the fact that where less than full three-dimensional positional order is present, there can be a graduation in rotational correlation from no order through short-range order to long-range order.

Protocol for Successive Shear Disordering Operations

If the chains themselves are to remain periodic and straight, there are four shear elements available for disordering: O_{31}, O_{32}, O_{12}, and O_{21}. However, once an element has disordered, let us say the element O_{12}, then it is difficult to visualize any subsequent shear disordering on other planes which do not contain the direction of the first disordering translations. Here this means the planes $_1$. Hence, we cannot consider, for example, the subsequent loss of order elements O_{21} or O_{31}, as the planes themselves will have been scrambled by the first operation. The consequence is that, for disorder of the shear type only, the combined disorder of O_{12} and O_{21} is not permitted. Later, it will be shown that the disordering of these two elements can exist in combination with the loss of O_{11} and O_{22}. Furthermore, disordering operations on successive elements which do not have either the first or the second subscript in common, will also result in a third loss at the same time. For example, the removal of O_{31} followed by the removal of O_{12} will also destroy O_{32} or, in other words, we cannot have the loss of O_{31} and O_{12} without the loss of O_{32} as well. The stages in the loss of shear order can be followed if the disordering operations occur in a sequence which means that the two sharing the common direction occur first.

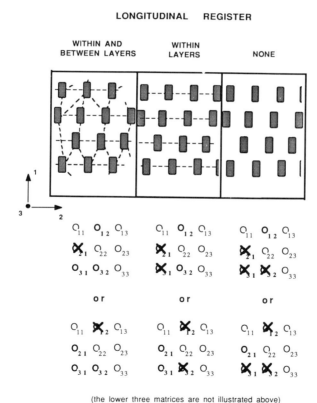

FIGURE 2.20. Three structures in which the shear correlation between layers of chains (horizontal on page) have been eliminated. The diagrams represent: full longitudinal register; longitudinal register within the horizontal layers; and no register.

The disording components involving the loss of longitudinal register expressed by O_{31}, O_{32} and the combination $O_{31}O_{32}$ have already been considered in Fig. 2.18 above. The next step is to consider the classes of structures resulting from the loss of the other elements. The loss of O_{12} or its equivalent O_{21} leads to a structure which still maintains full longitudinal register, even though it has been disordered on one set of planes by lateral shearing in a direction normal to the chain axis. It is illustrated in the first diagram of Fig. 2.20. The disordering of the equivalent combinations $O_{12}O_{32}$ or $O_{21}O_{31}$ leads to a series of layers containing the chain axis, which have two-dimensional crystalline order within them but no shear register whatsoever between them. The likelihood of such a structure existing has been questioned by Helfrich [31,35]. He argues that a molecularly thin crystal layer is unlikely to be stable at temperatures capable of providing totally thermal disorder between the layers, and that layer unregistering and two-dimensional melting of the layers go hand-in-hand. Although it is possible to envisage sheets of polymer molecules sufficiently strongly bonded so as to retain their integrity, even though all shear correlation between them is lost, as yet, no such structure has been clearly demonstrated, and there the matter must rest. If the existence of unregistered, two-dimensional sheets of chains is in doubt, then the third structure of Fig. 2.20, in which the longitudinal register in the layers is also absent, seems even less likely. It is, however, conceivable and results from the sequential loss of the order elements, $O_{31}O_{32}O_{12}$ or the equivalent $O_{32}O_{31}O_{21}$.

The loss of register on a single set of planes containing the chain axis, in each of the three structures of Fig. 2.20, is not compatible with axial symmetry and thus would not occur for chains which are not rotationally correlated.

Summary of Types of Shear Disorder

A modest regrouping of the various types of mesophase, obtained as a result of applying elements of shear disorder, enables the classification to be reviewed quite simply. The symbolism used in this and subsequent summaries, consists of a simplified matrix which is blocked out in stages to indicate the order elements lost. It should be noted that disordering elements sharing a common plane occupy the same column while those having the direction of disorder in common share the same row. Also, the final column which would represent disorder on the plane not containing the chain axes is not included in this classification. One reason for keeping account in this way is to ensure that no possible combinations of disorder are overlooked.

Summaries (i)–(iv)

(i) Full crystalline order

(ii) Loss of longitudinal shear register between all chains

(iii) Loss of shear register (longitudinal and/or lateral) between a single set of planes parallel to the polymer chains

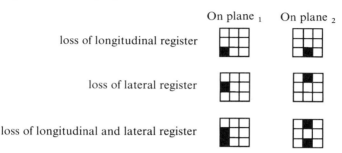

(iv) Loss of shear register (longitudinal and lateral) between a single set of planes within which the longitudinal register is also absent

 These four combinations are not permitted as they contain the loss of both O_{12} and O_{21} order elements (the top two diagonally opposed pairs)

 These two combinations are not possible as they will both generate a third disordering element making them the same as the two permitted ones above

Number Check. There are four different disordering shear elements and thus $2^4 (= 16)$ combinations. The sixteen combinations are shown above using the reduced scheme although six of them are not viable.

2.3.5 Mesophases Formed by Paracrystalline Disorder in One Lateral Dimension

Paracrystalline Disorder on a Single Set of Planes

The loss of the order represented by O_{11}, as a single operation, will have the effect of randomizing the spacings of a single set of planes, $_1$, without in-

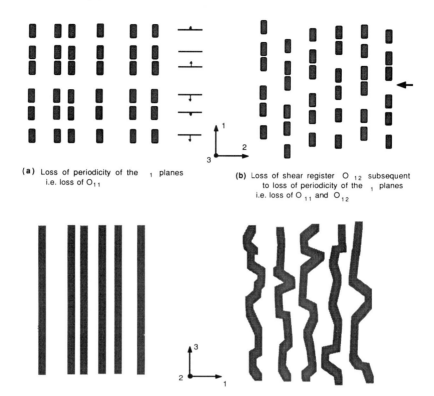

(a) Loss of periodicity of the $_1$ planes i.e. loss of O_{11}

(b) Loss of shear register O_{12} subsequent to loss of periodicity of the $_1$ planes i.e. loss of O_{11} and O_{12}

(c)   View of the chains shown in cross section in (b) from the direction of the horizontal arrow (note new axes).   On the left the chains are rigid and there is no loss of the shear order O_{13} . The chains on the right are not completely rigid and there is partial loss of O_{13} .

FIGURE 2.21. (a) Structure in which paracrystalline disorder has been introduced to convert the spacing of one set of layers (horizontal on page) from long-range order (periodic) to short-range order (aperiodic). (b) The loss of shear order between the layers which are vertical on the page, further to the introduction of paracrystalline disorder in (a). (c) Side view of the rod molecules in one of these (vertical) layers. In the left diagram, the disruption in spacing is assumed to be constant along the lengths of the rods, while in the right, there is limited shear disorder parallel to the planes normal to the molecules which leads to disturbance of the molecular trajectory.

fluencing the shear register between these planes in any way, as illustrated in Fig. 2.21(a) (and in Fig. 2.17(b)). The most useful next step is to add those combinations of shear disorder which do not disturb the planarity of the planes already disordered with respect to their spacing. We can generate a series of possible arrangements in which the register between the $_1$ planes is lost in either the $_3$ direction (losing O_{31} the longitudinal register) or the $_2$ direction (loosing O_{21}), or both. We could also remove O_{32} longitudinal register before O_{11}, or the sequences $O_{11}O_{31}$ or $O_{11}O_{21}O_{31}$ (N.B. the loss of

$$\begin{array}{ccc} \times_{11} & O_{12} & C_{13} \\ O_{21} & C_{22} & C_{23} \\ O_{31} & O_{32} & O_{33} \end{array} \qquad \begin{array}{ccc} \times_{11} & O_{12} & C_{13} \\ \times_{21} & O_{22} & C_{23} \\ O_{31} & O_{32} & O_{33} \end{array} \qquad \begin{array}{ccc} \times_{11} & O_{12} & C_{13} \\ O_{21} & O_{22} & O_{23} \\ \times_{31} & O_{32} & O_{33} \end{array}$$

$$\begin{array}{ccc} \times_{11} & O_{12} & C_{13} \\ \times_{21} & O_{22} & O_{23} \\ \times_{31} & O_{32} & O_{33} \end{array} \qquad \begin{array}{ccc} \times_{11} & O_{12} & C_{13} \\ O_{21} & O_{22} & O_{23} \\ O_{31} & \times_{32} & O_{33} \end{array} \qquad \begin{array}{ccc} \times_{11} & O_{12} & C_{13} \\ O_{21} & O_{22} & O_{23} \\ \times_{31} & \times_{32} & O_{33} \end{array}$$

$$\begin{array}{ccc} \times_{11} & O_{12} & C_{13} \\ \times_{21} & O_{22} & O_{23} \\ \times_{31} & \times_{32} & O_{33} \end{array}$$

In addition to these seven structures there will be the same number of equivalent structures based on the loss of O_{22}

FIGURE 2.22. The disordering combination which lead to one set of planes having aperiodic spacings, with various levels of shear disorder between these layers combined with the absence of longitudinal register in the planes. These structures are thought unlikely to occur and are not drawn.

$O_{32}O_{11}O_{21}$ also eliminates O_{31}). In these cases all longitudinal register would be lost within, as well as between, the unevenly spaced planes. The possible combinations of the above types of shear disorder with a single component of paracrystalline disorder are represented in Fig. 2.22 and in summary form below.

Summary (v)

(v) Loss of positional order in the spacing of a single set of planes (paracrystalline disorder) plus elements of shear disorder between the plane and the loss of longitudinal register within the planes (columns marked *)

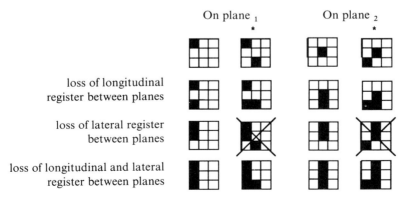

It is sensible at this stage to question the realism of these structures. For in each case disorder has been introduced into the spacing of a set of planes, at a level which will convert long-range order into short-range order, without any loss in planarity of the planes themselves. Whereas the reality of losing shear order between sheets which retain their planarity was questioned in the previous section, we are now faced with the even less likely arrangement of planes which melt, as far as their spacing is concerned, while maintaining their perfection as planes. While such a picture may be reasonable when considering discrete planar, covalently bonded molecules such as form discotic phases, it is not sensible if the planar entities are sheets of chain molecules, the integrity of which depends on comparatively weak intermolecular bonds.

Coupling of Paracrystalline Disorder with Shear Disorder to Give Translational Melting

The introduction of paracrystalline disorder in one dimension is more realistic if it is considered in relation to lattice rows rather than to a set of planes.

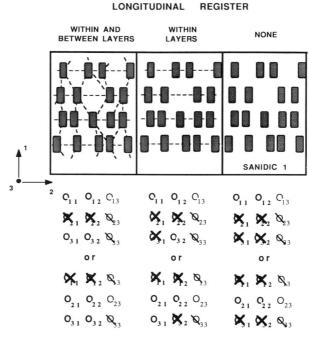

FIGURE 2.23. Three structures based on one paracrystalline disordering element (horizontal in the $_2$ direction in this case). The paracrystalline disorder is coupled with shear disorder with the disordering translations in the same direction. The result is that it is separate rows of chains which contain the aperiodicity, rather than the layers. The three diagrams represent stages in the loss of longitudinal register as before.

The spacing within each lattice row is disordered, to the level of short-range liquid-like order, and the register between the lattice points in successive rows is consequently lost. Organization of this type is shown in Fig. 2.21(b). It is the picture considered by Helfrich [31] who describes the process as *translational melting*. In terms of the formalism developed here, the implication is that translational disorder which removes O_{11} will at the same time remove O_{12} and O_{13}. However, this statement immediately begs a further issue, as we have so far considered the molecular chains to maintain their straightness and axial periodicity, implying that O_{13}, and also O_{23} and O_{33}, are immune to disorder. In the case of small molecule mesophases there are no special constraints on these three components of order. But, for polymeric phases the disorder is considered to involve the units of the chain rather than the complete chain itself, so that paracrystalline disorder normal to the molecule (such as the loss of O_{11}) can only involve loss of O_{13} if the chain no longer remains straight. Figure 2.21(c) illustrates this argument. A molecule in a mesophase will not be able to squirm to the extent of completely destroying shear register on planes normal to its axis, at least where separated by distances of the order of the axial repeat of the chain.

Once again we have a situation of intermediate order, loss of O_{11} will reduce O_{13}, and by implication O_{33} as a wandering chain will compromise its axial periodicity. However, the effects on O_{13} are in essence coupled to the loss of O_{11} and of limited extent, and will not be seen as playing a primary part in the classification. The limited loss of order elements on the plane normal to the chain axis is flagged by striking out the elements $O_{13}O_{23}O_{33}$ with a single bar, viz. $\not O_{13}$, as in Fig. 2.23 below, or as shaded patches in the summary maps such as those at the end of Sections 2.3.5 and 2.3.6.

Translational Melting in One (Lateral) Dimension

Figure 2.23 shows a sequence of structures generated by translational melting in the $_2$ direction eliminating O_{22}. In each case loss of O_{21} destroys the shear correlation between the horizontal layers. The sequence of three diagrams is for decreasing degrees of longitudinal register which is intact in the first structure, underlining that full longitudinal register, i.e., that between, as well as within the layers, is completely compatible with a single set of layers. In the structure shown in the middle diagram, O_{31} has been eliminated so that there is longitudinal register within but not between the layers. Both O_{31} and O_{32} are absent in the structure on the right where all longitudinal register has been eliminated. Each of these shear randomizations is best considered to be applied prior to the elimination of O_{22} and the consequential loss of O_{21}. It should be emphasized that we cannot consider longitudinal register between the defined layers, if the chains within the layers are not themselves in longitudinal register. In other words, the loss of $O_{31}O_{22}O_{21}$ would also remove O_{32}. Figure 2.23 also includes the disorder matrices for equivalent structures based on the loss of the element, O_{11}.

The summary (vi) below shows a wide range of combinations which are not permitted for the reasons already outlined.

Summary (vi)

(vi) Loss of positional order in spacing between single sets of planes, plus loss of lateral shear order on intersecting planes to give "translational melting"

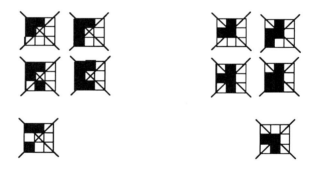

	On plane $_1$	On plane $_2$
Loss of spacing register		
plus: loss of lateral shear register on intersecting shear planes		
or		
plus: loss of lateral and longitudinal shear register on intersecting shear planes		
or		
plus: loss of shear register as above and loss of all longitudinal register		

The combinations below do not occur because they contain either $_{12}$, $_{21}$ pairs, or $_{21}$, $_{32}$ pairs without $_{31}$ or $_{31}$, $_{12}$ pairs without $_{32}$ or some combination of these.

The Sanidic 1 Structure

The *sanidic 1* structure with no periodicity of chain packing within the single set of layers, and no longitudinal register either within them or between them, was the first sanidic phase to be recognized [36]. The molecule was (10), and the wide-angle x ray diffraction patterns showed sharp peaks corresponding to the width of the molecules and diffuse maxima at higher angles, corresponding to the disordered packing in the direction of the thickness axis of the molecular lathe.

$$
\left[
\begin{array}{c}
\text{structure} \\
\end{array}
\right]_n
\tag{10}
$$

$$R = -C_{12}H_{25}$$

The representation of sanidic 1 in Fig. 2.23 (right-hand diagram) is complemented by a copy of the original diagram from the paper in Fig. 2.24. Note that the molecules are arranged into the equally spaced layers with their (longer) width axes normal to the layers, but that their spacing within the layers is irregular. We could envisage within the same classification an equivalent situation, except that it is the shorter thickness axes of the cross sections which are normal to the single set of periodic layers. Such an arrangement has been described by Duran et al. [37], which they call *pallisidic*, in the sense that the organization is a fence-like arrangement of planks. Recent high-resolution transmission electron microscopy by Voigt-Martin et al. [38], on the very similar sanidic molecule (11), has revealed the lateral positional order as layers parallel to the chain axes, as shown in Fig. 2.25. The spacing of 23 Å corresponds to the width of the molecular lathes (planks) which lie flat in the plane of the sample. For this reason, it is not possible to be sure that there is no positional order in the second lateral direction, i.e., normal to

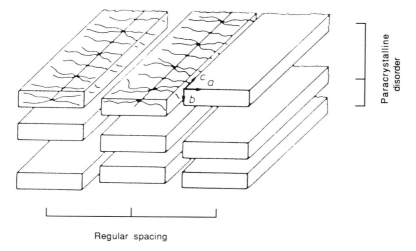

FIGURE 2.24. Reproduction of diagram of a *sanidic* structure showing the nature of the chain packing [30].

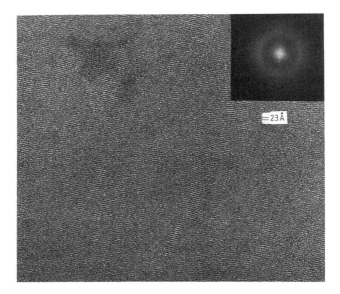

FIGURE 2.25. High-resolution electron micrograph of a sample of molecule (11) imaged in the sharp diffraction peaks corresponding to the layers containing the molecular chains. Low-dose conditions were used. Courtesy of Dr. I.G. Voigt-Martin.

the plane of the thin film specimen, however, there is as yet no direct evidence for it and the structure is viewed as sanidic 1 on the basis that the full width of the molecules is more like 40 Å, so that the layers could not be regularly positioned edge on to each other and give a spacing of 23 Å. The micrograph is also important in that it shows a degree of disorder within the layers, both with respect to them not being exactly straight, and that the layer-type order only extends over a finite distance and is limited in length in the direction of the chain axes to something of the order of the molecular length.

$$\tag{11}$$

$$R = -C_{12}H_{25}$$

2.3.6   Mesophases Resulting from Paracrystalline Disorder in the Two (Lateral) Dimensions, and Translational Melting in Two Dimensions

Two-Dimensional Paracrystallinity Resulting in the Retention of Layers

It is possible to conceive of the loss of the two translational order elements, O_{11} and O_{22}, together without the involvement of any shear disorder. However, the result, depicted in Fig. 2.26, is not really possible as a consequence of imperfect packing or thermal disorder for the spacings between the orthogonal planes are disturbed at random, while the chains are still deemed to lie exactly in the planes which remain perfectly delineated. Combinations of disorder which include the loss of both O_{11} and O_{22} are probably only of inerest where there is the loss of lateral shear disorder on one set of planes. The examples, shown in Fig. 2.27, are really equivalent to the series of mesophases shown in Fig. 2.23, but with the spacing between the intact planes no longer being periodic but subject to a paracrystalline distribution of spacings. The comparison of these structures raises the issue as to whether phase differentiation is possible on the basis of a level of paracrystalline disorder between a single set of planes. It is addressed separately in Section 2.3.8 below.

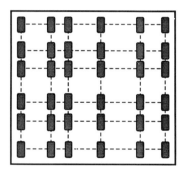

FIGURE 2.26. A rather contrived type of disorder with the loss of O_{11} and O_{22} elements only. It is considered to be extremely unlikely.

Two-Dimensional Melting

At this stage we return to the possibility of lateral shear disordering on intersecting planes, i.e., the combined loss of O_{12} and O_{21}. In purely shear terms, it is not really viable as the loss of O_{12} destroys the identity of the planes $_1$ on which the second shear disorder would take place. However, if the combined loss of O_{11} and O_{12} is seen as creating a translational melting element (after Helfrich), then it is possible to view this process as occurring in

LONGITUDINAL   REGISTER

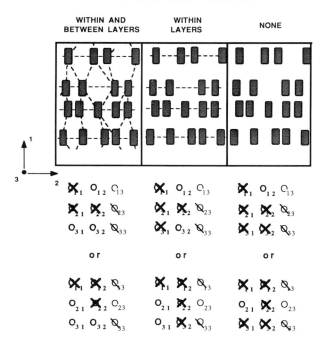

FIGURE 2.27. A series of structures developed from those in Fig. 2.23, but with the additional level of disorder that the spacings between the distinct planes are now themselves, additionally, subject to paracrystalline disorder.

two dimensions, destroying the two-dimensional lateral order and creating nematic-type packing. In fact, the mesophase has melted in two dimensions, and the only aspect of positional order which remains is that associated with longitudinal register which is either present in *both* lateral dimensions or completely absent. Figure 2.28 (top) shows the two possible structures.

We will focus on the so-called *biaxial nematic* structure first (top-right). The name stems from the optical properties of such a phase which shows two *optic axes*, as opposed to one in the case of a nematic. Of course, all the phases considered so far in which there is rotational correlation will in general also be optically biaxial. However, with the exception of the sanidics, and of the three-dimensional crystal, there is rather weak evidence for their existence in polymers. There is no long-range positional order of any type in the structure and yet the molecules are rotationally correlated about their chain axes. Biaxial nematics are far from common in main-chain liquid crystalline polymers. Probably the best-documented example is in the case of a sanidic molecule (12) in which the ratio of width to thickness of the cross section, for an idealized conformation, is of the order of 5 to 7 [30].

LONGITUDINAL   REGISTER

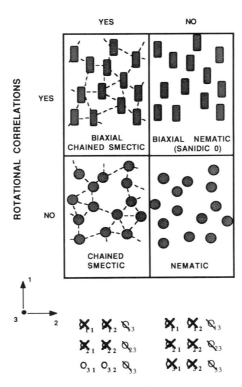

FIGURE 2.28. Structures (top) showing the effect of paracrystalline disorder in the two dimensions normal to the molecules. The presence of longitudinal register leads to what is called a *biaxial chained smectic*, its absence to a *biaxial nematic*. Both these structures are compatible with the absence of rotational correlations giving rise to the equivalents of *chained smectic* and *nematic*.

$$(12)$$

$$R = \!-\!O(CH_2CH_2O)_2CH_3$$

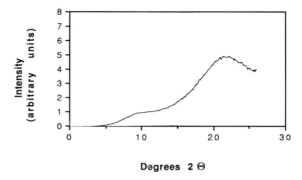

FIGURE 2.29. Wide angle x ray diffraction scan of a *sanidic 0* molecule showing the two very broad peaks which mark it as an example of a biaxial nematic [30].

At 160 °C and above, the wide angle diffraction pattern of the birefringent phase showed two diffuse halos, one of which corresponded to the approximate spacing of the molecules in the direction of their width, while the second, at a higher angle, represented packing in the direction of the laths' thickness (Fig. 2.29). The melt was birefringent and conoscopic images in the transmission polarixing microscope were consistent with biaxial optical symmetry.

The top left-hand diagram of Fig. 2.28 shows a structure in which longitudinal register is added onto the biaxial nematic. This generates layers which do not contain the chain axis. In one sense, the order is analogous to that in low-molar-mass *smectics*, but the difference is that the layers are not discrete in this case but contiguous through the molecular chains. The problem is what to call such a structure. It is perhaps not helpful to suggest another smectic subclass of the series S_A, S_B, etc., and any attempt to refer to a "polymer smectic" would be ambiguous as side-chain liquid crystalline polymers can equally form smectic structures, in which the layers are not necessarily connected through the backbones. Furthermore, the very term *smectic* implies a slippery feel which, in conventional smectics, is produced by the ease of shearing of the layers over each other, a process which is, of course, not possible when the layers are connected through the chains. For these reasons the term *biaxial chained smectic* is preferred. It would, of course, be possible to envisage subclasses equivalent to S_A and S_C, but that is not for now.

Each of these two structures is also compatible with axially symmetric chains or, more realistically, real chains without rotational correlation. The lower two diagrams of Fig. 2.28 show the equivalent structures classified, respectively, as *chained smectic* and *nematic*.

There is considerable evidence for longitudinal register and thus chained smectic phases in main chain polymers, consisting of alternating sequences of mesogenic units and aliphatic flexible spacers [39]. However, for molecules

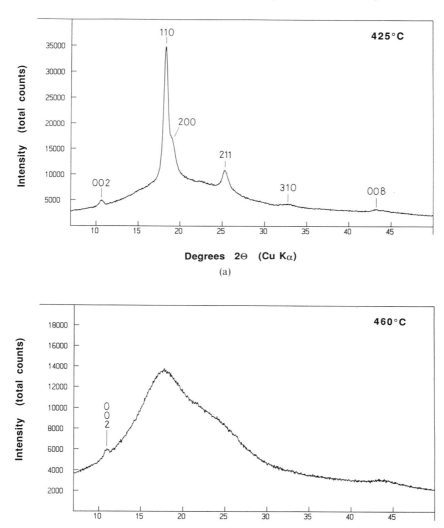

FIGURE 2.30. Wide-angle x ray diffraction scans of polyhydroxynaphthoic acid: (a) in the high-temperature crystal phase at 425 °C; and (b) above the crystal melting point at 460 °C. Note the 002 peak is still present in the mesophase. This indicates that it is an example of a *chained smectic* structure with layers normal to the chain axes.

in which the chain repeat lengths are of the order of their diameter, examples are more difficult to come by. Polyhydroxynaphthoic acid has been shown to have a chained smectic phase at temperatures above 435 °C [40,41]. Figure 2.30 shows the x ray diffraction scans in the crystalline and chained smectic phase ranges of this polymer.

Summary for Paracrystalline Combinations in Two Dimensions

Summaries (vii and viii)

(vii) Loss of positional order in spacing between single sets of planes, plus loss of lateral shear order on intersecting planes to give "translational melting"

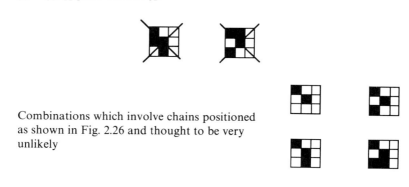

Loss of spacing register on both planes:

	On plane $_1$	On plane $_2$
plus: loss of lateral shear register on one set of planes		
or		
plus: loss of lateral and longitudinal register on one set of planes		
or		
plus: loss of shear register as above and loss of all longitudinal register		

The combinations below do not occur because they contain $_{21}$, $_{32}$ pairs without $_{31}$ or $_{31}$, $_{12}$ pairs without $_{32}$.

Combinations which involve chains positioned as shown in Fig. 2.26 and thought to be very unlikely

(viii) Loss of positional order in spacing in both lateral dimensions (paracrystalline disorder in two dimensions)

with longitudinal register

without longitudinal register

combinations incommensurate
with the symmetry

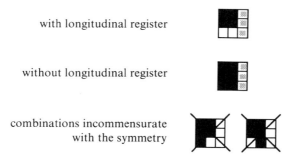

2.3.7   Audit of Possible Combinations

Keeping to one side the possibility of independent shear disorder on the plane which is normal to the chains, and maintaining the supposition that the chains remain essentially periodic along their axes, then all possible combinations of the disordering elements will be accounted for by the first two columns of the matrix, i.e., six elements. There will therefore be 2^6 (64) possible combinations and these have all been considered above. (There are 64 different diagrams in the summary sections.) Of these 64, 22 are considered non-viable either because a particular combination generates the loss of a further element of order, or because it is not possible to consider loss of shear disorder on a set of planes if the set has already been scrambled by disorder on an intersecting set. Of the remaining 42, 24 are distinct in the sense that planes $_1$ and $_2$ are equivalent.

2.3.8   Occurrence of Mesophase Structures

To identify mesophase structures with various degrees of disorder and to draw them is one thing, to be able to assess the probability of their occurrence in an actual polymer system is quite another. Already it has been pointed out that structures such as that illustrated in Fig. 2.26 are very unlikely indeed, and the three distinct combinations of such an arrangement with different degrees of longitudinal register (summary (vii)) are to be discounted.

It is also useful to address the issue of order between a single set of identifiable layers containing the chains. Where such layers can be observed, because, for example, there has been a loss of longitudinal register, or lateral shear order, or both, between them, then we have to decide whether the periodicity of their spacing is retained. In the treatment above, an attempt has been made to distinguish between perfect periodicity, on one hand, and the loss of this order as the loss of O_{11} or O_{22}, on the other. Such a distinction, however, is not strictly permissible as, with one-dimensional order, the

72   A.H. Windle

levels of order can vary continuously from merely short range to nearly (but never completely) perfect periodicity. In other words, there is no melting transition in one dimension, and the positional correlation function will have an algebraic decay. While the diffraction peaks from the layers of small molecule smectic liquid crystals often appear very narrow, high-resolution x ray diffraction measurements [42] have confirmed that the layer peaks are not

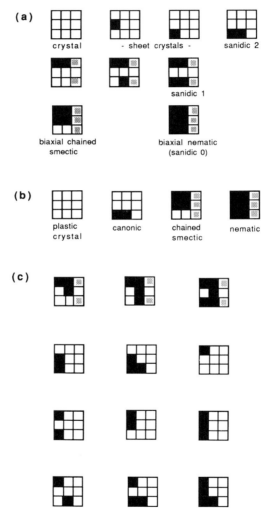

FIGURE 2.31. Summary of the levels of order in: (a) the nine mesophases seen as most likely to occur where there is rotational correlation between the chain axes; (b) the four most likely mesophases which contain rotational symmetry; and (c) the "second league" of twelve structures which are considered quite possible but, as yet, have not been observed experimentally.

totally sharp. There is also evidence that two dimensions are also insufficient to propagate perfect periodicity over long ranges. However, while noting that the disorder criterion is not strictly sufficient to differentiate phases in layer structures, it is still held to be useful in describing the general pattern of order.

Nevertheless, difficulties remain. The existence of paracrystalline disorder in the spacing of otherwise perfectly defined planes is seen to be very unlikely, while Helfrich has already argued that the complete absence of shear register between perfect planes is likely to cause the planes themselves to disorder. In terms of the overall classification, the first factor will suggest elimination of all those structures in which O_{11} is disordered without O_{12} (or, likewise, O_{22} without O_{21}). In fact, the ten distinct arrangements from summaries (v) and (vii). The second factor, the elimination of structures with shear disorder only in which O_{21} and O_{31} (or O_{12} and O_{32}) coexist. This would remove two distinct arrangements, one from summary chart (iii) and one from summary chart (iv).

We are therefore left with nine primary structures for the case of rotational correlation between the chain molecules. These are illustrated in Fig. 2.31(a), and may be seen as prime candidates for mesophase structures in main-chain, nonchiral liquid crystalline polymers. It should be noted that all the rotationally correlated structures observed experimentally are included within this sequence. The classification also suggests that variants of the sanidic 1 structure in which there are different degrees of longitudinal register should exist and may well be found, especially in molecules in which this register is encouraged by block sequences of fixed lengths but different chemical character.

Where there is no rotational correlation between the molecules, the number of options is reduced to four, Fig. 2.31(b), all of which are well known. The second league of possible structures is represented in Fig. 2.31(c). Although none has yet been observed experimentally, they are not completely unreasonable, and their discovery would be of great interest.

2.3.9 Disruption of Order along the Chain Axis

So far, the possibility of disorder elements in the third column of the matrix, i.e., those associated with planes normal to the chains, has been avoided, save for the discussion of the occurrence of derived disorder in these elements in Section 3.35, and its subsequent representation as a faint overstrike on the matrix or as a shaded patch within the reduced scheme. However, chain mesophases are possible in which the periodicity along the chain is completely absent and thus the element O_{33} lost. In most situations, this loss of axial order will result in the loss of longitudinal shear register between the chains, and thus of O_{31} and O_{32} as well. There are two main ways in which a chain may have no axial periodicity but remain sufficiently straight to contribute to a mesophase. The first, involves a pattern of conformational disorder which, while permitting the chain to run straight, severely disrupts the axial order. The approach was investigated by Pechhold [43] as a part of his "meander model" for nonmesogenic polymer melts and may be relevent to nematic phases, while Wunderlich and coworkers (for example [44]) have explored the possibility of conformational disorder occurring in systems in which there is full lateral positional order and are referred to as "Condis crystals." In the reduced scheme these possibilities could be represented as:

respectively

The second way in which chain periodicity can be disrupted is through the use of random copolymerization of two or more different units, especially where these units differ in length. As has been described in Section 2.1.2, such random copolymerization is used as a technique for reducing the crystalline melting point in commercial series of thermotropic polymers, and is typified by the molecules (5) and (6). The nematic phase of these materials is therefore describable by the matrix on the left above. The loss of chain periodicity, especially that associated with random copolymerization of two units, is not a state of total disorder, but can be considered as an example of paracrystallinity in which the positional correlation function has a strictly bimodal distribution [45,46]. An intriguing side issue of this particular form of axial disorder, is that it is possible to form crystals as a consequence of segregation of matched, yet aperiodic, sequences. Such a crystal would be classified as having lost O_{33} only:

The crystals consist of nonperiodic layers (NPL) in a scheme recently worked out for two-component random copolymers [47,48] where they form as a result of sequence segregation within restricted, planar regions of the polymer [49]. Such crystals are reminiscent of the "aperiodic crystals" envisaged by Schrödinger in his seminal treatise "*What is Life?*" [50].

References

1. M. Balauff, D. Wu, P.J. Flory, and E.M. Barrall II, Ber. Bunsenges. Phys. Chem. **881**, 52 (1984).

2. J. Mooney, *Synthesis and Characterization of Short Liquid Crystalline Chains*, Ph.D. Thesis, Cambridge University, 1990.

3. Hoechst Celanese Corporation, Summit, New Jersey.

4. A.M. Donald and A.H. Windle, *Liquid Crystalline Polymers* (Cambridge University Press, Cambridge, 1992).

5. L. Onsager, Ann. N.Y. Acad. Sci. **51**, 627 (1949).

6. J.D. Parsons, Phys. Rev. A **19**, 1225 (1979).

7. P.J. Flory and G. Ronca, Mol. Cryst. Liq. Cryst. **54**, 311 (1979).

8. W. Maier and A. Saupe, Z. Naturforsch. A **4**, 882 (1960); **1**, **15**, 287 (1959).

9. B. Jung and M. Depner, Makromol. Chem. **192**, 1667 (1991).

10. O. Kratky and G.F. Porod, Recl. Trav. Chim., Pays-Bas **68**, 1106 (1949).

11. P. Coulter and A.H. Windle, Amer. Chem. Soc., Polymer. Prepr. **30**, 67 (1989).

12. S.E. Bedford, K.Yu, and A.H. Windle, *The Conformations of Flexible Molecules in Fluid Phases*, Southampton, Faraday Symposium No. 27 (1991).

13. W. Kuhn, Kolloid. Z. **76**, 2586 (1936).

14. P.J. Flory, Macromolecules **11**, 1141 (1978).

15. R.R. Matheson and P.J. Flory, Macromolecules **14**, 954 (1981).

16. A.C. Griffin and S.J. Havens, J. Polymer. Sci., Polymer. Phys. Ed. **19**, 951 (1981).

17. G.J. Vroege and T. Odijk, Macromolecules **21**, 2848 (1988).

18. M. Warner, in *Materials Research Society Symposium Proceedings*, Boston (1988).

19. A.H. Windle, C. Viney, R. Golombok, A.M. Donald, and G.R. Mitchell, Faraday Discuss. Chem. Soc. **79**, 55 (1985).

20. N.J. Alderman and M.R. Mackley, *ibid*, p. 149.

21. C. Viney and A.H. Windle, J. Mater. Sci. **17**, 2661 (1982).

22. T. Odijk, Liq. Cryst. **1**, 553 (1986).

23. V.G. Taratuta, F. Lonberg, and R.B. Meyer, Phys. Rev. A (Rap. Comm.) **37**, No. 5, 1831 (1982).

24. J.P. Straley, Phys. Rev. A **8**, 2181 (1973).

25. P.G. de Gennes, Mol. Cryst. Liq. Cryst. Lett. **34**, 177 (1977).

26. R.B. Meyer, in *Polymer Liquid Crystals*, edited by A. Ciferri, W.R. Krigbaum, and R.B. Meyer (Academic Press, New York, 1982).

27. J.I. Jin, S. Antoun, C. Ober, and R.W. Lenz, British Polymer. J. **12**, 132 (1980).

28. G.W. Gray and J.W. Goodby, *Smectic Liquid Crystals* (Leonard Hill, Glasgow, 1984).

29. A.J. Leadbetter, in *Thermotropic Liquid Crystals*, edited by G.W. Gray (Wiley, Chichester, 1987).

30. M. Ebert, O. Herrmann-Schonherr, J.H. Wendorff, H. Ringsdorf, and P. Tschirner, Makromol. Chem., Rapid Commun. **9**, 445 (1988).

31. W. Helfrich, in *Liquid Crystals*, edited by S. Chandrasekhar (Heyden, London, 1980).

32. B.K. Vainshtein, *Diffraction from Chain Molecules* (Elsevier, Amsterdam, 1966).

33. F.C. Frank, in *Liquid Crystals*, edited by S. Chandrasekhar (Heyden, London, 1980).

34. S. Hanna and A.H. Windle, Polymer. Commun. **29**, 235 (1988).

35. W. Helfrich, Phys. Lett. **58A**, 457 (1976).

36. O. Hermann-Schonherr, J.H. Wendorff, H. Ringsdorf, and P. Tschirner, Makromol. Chem., Rapid Commun. **7**, 791 (1986).

37. R. Duran, D. Guillon, P. Gramain, and A. Skoulios, Makromol. Chem., Rapid Commun. **8**, 181 and 321 (1987).

38. I.G. Voigt-Martin, P. Simon, R.W. Garbella, H. Ringsdorf, and P. Tschirner, Macromol. Chem., Rapid Commun. **12**, 285 (1991).

39. P. Meurisse, C. Noel, L. Monnerie, and B. Fayolle, British Polymer. J. **13**, 55 (1981).

40. G. Schwarz and H.R. Rvicheldorf, Macromolecules **24**, 2829 (1991).

41. S. Hanna and A.H. Windle, Polymer **33**, 2825 (1992).

42. J. Als-Nielsen, J.D. Litster, R.J. Birgeneau, M. Kaplan, C.R. Safinya, A. Lindegaard-Andersen, and B. Mathiesen, Phys. Rev. B **22**, 312 (1980).

43. W. Pechhold, J. Polymer. Sci. C, Polymer. Symp. **32**, 123 (1971).

44. B. Wunderlich and J. Grebowicz, in *Liquid Crystal Polymers*, II/III, edited by M. Gordon and N.A. Platé (Springer-Verlag, Berlin, 1984).

45. G.R. Mitchell and A.H. Windle, Colloid. Polymer. Sci. **263**, 230 (1985).

46. J. Blackwell, A. Biswas, and R.C. Bonart, Macromolecules **18**, 2126 (1985).

47. S. Hanna and A.H. Windle, Polymer **29**, 207 (1988).

48. R. Golombok, S. Hanna, and A.H. Windle, Mol. Cryst. Liq. Cryst. **155**, 281 (1988).

49. S. Hanna, T.J. Lemmon, R.J. Spontak, and A.H. Windle, Polymer **33**, 3 (1992).

50. E. Schrödinger, *What is Life?* (Cambridge University Press, Cambridge, 1944).

3

Molecular Architecture and Structure of Thermotropic Liquid Crystal Polymers with Mesogenic Side Groups

V.P. Shibaev, Ya.S. Freidzon, and S.G. Kostromin

3.1 Introduction

Of the variety of scientific trends developed in the last ten to fifteen years, within the field of physical chemistry of polymers, the design and investigation of liquid crystal (LC) polymers [1–15] has been the one growing most actively and fruitfully.

Theoretically, the possibility for the anisotropic LC phase to be formed in a solution of long rigid rods (lyotropic LC systems) was, for the first time, demonstrated in the 1950s by Onsager [16] and Flory [17], and was then experimentally verified in the studies with solutions of polypeptides whose macromolecules are known to acquire, in certain solvents, a helical conformation [18] spatially approximated by rods. In spite of the remoteness of these events, for a relatively long time afterward the studies of LC lyotropic systems were confined to purely academic research. However, at the beginning of the 1970s, aromatic polyamides, used to fabricate ultrastrong high-modulus thermostable fibers like "Kevlar," were shown to form LC solutions—a discovery that has given a mighty impetus to investigations aiming at new LC macromolecular systems.

On the other hand, a dashing introduction into the market of low molecular weight liquid crystals, such as digital electronic indicators in scientific instrumentation, consumer goods, and information processing devices, has also affected the interests of polymer scientists, shifting them to thermotropic LC polymers displaying LC phases in the polymer melt. These studies resulted in the creation of the so-called mesogenic LC polymers incorporating mesogenic groups (simulating the structure of low molecular weight liquid crystals), either in the main chains of the macromolecules ("main-chain LC polymers") or as pendant side branches ("side-chain" or "comb-shaped" LC polymers). Hereafter, we will use the latter name.

In these cases, formation of a mesophase is governed by the interaction between "rigid" fragments and is essentially dependent on the polarizability anisotropy of mesogenic groups; the macromolecules per se retain relative flexibility, as is evident from the magnitude of Kuhn's segment varying for mesogenic polymers of different chemical structure from 60 Å to 100 Å [19]. It is distinctly this feature of their molecular structure that stipulates the ambiguity of their nature: on the one hand, mesogenic polymers display the properties inherent to polymers and these are associated with the effect of the backbone, while, on the other hand, these polymers reveal LC properties which are due to the mesogenic groups.

Such dual performance of LC polymer macromolecules is most explicitly manifested by comb-shaped polymers incorporating mesogenic groups chemically attached to the main chain, with its one terminus and showing sufficient autonomy. Quite naturally, when elaborating on the structural organization of such systems, the role of each of the structural components and their mutual effects in forming the mesophase appear to be the key aspects.

One of the approaches dealing with investigation of the structural organization of comb-shaped LC polymers involves the characterization of the mode and type of LC ordering pursued by polymer side groups. As a rule, principles, methods, classification, and even the notation elaborated for low molecular weight compounds are used for the purpose, somehow neglecting the specific features introduced into their structure by the polymeric nature of the molecules of these substances. This approach provides the answer to the basic question concerning the structural type of LC polymer mesophases.

However, the macromolecular nature of comb-shaped molecules reveals itself even at this first stage of examining the structural organization. The most vivid example is the observed dependence of the mesophase type on the average molecular weight and the width of molecular weight distribution (polydispersity) (see Section 3.3).

Within the framework of the above-mentioned approach, using the methods developed for low molecular weight liquid crystals, profound studies of the structural organization of various LC phase types were also performed. These were, at first, aimed at the measurement of the dynamic order parameter (NMR, IR-dichroism),* molecular mobility (dielectric relaxation technique), and determination of the structural correlation functions and electron density profiles of macromolecular fragments packing derived from quantitative x ray experiments, etc.

Usually, these parameters are sufficient for a complete description of low molecular weight LC compounds. However, this is not the case with LC polymers. The placement within the mesophase of the polymer main chain, as of a whole macromolecule as well as of its fragments, at the segmental scale appear to be the questions yet unclear. What are the routes along which the

---

* See Chapter 6.

relevant answers are sought? At first, the placement of the main chain is appraised via indirect observations involving the features, revealed in optical, thermodynamic, structural studies of LC polymers, that are uncommon for low molecular weight liquid crystals. Certain inferences may rely on the comparison of various homologous series of LC polymers comprising one and the same side fragment and the main chains of different chemical structure. However, it is the direct methods developed specifically for the investigation of the polymer structure that provide the most valuable information. Among these methods, small-angle neutron scattering, which makes it possible to judge the conformation (shape) of the macromolecule as a whole, appears to occupy the leading position. Below we will consider, in detail, the results of structural studies of comb-shaped polymers distinguishing between x ray data, pertaining, as a rule, to the packing of mesogenic side groups, and small-angle neutron scattering disclosing conformation of main chains. However, let us first briefly survey some general aspects concerning the molecular structure of comb-shaped polymers, and some specific features of their structural and thermal properties related to their macromolecular nature.

3.2   Molecular Architecture of Liquid Crystal Polymers

The first thermotropic LC polymers were of the comb-shaped structure. The basic principle for the design of such systems by chemical attachment of the molecules or the fragments of mesogenic groups of the low molecular weight liquid crystals to the aliphatic side chains, performing as spacers (Fig. 3.1(a)) [20–23], of comb-shaped polymers is by today a generally adopted synthetic procedure leading to thermotropic LC polymers with mesogenic side chains. The flexible aliphatic (or any other) spacer distancing mesogenic groups enables a high degree of independence for the mesogenic groups with regard to the backbone making possible their cooperative interaction resulting in the formation of the mesophase.

The majority of comb-shaped polymers were synthesized using this approach. As regards the structure of such systems, it is overwhelmingly versatile. In the first studies published, relatively simple polymeric systems (Fig. 3.1(a)) were reported; these systems comprised rod-like mesogenic groups, the length of the spacer and the chemical nature of the main chain having been the parameters varied. However, the series of reports that followed described comb-shaped polymers of sufficiently complex and even exotic structures.

Figure 3.1(a–j) shows the molecular structure of the major types of comb-shaped LC polymers. As is seen, beside the branched (comb-shaped systems) polymers proper containing one (Fig. 3.1(a)) or two (Fig. 3.1(b)) mesogenic

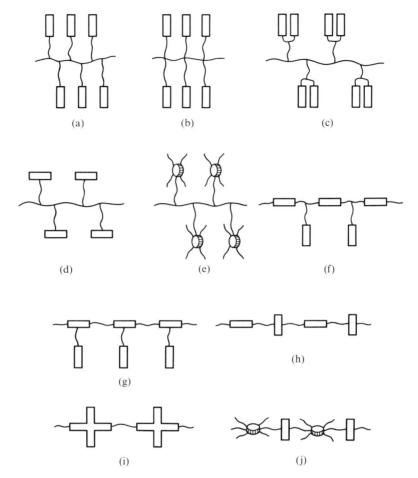

FIGURE 3.1. Schematic representation of macromolecules of LC polymers with mesogenic side groups.

groups within a monomeric unit as well as laterally linked and paired meso-gens (Fig. 3.1(c, d)), there exist the polymers with disc-shaped side groups (Fig. 3.1(e)) and complex macromolecules incorporating mesogenic groups both within the main and the side chains (Fig. 3.1(f–h)). Even more complex LC polymers with cross-shaped mesogenic groups (Fig. 3.1(i)), as well as the polymers comprising macromolecules with alternating laterally linked rod-like and disc-shaped groups (Fig. 3.1(j)), were reported [11,12].

Board-shaped polymers (Fig. 3.2) represent a quite peculiar type of comb-shaped polymers. Their main chains are that of aromatic polyamides (I) or polyesters (II) whereas long aliphatic pendant groups are attached to the benzene rings within the backbone [24–25]:

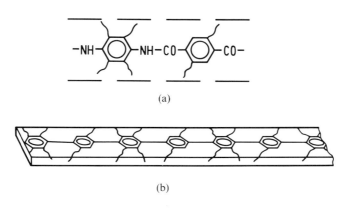

$$(I)$$

or

$$(II)$$

where R is the alkyl group.

As is seen from Fig. 3.2, in this case the arrangement and interaction of the side groups "suppress" the specific features associated with the aliphatic side

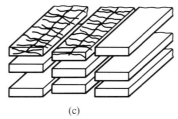

(a)

(b)

(c)

FIGURE 3.2. Formation of board-like structures by highly substituted aromatic polyamides [25].

chains, and the macromolecules as a whole are packed in layers (Fig. 3.2(b, c)). The packing of such flat board-shaped molecules can proceed by different routes, leading to the formation of so-called sanidic ordered and disordered mesophases; these were treated in detail in [25].

New metal-containing polymers whose macromolecules contain ferrocene as the pendant side group [26] are also of undoubted interest:

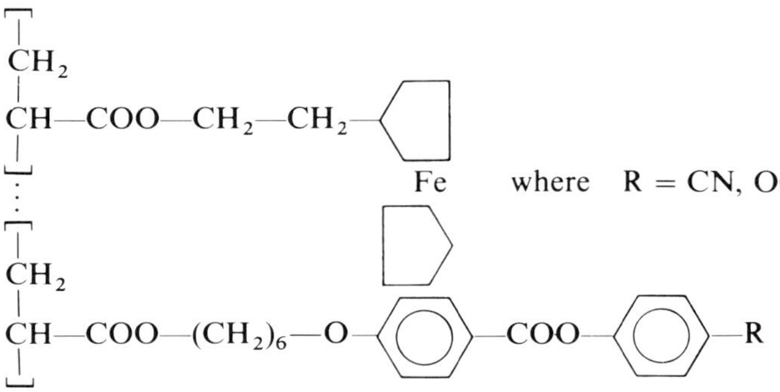

where   R = CN, OCH$_3$ (III)

Polymers of this type, containing up to 10% ferrocene, can form either the nematic or the smectic modifications depending on whether the polymer is in the "initial" (III) or the oxidized form. These new LC compounds were called the redoxactive ionomers and are of unquestionable applied interest.

Another type of ionic thermotropic LC polymers was recently obtained by an interaction between anionic polymer backbones and cationic mesogenic molecules (Fig. 3.3(a)); the same type of ionic polymers can be obtained by polymerization of the complex between cationic and anionic units (Fig. 3.3(b)) [27]. Examples of such ionic polymers are shown below:

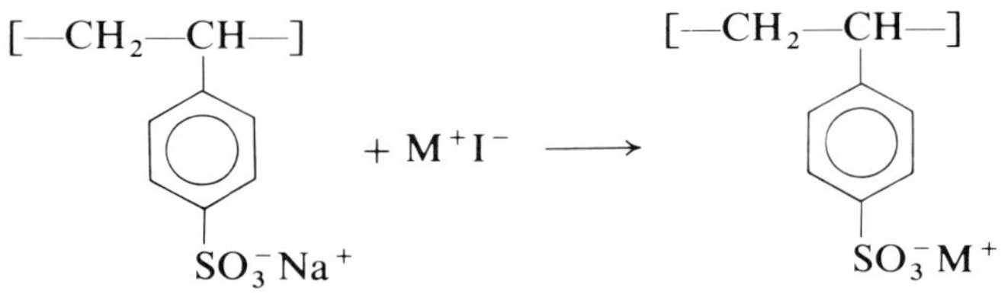

(IV)

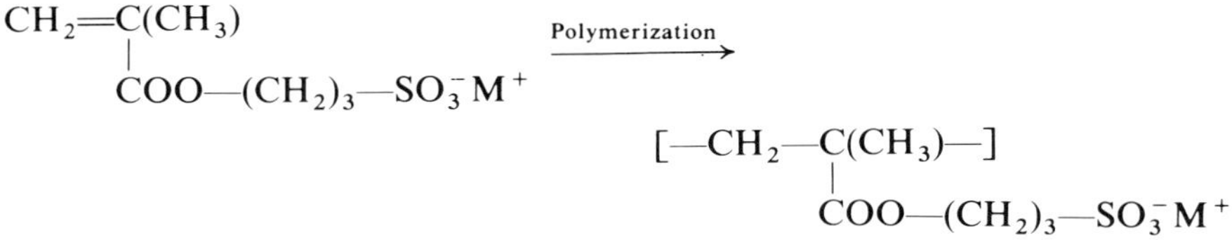

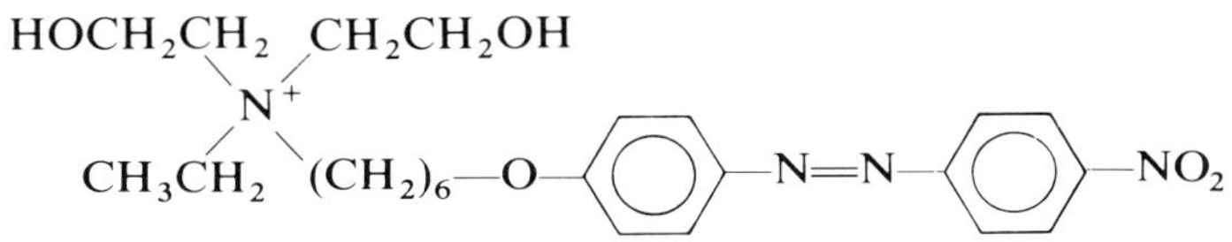

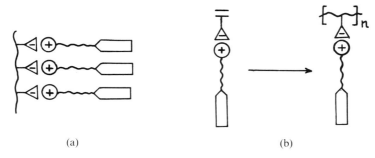

FIGURE 3.3. Schematic molecular structures of thermotropic LC ionic polymers, obtained by polymer analogous ion-exchange reaction (a) and polymerization (b) [27].

The first examples of graft LC polymers have also been synthesized by copolymerization of polystyrene (PS) macromonomers and methacrylic mesogens [28]. In this case, the nematic LC state is even realized at a high concentration of polystyrene (up to ~40%): besides it was found that LC copolymers have two glass transition points, corresponding to the backbone and PS segments existence.

The schemes depicting the macromolecules of comb-shaped LC polymers (Fig. 3.1) and the above listed data serve to illustrate the fact that LC polymers are most likely to occupy one of the leading positions in the field dealing with the design of new macromolcular structures.

Table 3.1 lists the various types of main chains and spacers assembled in LC comb-shaped polymers synthesized by today. As is seen, comb-shaped polymers encompass a sufficiently large number of the major classes of polymers as of organic, so as of organoelement and even inorganic (polyphosphazenes) types.

As regards mesogenic groups, actually any mesogenic group (or actually any low molecular weight LC compound) can be more or less easily incorporated within a comb-shaped polymer as a pendant side group.

At the same time, in spite of the versatility of the types of molecular structure of LC polymers whose overall number perhaps exceeds thousands, there is only around several tens of studies devoted to the detailed analysis of the

TABLE 3.1. Chemical structure of main chains and spacers of synthesized liquid crystal comb-shaped polymers.

Main chains	Polyacrylic, polymethacrylic, polymethylsiloxane, polyether, aromatic and aliphatic polyesters, polystyrene, polyolefinic, polyitaconic, polyphosphasene.
Spacers	aliphatic, oxyaliphatic, siloxane.

structure of such complex LC polymers as, for instance, those shown in Fig. 3.1(b–g) and Fig. 3.2.*

A major piece of data was obtained for LC polymers of the simplest structure, incorporating a rod-like mesogenic fragment within each monomeric unit. In other words, today the art of the synthesis of LC polymers is far ahead of the studies concerning their structure and physico–chemical properties, not to mention the ways of their application.†

Such a situation is probably due to the fact that the synthesis of comb-shaped LC polymers with strictly defined molecular characteristics is still a complicated task, and the yield of synthetic procedures is often too low to produce the amount of polymer sufficient for a complex structural and physical investigation to be undertaken. In their turn, such studies require special complex instrumentation managed by physicists. In this regard only a helpful collaboration of both synthetic chemists and physicists can lead to final success.

3.3   Specific Features of Comb-Shaped Polymers Associated with Their Macromolecular Nature

Despite the fact that the mesophase in comb-shaped polymers is formed by the mesogenic groups and the type of mesophase is predominantly (although not always) determined by their chemical nature, the polymer backbone contributes essentially to the physico–chemical behavior of LC polymers.

This is manifested in the slowing-down of all relaxation processes taking place in LC polymers, as well as in the existence of a large number of non-equilibrium states, which can be falsely interpreted as the equilibrium ones. For a LC polymer to get to the equilibrium state, prolonged annealing of the polymer sample is necessary under the conditions favoring fast relaxation of the system (usually at temperatures above T_g). Indeed, the disguised optical textures reported for a number of LC comb-shaped polymers [30–33] were transformed into distinctly shaped optical patterns similar to the textures of low molecular weight liquid crystals.

When speculating on the role of the backbone in comb-shaped polymers, the first feature to be emphasized is that its main effect involves an essential increase of the mesophase thermostability as compared to the low molecular weight liquid crystals. Comparison of transition temperatures of low molecular weight alkoxycyanobiphenyls and of comb-shaped polyacrylates,

---

* Chemical formulas of LC polymers and the methods for their synthesis are described in [11,12].
† Concerning the applied aspects, we can find the relevant information in a special issue of *Mol. Cryst. Liq. Cryst.* [29] dedicated to the topic, or in the review by Shibaev and Belyaev, "The Prospects in Applying Functional LC Polymers and Composites" [14].

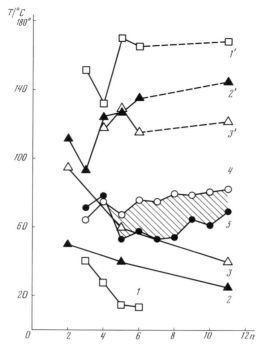

FIGURE 3.4. Glass transition (1–3), clearing (1′,3′,4′) and melting temperatures (5) of polysiloxanes, polyacrylates, polymethacrylates, and alkoxycyanobiphenyls as a function of carbon atoms (n) in an aliphatic fragment

1,1′-[—Si(CH$_3$)—O—]
        |
     (CH$_2$)$_n$—O—⟨○⟩—⟨○⟩—CN

2,2′-[—CH$_2$—CH—]
              |
           OCO(CH$_2$)$_n$—O—⟨○⟩—⟨○⟩—CN

          CH$_3$
          |
3,3′-[CH$_2$—C—]
          |
       OCO—(CH$_2$)$_n$—O—⟨○⟩—⟨○⟩—CN

4,5-C$_n$H$_{2n+1}$O—⟨○⟩—⟨○⟩—CN

TABLE 3.2. Influence of molecular parameters of polyacrylic polymer

$$[-CH_2-CH-]$$
$$\underset{OCO-(CH_2)_5COO-\langle\bigcirc\rangle-OOC-\langle\bigcirc\rangle-OC_4H_9}{|}$$

on its phase transitions [35].

Sample	$\overline{M}_w \cdot 10^{-4}$	$\overline{M}_w/\overline{M}_n$	Phase transitions (°C)
Fraction	3.95	1.2	S_F 68 RN122 S_A135 N148 I
Unfractionated	4.10	2.9	S_F 70 N 140 I

* RN is the nematic reentrant phase.

polymethacrylates, and polysiloxanes containing cyanobiphenyl mesogenic groups [34] (Fig. 3.4) illustrates the statement. As is seen from the figure, in LC polymers the mesophase spans a much broader temperature range than in the case of a low molecular weight LC analogue.

When comparing polymeric and low molecular weight liquid crystals we should bear in mind the following important trait. Owing to the specific features associated with the synthetic procedures of polymer production (chain polymerization, polycondensation) polymers are always a mixture of macromolecules of different length, i.e., strictly speaking, they are multicomponent systems. The macromolecules distribution function over their length, which is primitively described by the ratio between two different average molecular weights, the weight average $\overline{M}_w$ and the number average $\overline{M}_n$, may appear rather broad ($\overline{M}_w/\overline{M}_n$ may reach several units in magnitude). Table 3.2 illustrates the effect of molecular weight distribution width on the thermodynamic parameters of LC polymers.

As is seen, the two samples with close values for $\overline{M}_w$ but differing in polydispersity are essentially different with regard to the mesophases formed and the relevant clearing points. This example also reflects a general regularity emphasized in a number of studies [33,36,37], viz., the samples with a broader molecular weight distribution (MWD) exhibit lower transition temperatures and a broader transition interval, which may sometimes reach 20°–30° [36]. Moreover, when the MWD is broad not all LC phases characteristic of the given polymer are liable to be observed. Therefore, sufficiently narrow fractions ($\overline{M}_w/\overline{M}_n < 1.2$–1.3) should be used to ensure correct results in the studies of LC polymers.

Separation and investigation of such sufficiently narrow fractions spanning a wide range of average molecular weights made it possible to trace the effect of polymer chain length on the properties of mesophases. Figure 3.5 depicts the plots of transition temperature versus the average degree of polymerization (DP) for polymers V–VII.

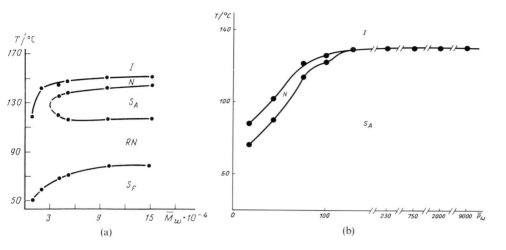

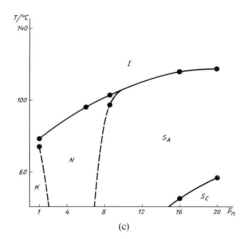

FIGURE 3.5. Phase transition temperatures of polymers

(a) $[-CH_2-CH-]$
$\quad OCO-O-(CH_2)_5-COO-⬡-OOC-⬡-OC_4H_9;$

(b) $[-CH_2-CH-]$
$\quad OCO-(CH_2)_5-O-⬡-⬡-CN;$ and

(c) $[-CH_2-CH-]$
$\quad O-(CH_2)_6-O-⬡-⬡-CN$

as a function of the degree of polymerization [35,38]

$[-CH_2-CH-]$
$COO-(CH_2)_5-COO-\langle\bigcirc\rangle-OOC-\langle\bigcirc\rangle-OC_4H_9$ (V)

$[-CH_2-CH-]$
$COO-(CH_2)_5-O-\langle\bigcirc\rangle-\langle\bigcirc\rangle-CN$ (VI)

$[-CH_2-CH-]$
$O-(CH_2)_6-O-\langle\bigcirc\rangle-\langle\bigcirc\rangle-CN$ (VII)

For long macromolecules (Fig. 3.5) the dependence is either too weak or not observed at all. For short chains (oligomers) the dependence becomes conspicuous. For instance, in the majority of cases, mesophase types of the polymer and its low molecular weight analogue (liquid crystal, corresponding to the structure of the polymer side group) are different; in the "oligomeric region" the transition from one type to another takes place. For all cases depicted in Fig. 3.5 the transition from the only nematic mesophase characteristic of the low molecular weight analogue to the S_A phase (frequently along with other phases, Fig. 3.5(a)) is observed. In order to trace the onset of the S_A phase formation we have to use oligomers with a very low DP. This is quite a difficult task in the case of polyacrylates synthesized by radical polymerization (Fig. 3.5(a, b)); however, it is feasible in the case of polyvinyl ethers obtained by cationic polymerization [38]. The fractions of these polymers allow us to monitor the initial stage of the transition from the monomer to oligomers with very short chains (Fig. 3.5(c)). For instance, an oligomer with $\bar{P}_n = 7$ loses the ability to crystallize, retaining however the same mesophase type (N-type) as the monomer. Already at $\bar{P}_n = 11$ the smectic A phase, specific for the polymer, arises. Hence, the mesophase type that would be retained for a polymer on the further growth of DP already starts forming for such short chains.

An important consequence ensues from the data presented. When elaborating on the assignment of a set of phases and definite phase transition temperatures to a distinct polymer, sufficiently narrow fractions of a polymer have to be examined, yet with the average degree of polymerization exceeding a certain threshold value above which the clearing temperature is independent of DP (Fig. 3.5). Only such data can be used to establish correlations between the chemical structure of comb-shaped polymers and their mesomorphic properties.

It is of no doubt that experimental studies in this field, and the ensuing accumulation of relevant data, are necessary stages in the investigations of the thermal properties of comb-shaped LC polymers. The outcome of this work should be the compilation of LC polymers, with strictly identified physical characteristics, within a handbook similar to *Flussige Kristallen in Tabellen* [39] widely used by researchers working in the field of low molecular weight liquid crystals.

Neglecting the reliably established dependences between the molecular weight characteristics of LC polymers, and their thermal and structural properties, may lead to serious errors in assessing thermostability and even the mesophase type. Another, no less important, feature associated with the polymer chain mediated phenomena in LC polymers is related to the effect of the backbone flexibility on mesophase thermostability, i.e., on the clearing point.

Table 3.3 lists the values for the glass transition and clearing temperatures of cholesterol-containing polymers, with approximately equal spacer lengths and close DP values (except for polymer 5), but an essentially different structure of the main chain (and, naturally, different chemical links between the backbone and the spacer). As is seen, T_{cl} values decrease with increasing

TABLE 3.3. Influence of main chain flexibility on the temperature interval of the LC state for the series of cholesterol-containing homopolymers [40].

No.	Polymer	T_g (°C)	Phase transitions (°C)
1	CH$_2$ | CH—CO—NH(CH$_2$)$_5$COO— [Chol] *	165	S_A 225 I
2	CH$_2$ | CH$_3$—C—CO—NH—(CH$_2$)$_5$—COO— [Chol]	130	S_A 220 I
3	CH$_2$ | CH$_3$—C—CO—NH—(CH$_2$)$_5$—COO— [Chol]	85	S_A 190 Chol 210 I
4	CH$_2$ | CH$_3$—C—COO—(CH$_2$)$_5$—COO— [Chol]	55	S_A 218 I
5	O | CH$_3$—Si—(CH$_2$)$_3$—COO— [Chol]	45	S 115 I

* Chol is cholesterol.

TABLE 3.4. Phase transitions of LC comb-shaped polymers with identical mesogenic groups, close spacer length, but different main chains [31,34,37,41].

| | Mesogenic group and spacer | | | | |
| Main chain | $-(CH_2)_2-O-$⬡$-COO-$⬡$-OCH_3$ | | | $-(CH_2)_6-O-$⬡$-$⬡$-CN$ | |
	T_g (°C)	Phase transitions (°C)		T_g (°C)	Phase transitions (°C)
Polymethacrylic $[-CH_2-C(CH_3)-]$ $\mid$ OCOR	96	LC* 121 I		55	S 115
Polyacrylic $[-CH_2-CH-]$ $\mid$ OCOR	47	LC 77 I		35	S_A 133 N 136
Polymethylsiloxane CH$_3$ $\mid$ $[-Si-O-]$ $\mid$ R	15	LC 61 I		14	S 166

| | Mesogenic group and spacer | |
| Main chain | $-(CH_2)_6-O-$⬡$-$(pyrimidine N=, N)$-(CH_2)_7-CH_3$ | |
	T_g (°C)	Phase transitions (°C)
Polymethacrylic $[-CH_2-C(CH_3)-]$ $\mid$ OCOR	29	S_A 96 I
Polyacrylic $[-CH_2-CH-]$ $\mid$ OCOR	7	S_A 97 I
Polymethylsiloxane CH$_3$ $\mid$ $[-Si-O-]$ $\mid$ R	8	S 110 I

* LC is the liquid crystalline phase. The mesophase type is not determined.

backbone flexibility (glass transition temperatures decrease in the series from polymer 1 to polymer 5) [40].

A similar dependence is observed for polymers with a main chain structure similar to that of the polymers listed in Table 3.3 but incorporating the other type of mesogenic group (Table 3.4, on the left). However, the two columns on the right of Table 3.4 is evidence of the reverse mode; indeed, clearing points were found to increase with increasing backbone flexibility when cyanobiphenyl and pyrimidine mesogenic fragments were attached to the same backbones (see Table 3.4).

The contradictory data of Tables 3.3 and 3.4 (two last columns) can be accounted for by what different types of mesophases are formed in LC polymers bearing identical mesogenic groups, but differing in the structure of the backbone (we say these polymers comprise a "homologous series"); the other reason, already mentioned above, may be the variance of the molecular weight characteristics.

In any case, at the present state of the art, we can hardly count on making strict conclusions concerning the role of backbone flexibility with regard to the effects it imposes on the thermostability of the mesophase, thus disguising the targeted choice of monomers with the required temperature range of the mesophase. Studies of homologous series with strictly defined molecular weight characteristics, like those discussed above, are still one of the most vital and urgent tasks in the field involving synthesis and investigation of comb-shaped LC polymers.

3.4 Mesophase Types of Comb-Shaped Liquid Crystal Polymers. Problems of Classification

At present, quite a number of mesophases of various structural types has been observed in comb-shaped LC polymers. Indeed, bearing no pretense to give a complete list of the relevant publications, we still feel it appropriate to support the statement by naming some of the publications reporting on nematic [30,41,42], smectic A [43–47], smectic C [48–52], smectic B [53–56], smectic F [35,52,57], smectic G [57], smectic E [58,59], and smectic J [60] phases. The reentrant nematic phase was also discovered [35,61–63]. For chiral homo- and copolymer cholesterics, chiral S_C^* and blue phases were observed. Contemporaneously, there are no virtual reasons to prevent the manifestation of any definite structural type of LC ordering known for low molecular weight compounds in comb-shaped polymers. Moreover, there is a number of examples of a mesophase type having been discovered in polymers, before it was observed in "conventional" low molecular weight liquid crystals. For instance, in low molecular weight systems the chiral S_A phase [64] was first reported only four years after [65] it had been detected in cholesterol-containing polymers (see Section 3.7). The second example involves the so-called nematic B phase first reported in [66]; this one has not yet been observed in low molecular weight liquid crystals.

The simplest and most commonly used method for detection of the mesophase type is polarization optical microscopy. High viscosity of polymers appears to be an essential impediment to this technique, preventing quite often the formation of large-size textures. By prolonged (for several hours and sometimes more) annealing of a sample at temperatures slightly lower than T_{cl} (or the temperature of the transition between the two mesophases), characteristic textures allowing reliable identification of mesophases were obtained; the mesophases thus distinguished were marble and Schlieren textures for the N phase, fan-like and confocal for the S_A phase and broken fan-like for the S_C phase.

For the cholesteric mesophase the confocal texture and the "oily strikes" texture were observed, both exhibiting selective light reflection. However, application of this method is not always a success with polymers having relatively stiff chains (polymethacrylates) as well as with polymers of sufficiently large molcular weight; in these cases characteristic textures quite often failed to be obtained.

For low molecular weight liquid crystals the concept of miscibility [67] is also used to identify LC phases. According to these concepts, continuous and unlimited mutal miscibility of the two mesophases displayed by the two substances indicates that these mesophases are of the same structural type. This principle appears to offer the sufficient but not the necessary condition for identification of mesophases; indeed, even for low molecular weight liquid crystals, complete compatibility between identical mesophases was reported to be lacking in quite a number of examples. This is often ascribed to the effects related to the polarity, association, and steric interaction of molecules in mesophases [68,69]. For polymers, application of this concept is even more restricted because of the thermodynamic and kinetic features of the dissolution of a polymer, these features being associated with incommensurability of the molecular sizes of a polymer and a low molecular weight solvent.*

One of these features, in general characteristic for all polymers, not only the liquid crystalline ones, is the existence of large phase separation regions. For the mixtures of comb-shaped LC polymers and low molecular weight liquid crystals such biphasity was reported in [70,71]. The regions covering the two coexisting nematic phases and the two coexisting isotropic phases were discovered. A detailed investigation of the structure of low molecular weight liquid crystals, as related to their compatibility with liquid crystalline polysiloxanes, disclosed [72] that chemical affinity appeared to control compatibility to a larger extent than the mere coincidence of mesophase types.

The molecular weight of a polymer component appears to be one of

---

* See also Chapter 4.

the basic features responsible for compatibility. In studies of the binary mixtures of cyanobiphenyl-containing polyacrylates of various molecular weights, with cyano-substituted low molecular weight liquid crystal, complete compatibility in structurally alike mesophases was found to be characteristic only for oligomers with the DP less than 200. At larger DP compatibility in the nematic phase ceases to exist, whereas at $DP > 10^3$ phase separation starts in the S_A phase [73]. Therefore, identification of mesophases relying on miscibility with low molecular liquid crystals can probably be successful only in the case of LC oligomers.

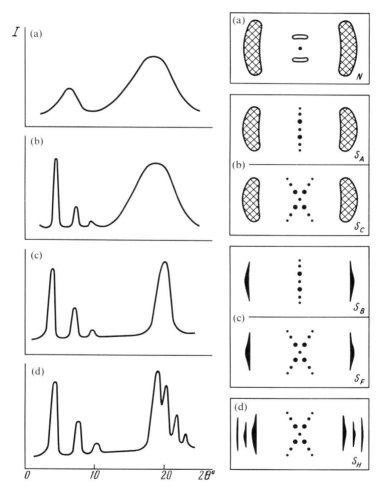

FIGURE 3.6. Schematic representation of typical x ray diffraction curves and patterns corresponding to the different types of mesophases of comb-shaped polymers [74].

X ray diffraction is the method that is most important for the identification of mesophases in LC polymers. x ray diffraction patterns of oriented, and especially of nonoriented, samples enable a sufficiently reliable identification of mesophases relying on the same principles as elaborated on for low molecular weight LC compounds. Figure 3.6 shows the schematic representations of x ray diffraction patterns corresponding to the different mesophases types that were observed in comb-shaped polymers.

Reflections at wide angles arise from the packing of mesogenic fragments, whereas the small-angle maxima are related to their longitudinal (along the LC director) packing. From the profiles of these reflections the correlation function $\gamma(x)$ can be computed

$$\gamma(x) = \frac{\int s^2 I(s) \cos(sx)\, dx}{\int s^2 I(s)\, ds}, \tag{3.1}$$

where $s = 2(\sin\theta)/\lambda$ and θ is half the scattering angle.

The packing perfection along any direction can be described by the correlation length ξ, which is determined from the die-away pattern of the correlation function along certain coordinate

$$\gamma(x) = \gamma_0 \exp(-2x/\xi). \tag{3.2}$$

Hence, such an analysis of wide- and small-angle reflections not only makes it possible to identify the mesophase type, but also enables a quantitative assessment of the quality (regularity) of the mesogenic groups arrangement along $(\xi_\parallel)$ the LC ordering director, and perpendicular $(\xi_\perp)$ to it. Let us examine the ratio between these quantities for some mesophase types of LC polymers.

The nematic phase of comb-shaped LC polymers is characterized by diffuse scattering at wide angles being concentrated at the equator with regard to the texture axis, and weak diffuse scattering at $2\theta \approx 1\text{--}6°$ at the meridian (Fig. 3.6(a)). This is indicative of weak correlation in packing of the polymer side groups in longitudinal ($\xi_\parallel \approx 2\text{--}6$ nm) and lateral ($\xi_\perp \approx 0.3\text{--}1.0$ nm) [74] directions.

However, recently in the acrylic polymer

$$[-CH_2-CH-]$$
$$COO-(CH_2)_5-COO-\langle O\rangle-OOC-\langle O\rangle-OCH_3 \quad (VIII)$$

a new type of nematic mesophase, as yet unknown for low molecular weight liquid crystals, was discovered [66]. For polymer (VIII) at temperatures below 60° the x ray patterns are quite unusual for the nematic liquid crystals: at wide scattering angles a single sharp reflection corresponding to the in-

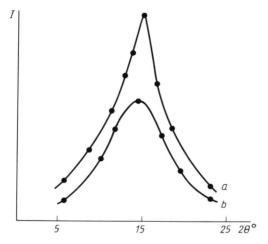

FIGURE 3.7. X ray diffraction curves of polymer (VIII) at (a) 25 °C and 70 °C (b) [66].

terplanar distance of 0.44 nm is observed, the small-angle reflexes lacking (Fig. 3.7(a)).

Such a pattern indicates that a nematic structure, with the ordered hexagonal arrangement of mesogenic groups displaying no periodicity along their long axes, is formed in this polymer (Fig. 3.8). The correlation length $\xi_\perp$ in the direction perpendicular to the long axes of mesogenic groups was calculated

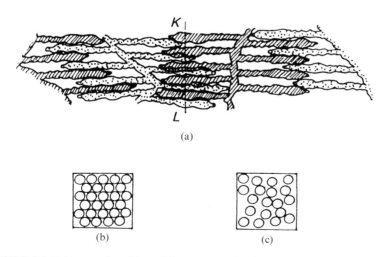

FIGURE 3.8. Schemes of packing of the macromolecules in nematic phases. (a) Packing of mesogenic groups in (b) N_B and N_A (c) phases in a plane KL, perpendicular to long axes of the side chains [66].

to be 7.0 nm. At temperatures above $60°$ a sharp reflection transforms to a diffuse halo (Fig. 3.7(b)) characteristic of a common nematic structure; the correlation length $\xi_\perp$ is only 1.5 nm. To distinguish between these two phases a conventional nematic phase was denoted as the N_A phase, whereas the hexagonally ordered one was denoted as the N_B phase. It is quite notable that the acrylic polymer with the like structure differing only in the orientation of the ester link between the two benzene rings

$$[-CH_2-CH-]$$

COO—$(CH_2)_5$—COO—⟨O⟩—COO—⟨O⟩—OCH$_3$ (IX)

forms only the N_A phase.

As in the case of low molecular weight compounds, small-angle reflections are related to the presence of sibotactic groups and small regions displaying layered smectic packing of the orthogonal type [75].

Nonordered smectic phases of A and C types are characterized by sharp small-angle maxima (usually with 1–2 reflection orders) proving the existence of extended layers (Fig. 3.6(b)). For the S_A phase these maxima are positioned at the meridian, whereas the S_C phase gives a four-dot reflection. According to [74] the values for $\xi_\parallel$ of these mesophases are within the 6–100 nm range. At the same time a diffuse reflection is retained at wide angles, whose width is not largely different from that of the nematic ($\xi_\perp \approx 0.4$–1.0 nm); however, it is more compact with regard to the azimuthal distribution (the order parameter of smectic A and C mesophases is usually larger than that of the nematic phases).

Hexatic smectic phases of B and F types with ordered layers are described by a single reflection at wide angles, with the intensity substantially higher and the half-width substantially less ($\xi_\perp = 2$–4 nm) than for the smectic phases with disordered layers. The small-angle scattering patterns are actually the same as for S_A and S_B phases, respectively (Fig. 3.6(c)). Davidson and Levelut [56] emphasize the substantial difficulties in the correct assignment of the polymeric S_B to one of the variants known for low molecular weight liquid crystals [76], either the hexatic smectic B phase or the B phase, which is a true but disordered crystal.* These impediments are natural for polymers. Indeed, it is difficult to obtain an equilibrium structure in a polymer (because of the high viscosity and the defectness of its structure arising from the polymer backbone); for the polymers investigated no truly crystalline phase was detected (therefore it is impossible to compare with the S_B phase observed).

An even greater uncertainty due to this cause arises when investigating polymer mesophases of E, G, H, and I types, which are usually considered as

---

* Closely related to these structural types are the structures of comb-shaped polyacrylates, polyvinyl ethers, and esters with long alliphatic substituents (containing no mesogenic groups), all forming the so-called rotational crystalline phase of hexagonal type [5]).

disordered crystals [76]. These phases (Fig. 3.6(d)) give rise to several sharp reflections at wide angles ($\xi_\perp = 4-10$ nm). Only for polymers investigated in [60], both the S_I mesophase and the truly crystalline phase were observed, enabling a sufficiently reliable classification of this mesophase. In other cases, only a single ordered phase is observed; hence the doubt that the sample is representative of a defect semicrystalline polymer sample, whose structure and degree of crystallinity are essentially dependent on thermal prehistory and orientation procedure.

When describing the scattering patterns arising from various mesophase types of LC polymers we confined ourselves to listing only the "major" reflections, i.e., the most intensive maxima at wide and small angles. However, besides these reflections, similar to the corresponding reflections from the low molecular weight mesophases, the x ray patterns often display a large number of diffuse reflections. As an example, Fig. 3.9 [77] shows the x ray diffraction pattern of the S_A phase.

The large number of "alien" (not specific for the low molecular weight liquid crystals) diffuse reflections which are not associated directly with the packing of mesogenic fragments (the essential limitations on the applicability to polymers of the miscibility concept), on which mesophase classification in low molecular weight liquid crystals rests, provoked Diele et al. [60] to sug-

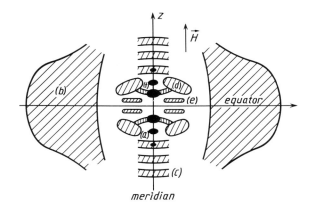

FIGURE 3.9. Schematic representation of the x ray diffraction pattern of the polymer

$$[-CH_2-\overset{\overset{\displaystyle CH_3}{\displaystyle |}}{\underset{\underset{\displaystyle COO-(CH_2)_6-O-\langle\bigcirc\rangle-COO-\langle\bigcirc\rangle-OC_4H_9}{}}{C}}-]$$

(a) Bragg spots; (b) large-angle diffuse crescents; (c) diffuse lines; (d) diffuse spots; (e) diffuse streaks; and (f) moustaches. (The inner circle going through the first Bragg spots is due to the small disorientation of the sample [77].)

gest specific notation for the mesophases in LC polymers, viz., P_N, P_{S_A}, P_{S_B}, etc. The symbol letter P is added to the conventional notation to denote that the mesophase is polymeric but not the true one. We might agree with the suggestion had it not been for the problem concerning the criterion that enable distinguishing between the N and P_N, S_A and P_{S_A} phases, respectively. If it is the lack of complete compatibility between polymer and low molecular weight liquid crystals in structurally alike mesophases that is used as this criterion, then, as emphasized above, it is not the "fault" of the structure of the polymer mesophase. Immiscibility is at first determined by the chain structure of polymer molecules. It is worth mentioning that for a large class of polymer liquid crystals, viz., LC oligomers, the miscibility concept appears to be valid. Moreover, in different phases the threshold for immiscibility lies at different molecular weight values of a polymer. Hence, this criterion appears to be rather ambiguous.

The criterion relying on the emergence of additional diffuse reflections in the mesophase x ray diffraction patterns also seems to be somehow indefinite. Indeed, along with Bragg's sharp maxima, reflections of this type were also registered for low molecular weight liquid crystals, especially those of a complex chemical structure (three benzene rings containing liquid crystals with polar substituents) displaying a number of feasible molecular packings [78]. In particular, the long spacing d of the polymeric S_A phase, presumed to reflect the undulations of the smectic interface [79,80], was also observed for low molecular weight compounds.

A number of authors [46,56,77,80,81] have undertaken attempts to offer a reasonable explanation for the diffuse reflections shown in Fig. 3.9. Some of the reflections (for instance, reflection c) are associated with the packing modes of mesogenic groups, others (reflection f) are related to dislocations arising from the polymer backbone. Hence, up to date, there are no adequate grounds for introducing special notation to denote polymer mesophases. At the same time, we believe that the efforts spared, on elaborating structural classification specifically for polymers, stimulate the development of this field and might be successful in the future.

3.5   Effect of Mesophase Type on the Structure of Polymers Oriented by Uniaxial Drawing

Studies with oriented polymer samples, prepared by drawing the fibers from LC polymer melts, disclosed that the structure of such fibers is determined by the mesophase type of the melt used to produce the fiber [82]. Uniaxial drawing of nematic polymers leads to the fact that the wide angle maximum in the x ray patterns splits into equatorial diffuse arcs (Fig. 3.10). This holds true for the N_A phase as well as for the N_B phase. The polymers in smectic S_A and S_C phases are also easily drawn into fibers. Figure 3.11 depicts the x ray

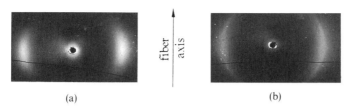

(a)   (b)

FIGURE 3.10. X ray diffraction patterns of uniaxially oriented polymers VIII and IX in N_B (b) and N_A (a) phases [82].

patterns of such polymers, which are in good agreement with the general scheme of Fig. 3.6.

Polymers in the "low mobility" ordered S_B and S_F phases cannot be oriented by the drawing process. Their orientation is plausible only when starting at higher temperatures from the two-dimensional ordered or nematic phases. In this case, on cooling, the S_A, S_C, or a nematic phase of the fiber prepared, undergoes the transition to the S_B or S_F phase. X ray patterns of such polymers are exemplified in Fig. 3.12. As is seen, in samples obtained by orientation in either of the smectic phases, small-angle reflections are always split into dots lying at the equator, whereas the wide-angle reflection is split into two meridional arcs (for phases with orthogonal position of the side groups S_A and S_B) or into four arcs (for phases with a tilted arrangement of the side groups S_C and S_F). The x ray pattern of the polymer forming the

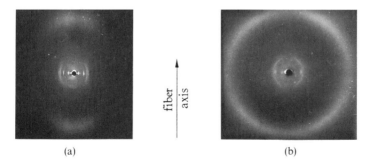

(a)   (b)

FIGURE 3.11. X ray diffraction patterns of uniaxially oriented polymers in S_A (a) and S_C (b) phases.

(a) $[-CH_2-C(CH_3)-]$
           |
           $OCO-(CH_2)_{11}-O-\langle\bigcirc\rangle-\langle\bigcirc\rangle-CN$ S_A 121 °C   I

(b) $[-CH_2-CH-]$
           |
           $OCO-(CH_2)_{11}-O-\langle\bigcirc\rangle-\langle\bigcirc\rangle-CN$ S_A 145 °C   I
                                                                                    S_C 30 °C

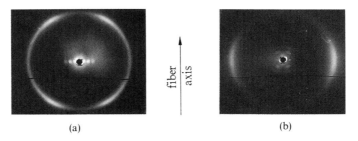

FIGURE 3.12. X ray diffraction patterns of uniaxially oriented acrylic polymers in the S_F phase

(a) $[-CH_2-CH-]$
      $OCO-(CH_2)_5COO-\bigcirc-COO-\bigcirc-OC_4H_9$

      S_F 45 °C S_C 145 °C   I

(b) $[-CH_2-CH-]$
      $OCO-(CH_2)_5COO-\bigcirc-OOC-\bigcirc-OC_4H_9$

      S_F 70 °C N_A 140 °C   I

The oriented samples of polymers (a) and (b) were prepared by stretching from the S_C and N_A phases, respectively [82].

smectic S_F phase, on cooling of the fiber obtained in the nematic phase, shows that as a result of the transition a wide-angle reflection is split into equatorial arcs, while the small-angle ones are split into four dot reflections (Fig. 3.12). An additional splitting of one of the small-angle reflections observed in cyanobiphenyl polymer is associated with the coexistence of two types of the layer packing of mesogenic groups (Fig. 3.11(b)).

Such an appearance of x ray patterns is related to the specific features of the orientation process manifested by polymers with mesogenic side groups in different types of mesophases.

3.5.1   Nematic Polymers

Orientation of polymers in the nematic phase is accompanied by arrangement of the side groups along the fiber axis. Such behavior of nematic polymers in a mechanical field appears to be a manifestation of the specific features of the deformation process of a polymer in the liquid crystalline state displaying regions with uniformly oriented mesogenic groups (the so-called domains). It is well known that in low molecular weight liquid crystal appli-

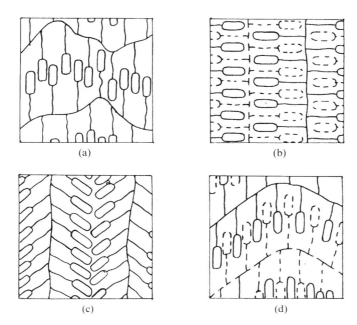

FIGURE 3.13. Schemes of macromolecular packing in oriented polymers in N_A and N_B (a), S_A and S_B (b), S_C and S_F (oriented in S_C) (c), and S_F (oriented in N_A) (d) phases [82].

cations of various fields (electric, magnetic, mechanical) causes orientation of the liquid crystal domains. In the polymers considered, the domains are formed by mesogenic side groups whose long axes are aligned during orientation along the fiber axis. Assuming such an orientation pattern of the mesogenic groups, the main chains of the macromolecules are more liable to be in a disordered conformation. Figure 3.13(a) shows a schematic presentation of the structure of an oriented nematic polymer.

It was also discovered [83,84] that, for nematic polymers with short aliphatic spacers, orientation results, in that the wide angle maximum splits into equatorial arcs. Zentel and Strobl [85] attempted to give an explanation for the observed phenomenon. They reject the possibility of mesogenic groups being aligned along the fiber axis, presuming that in this case the main chain should inevitably be perpendicular to this direction. What they suggest as an explanation is that the main chains are parallel, whereas mesogenic groups are perpendicular to the fiber axis, whereas macromolecules on the whole may be considered as a flat "ribbon-like" structure. The x ray diffraction pattern arises as a reflection from the parallel "ribbons." However, this explanation is inconsistent with the experimental fact (reported in [85] and observed by us) that the wide angle maximum corresponds to the distance between the nearest mesogenic groups.

3.5.2 Smectic Polymers

The sliding of layers with respect to each other is the easiest mode involved in the orientation of smectic polymers. The obvious consequence is that smectic layers are aligned along the force vector. Orientation of mesogenic groups is determined by the type of the smectic mesophase: in the S_A phase mesogenic groups are positioned normal to the smectic layers and, consequently, to the fiber axis, whereas in the S_C phase they are tilted (Fig. 3.13(b, c)).

The arrangement of smectic layers and mesogenic groups in ordered S_B and S_F phases depends on which was the precursor phase from which the fiber was formed. For instance, polymer (X) is subjected to orientation in the S_A phase, therefore, in the oriented sample, which is in the S_B phase, the smectic layers are aligned along the fiber axis while the mesogenic groups are perpendicular to the layers

$$[-CH_2-\underset{\underset{COO-(CH_2)_{11}-O-\bigcirc-CH=N-\bigcirc-C_4H_9}{|}}{\overset{\overset{CH_3}{|}}{C}}-] \qquad (X)$$

$$S_B\ 86\ ^\circ C \quad S_A\ 140\ ^\circ C \quad I$$

$$[-CH_2-\underset{\underset{COO-(CH_2)_5-COO-\bigcirc-COO-\bigcirc-OC_4H_9}{|}}{CH}-] \qquad (XI)$$

Polymer (XI) can be oriented in the S_C phase; in this case, the smectic layers are aligned along the fiber axis, whereas the mesogenic groups are at an angle of 34° to the fiber axis. In the course of the transition to the S_F phase an ordering of mesogenic groups in layers takes place, their orientation being retained (Fig. 3.13(c)). Polymer (V) is oriented in the nematic phase where the mesogenic groups are parallel to the fiber axis. Transition to the S_F phase is accompanied by displacement of mesogenic groups along their long axes, their orientation being retained, and formation of layers which are positioned at an angle to the fiber axis (Fig. 3.13(d)).

The splitting of a wide-angle maximum into meridional arcs, and of the small-angle reflections into the equatorial ones was also reported for smectic polymers by Zentel and Strobl [85]. They also discovered that resulting from the orientation of nematic polymers with long aliphatic spacers was the splitting of the wide-angle maximum into meridional reflections and the emergence of small-angle equatorial reflections, proving the formation of a sibotactic nematic mesophase. We believe that it is the existence of layer structures in these polymers that leads to the orientation of such samples, which proceeds via the same modes as orientation of all smectic polymers.

Hence, relying on the studies of a large number of uniaxially drawn polymers with mesogenic side groups, we may infer that the mode of ordering of the mesogenic side groups in such polymers is controlled by the type of mesophase from which the oriented species was prepared.

3.6 Placement of the Main-Chain in Smectic and Nematic Phases of Comb-Shaped Liquid Crystal Polymers

As was shown above, the chemical nature of the main chain and its length substantially affect both the temperatures of phase transitions and the type of mesophase.

The task of the structural investigation is to trace the placement of the main chain in the overall packing, to "visualize" it amongst the relatively bulky side groups. It is worth noting that the number of studies devoted to the subject is yet very small. Therefore, today it is hardly possible to proceed from the structural data to clarification of the known relations between the properties of the mesophase and the chemical structure of the main chain.

The major problem in applying x ray analysis techniques for investigation of the packing of main chains involves the necessity to identify the scattering arising from these regions, since its intensity amounts to only several percent of the overall polymer scattering intensity.

Lipatov et al. [86] succeeded in increasing the fraction associated with the main chain by introducing "heavy" atoms into the backbone. For instance, bromoacrylates

$$[-CH_2-CBr-]$$
$$COO-\bigcirc-COO-\bigcirc-OC_4H_9 \qquad (XII)$$

were shown to give a conspicuously higher intensity of the small-angle maximum (Fig. 3.14), as compared to conventional polymethacrylates

$$[-CH_2-C(CH_3)-]$$
$$COO-\bigcirc-COO-\bigcirc-OC_4H_9 \qquad (XIII)$$

proving that bromine atoms are predominatly localized within the planes parallel to the smectic layers. In order to determine the ordering pattern of the main chains within these pseudolayers, the wide angle maximum for polymethacrylate was substracted from that for polybromoacrylate. The resultant curve represented a diffuse maximum proving the liquid-like packing of the main chain fragments in the S_A phase (Fig. 3.14).

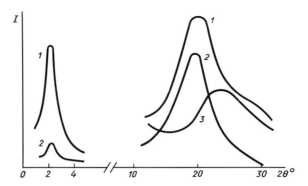

FIGURE 3.14. X ray diffraction curves of polymers (1) (XII), (2) XIII and their differential curve 3 [86].

The same inference appeared to be valid for polysiloxane in the S_A phase [87]

$$[-\underset{\underset{(CH_2)_3-O-(CH_2)_2-O-\langle\bigcirc\rangle-CN}{|}}{\overset{\overset{CH_3}{|}}{Si}}-O-]$$ (XIV)

Further development of the concepts concerning the shape of comb-shaped LC polymer macromolecules is associated with the application of the small-angle neutron scattering technique. In fact, by using mixtures of partially deuterated polymers oriented in the magnetic field

$$[-CD_2-\underset{COO-(CH_2)_6-O-\langle\bigcirc\rangle-COO-\langle\bigcirc\rangle-OC_4H_9}{\overset{|}{C}(CD_3)-}]$$ (XV)

$$[-CD_2-\underset{COO-(CH_2)_6-O-\langle\bigcirc\rangle-COO-\langle\bigcirc\rangle-OCH_3}{\overset{|}{C}(CD_3)-}]$$ (XVI)

$$[-\underset{CD_2-CHD-CD_2-O-\langle\bigcirc\rangle-\langle\bigcirc\rangle-CN}{\overset{\overset{CH_3}{|}}{Si}}-O-]$$ (XVII)

with nondeuterated analogues, it was demonstrated [88–92] that the conformation of macromolecules within mesophases was anisotropic. The macromolecules could be simulated by ellipsoids with axes (projections of

TABLE 3.5. The values of the radii of gyration, R_g, $R_\parallel$, and $R_\perp$ (in Å) of labeled LC polymers in the different mesophase types.

Polymer	Isotropic melt (R_g)	Nematic		Smectic A		Reentrant nematic		Ref.
		$R_\perp$	$R_\parallel$	$R_\perp$	$R_\parallel$	$R_\perp$	$R_\parallel$	
(XV)	61	65	59	86	22	—	—	[89]
(XVI)	92	120	94	—	—	—	—	[90]
(XVII)	13	—	—	16	10	—	—	[91]
(XVIII)	52	49	48	51	31	38	58	[92]

the gyration radius of a polymer coil) parallel ($R_\parallel$) and perpendicular ($R_\perp$) to the mesophase director.

As proved by the data of Table 3.5, in nematic and S_A phases, the ellipsoid is oblate ($R_\perp > R_\parallel$), the degree of the coil anisotropy ($R_\perp/R_\parallel$) in the smectic A phase reaching the value of 4. How can we imagine the packing of the main chain in such a coil? Relying on a generalized and comprehensive treatment of the neutron and x ray scattering studies, Pepy et al. [93] suggested the model delineated in Fig. 3.15(a); the main chain is seen to be placed in a single smectic layer repeatedly, and regularly passing across it. Having generalized the data of [89] and [90], the authors of [91,94] suggested another model shown in Fig. 3.15(b). The model stands on two basic assumptions: local stiffness of macromolecules is anisotropic, i.e., it is different along the layer and in the normal-to-the-layer direction; the main chains segregate from out of the smectic layer formed by the side groups into a separate phase (microphase) and form a two-dimensional layer (quasi-two-dimensional coil).

The main chain may pass from one layer into another, although this process is retarded by the chain stiffness anisotropy. Such chain migration results in the arrival of structural defects, "splitting" a macromolecule into a number of quasi-two-dimensional subcoils belonging to different smectic layers. Mathematical apparatus developed to describe the model makes it possible to assess the number of such subcoils and the magnitude of the Kuhn's segment for a two-dimensional coil. The magnitude of this parameter ($b_\perp = 63$ Å) in the S_A phase (polymer (XVI)) appears to be substantially larger than in the isotropic phase ($b_\perp = 39$ Å) indicating that the chain stiffness in the two-dimensional subcoil of the S_A phase is substantially higher.

For the polymer

$$[-CD_2-CD-]$$
$$COO-(CH_2)_6-O-\langle\bigcirc\rangle-\langle\bigcirc\rangle-CN \quad (XVIII)$$

forming the N, S_A, and reentrant nematic phases [92] the data on neutron scattering demonstrate that the changes in the packing of mesogenic groups

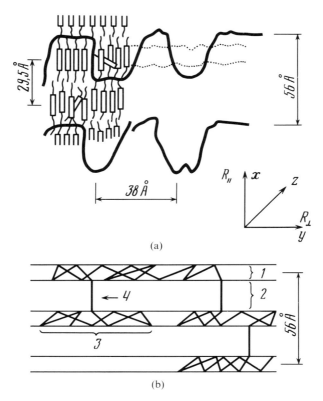

(a)

(b)

FIGURE 3.15. Proposed conformational models of the backbone of comb-shaped macromolecules in the smectic mesophase of polymers (3.XV) [93] and (XVII) [91]. (1) Intermediate layers, in which the backbones are disposed; (2) mesogenic groups layers; (3) quasi-two-dimensional subcoils; and (4) tie segments.

accompanying various phase transitions may affect the ratio between the axes of the macromolecular coil so dramatically as to change the very pattern of the ellipsoid from the oblate to the prolate one.

Indeed, in the course of the $S_A \rightarrow RN$ transition, the macromolecular coil sustains a transformation from the oblate ($R_\perp/R_\parallel \approx 1.6$) to the prolate ($R_\perp/R_\parallel \approx 0.6$) shape reflecting, as presumed by [92], the transition of mesogenic groups from orthogonal position in the smectic layers (S_A phase) to the tilted one in sibotactic groups (RN phase).

Contemporaneously, widespread application of the neutron scattering technique is somehow retarded by the scarcity of deuterated polymers. However, the unique information provided by the method would apparently promote this highly promising trend.

3.7   Structure of Cholesteric Polymers

Homo- and copolymers comprising cholesterol fragments as mesogenic groups represent the most common type of cholesteric polymers. Major representatives of these two groups of cholesteric polymers are listed in Tables 3.3 and 3.6. As is seen, for all homopolymers and the majority of copolymers, the smectic mesophase appears to be most common. Hence, we will start our introspect into the structure of such polymers by considering first the smectic mesophase in homopolymers; such an approach is supported by the fact that the cholesteric mesophase inherits many of the features of the "precursor" smectic structure.

3.7.1   Structure of Cholesterol-Containing Homopolymers

Smectic Mesophase

A detailed study of the structure of oriented and nonoriented polymer samples, some of them are summed up in Table 3.3, demonstrated that at temperatures below the transition to the cholesteric phase the existing mesophase is of the S_A type characterized by the fact that the macromolecular backbones lie flat within the smectic layers, whereas the mesogenic side groups are positioned normal to the plane [95,96]. The type of side group packing depends on the length of the methylene spacer and on the chemical nature of the bond linking the side group to the backbone.

For instance, in polymers with an amide link and a long spacer ($n > 7$) an antiparallel packing of the side groups is observed, whence the methylene fragments of one branching are surrounded by cholesterol fragments of the other branches and vice versa (Fig. 3.16(a)) [95]. The same packing holds true for polymers with the ester link and a spacer with $n > 11$ [96–98]. In these cases, the thickness of a smectic layer corresponds to the length of the side branching. This is the so-called one-layer packing. As is seen from the figure, such a structural model is feasible only when the length of the methylene sequence is sufficient to let the "rigid" cholesterol fragment align with the axis of the methylene chain. In polymers with no spacer (polycholesteryl acrylate and methacrylate) or with the short one ($n > 5$) there is no chance for the cholesterol fragment to accommodate itself in antiparallel packing, hence, the observed arrangement is that of parallel packing (Fig. 3.16(c)). In this case, the thickness of the smectic layer is double the length of the side group, i.e., the packing is bilayer. A similar structure with a bilayer parallel arrangement of the side groups was reported in poly(cholesteryl-*p*-acryloyl oxybenzoate) [99,100]. In polymers with the amide bond and a methylene spacer of 5–7 methylene units, as well as in polymers with the ester bond and 9–11 methylene units, long spacer bilayer parallel packing with partial overlapping of the alkyl "tails" at the seventeenth carbon atom is most common (Fig.

TABLE 3.6. Cholesterol-containing copolymers.

Series	Structure of copolymer	Content of chiral units (mol. %)	Phase transitions (°C)
1.1	$\mathrm{CH{-}COO{-}(CH_2)_5{-}COO{-}\bigcirc{-}COO{-}\bigcirc{-}OCH_3}$	17	Ch 121 I
1.2	$\mathrm{CH_2}$	21	Ch 110 I
1.3	⋮	24	Ch 127 I
1.4	$\mathrm{CH{-}COO{-}(CH_2)_{10}{-}COO{-}Chol}$	28	Ch 118 I
1.5	$\mathrm{CH_2}$	38	Ch 115 I
2.1	$\mathrm{CH{-}COO{-}(CH_2)_5{-}COO{-}\bigcirc{-}COO{-}\bigcirc{-}OC_3H_7}$	28	S_A 110 Ch 118 I
2.2	$\mathrm{CH_2}$	31	S_A 98 Ch 110 I
2.3	⋮	40	S_A 80 Ch 96 I
2.4	$\mathrm{CH{-}COO{-}(CH_2)_{10}{-}COO{-}Chol}$ / $\mathrm{CH_2}$	55	S_A 85 Ch 104 I
3.1	$\mathrm{CH{-}COO{-}(CH_2)_5{-}COO{-}\bigcirc{-}OOC{-}\bigcirc{-}OCH_3}$	21	N_B^* 40 Ch 120 I
3.2	$\mathrm{CH_2}$	34	Ch 113 I
3.3	⋮	41	Ch 121 I
3.4	$\mathrm{CH{-}COO{-}(CH_2)_{10}{-}COO{-}Chol}$ / $\mathrm{CH_2}$	55	Ch 136 I
4.1	$\mathrm{CH{-}COO{-}(CH_2)_5{-}COO{-}\bigcirc{-}OOC{-}\bigcirc{-}OC_4H_9}$	17	S_F^* 50 Ch 125 I
4.2	$\mathrm{CH_2}$	20	S_F^* 50 Ch 120 I
4.3	⋮	34	S_A 70 Ch 121 I
4.4	$\mathrm{CH{-}COO{-}(CH_2)_{10}{-}COO{-}Chol}$	43	S_A 92 Ch 119 I
4.5	$\mathrm{CH_2}$	49	S_A 70 Ch 110 I

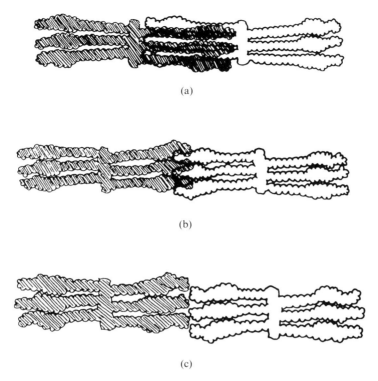

(a)

(b)

(c)

FIGURE 3.16. Schematic representation of the packing of macromolecules of the cholecteric homopolymers: (a) single-layer antiparallel packing; (b) intermediate packing with partial overlapping of cholesterol groups; and (c) double-layer parallel packing. The shaded molecules lie in the plane parallel to the plane of the figure.

3.16(b)). At the same time, a one-layer antiparallel packing is also observed, so that both packing types actually coexist.

Analysis of the phase behavior of cholesterol-containing polymers has shown that only those with the ester link and a spacer of 5–11 methylene units long form the cholesteric mesophase. For polymers with a spacer of five methylene units, the temperature range for the cholesteric mesophase is so high that the polymer is subject to thermal decomposition. Therefore, temperature-dependent structural changes were most scrupulously examined in polymers (XIX) and (XX) [65]

$$[-CH_2-CH-]_n$$
$$COO-(CH_2)_{10}-COO-$$

$$CH_3$$
$$CH_3$$
$$CH-(CH_2)_3-CH-CH_3$$
$$CH_3$$

(XIX)

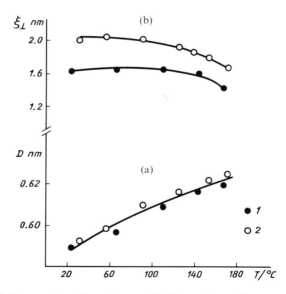

(XX)

Increasing the temperature to T_{Ch-1} does not affect the general features of the mesophase structure in polymers (XIX) and (XX), however, some structural rearrangement takes place. As is known, the half-width of the diffraction maxima in the x ray patterns of polymers is directly related to the perfection in the short-range order. Assuming that the profile of the wide-angle maximum abides by the Lorentz shape, we can assess the correlation length in the plane of the smectic layer ξ by using the relationship $\xi = \lambda/(\beta - \beta_0)$, where λ is the x ray radiation wavelength, β is the half-width of the diffraction maximum, and β_0 is the half-width of the instrumental function. At 25 °C the value for ξ was calculated to be 2.0 nm and 1.6 nm for (XX) and (XIX), respectively.

When the temperature is elevated the mean intermolecular distances corresponding to side packing increase, whereas the correlation length slightly decreases (Fig. 3.17). The increase in mean intermolecular distances is due to thermal expansion of the samples; within the precision of the experimental procedure it is equal for both samples.

From the half-width of the small-angle maxima the correlation length of

FIGURE 3.17. Temperature dependences of (a) interplanar distance and (b) correlation length for polymers (XIX) (1) and (XX) (2).

the longitudinal packing of mesogenic groups $\xi_\parallel$ was estimated. For both polymers, (XIX) and (XX), the magnitude of $\xi_\parallel$ is not less than 30 nm. On increasing the temperature up to transition T_{Ch-1} small-angle maxima sustain certain broadening and their intensity is somewhat redistributed, indicating that the contribution of the packing with partially overlapping mesogenic groups is reduced.

It is worth mentioning that, besides the cholesteric mesophase, low molecular weight cholesterol esters with sufficiently long alkyl substituents also show the smectic S_A phase. Wendorff and Price [101] observed identical packing patterns of the alkyl groups in the S_A phase in cholesterol-containing compounds, cholesteryl nonanoate and cholesteryl myristate, which are closely related in the structural aspect to the molecules of cholesteryl containing homopolymers; all showed a characteristic antiparallel packing. At the same time, other packing patterns are displayed by polymers with short spacers. Thus, the packing pattern of the side groups of cholesterol-containing homopolymers in the smectic phase is governed by the spacer length.

Cholesteric Mesophase

For a cholesteric mesophase, the small angle x ray scattering pattern, although changed, as compared to that for the smectic phase, retains sufficiently intensive reflections (Fig. 3.18) evidencing the layer packing with a one-layer antiparallel positioning of the cholesteric side groups.

In low molecular weight liquid crystals the cholesteric mesophase is, as a rule, considered as the twisted nematic phase [102]. An essential distinction

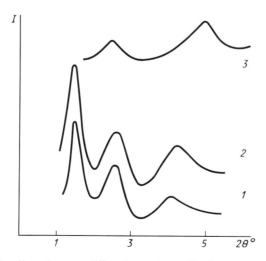

FIGURE 3.18. Small-angle x ray diffraction patterns for the polymer (XX) in S_A (1,2) and Chol(3) phases at (1) 20 °C, 90 °C (2), and 140 °C (3).

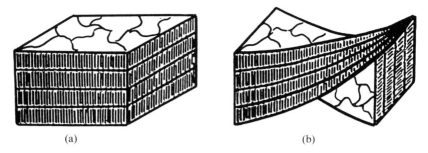

FIGURE 3.19. Schemes of the packing of mesogenic groups in (a) S_A and cholesteric (b) phases of side-chain liquid crystal polymers.

of the cholesteric mesophase in polymers pertains to explicit layer ordering; this feature prevents as from considering it as the twisted nematic. According to the model first proposed in [65,103], the cholesteric mesophase of a polymer is interpreted as the S_A smectic phase in which the layers are twisted along the axis normal to the long axis of mesogens, so that any cross section perpendicular to the axis of twisting shows a structure specific to the S_A phase (Fig. 3.19). Obviously, the twisting of the layers should lead to a certain distortion of these.

Coexistence of the helicoidal structure with the layer ordering may be delineated by way of the following speculation. Consider the length of the nondistorted fragment of the smectic layer 3.5 nm thick (this value corresponds to the layer thickness in homopolymers (XIX) and (XX)). Assuming the wavelength of light selectively reflected by the cholesteric structure to be $\lambda = 300$ nm, the helix pitch P calculated by the formula $P = \lambda/n$ (where n is the average refractive index equaling 1.5) is 200 nm. The angle by which the neighboring molecules are twisted around the helix axis may be calculated (Fig. 3.20). Indeed, assuming a cholesterol molecule to be 0.6 nm thick and the helix pitch 200 nm, there are $200:0.6 = 333$ molecules to a pitch; the angle is then $\alpha = 360:333 = 1.08°$. Since the projection of the molecule onto

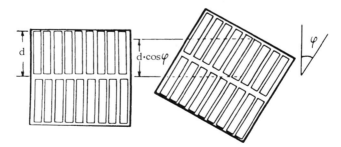

FIGURE 3.20. The size of a layered structure in the cholesteric mesophases of a polymer.

the normal to the plane of the layer decreases when the molecule is rotated, the distortion of the layer may be presumed to commence when the difference between the thickness of the smectic layer and the projection of the long axes of the mesogenic groups onto the normal exceeds the experimental error for determination of the interlayer distance (± 0.1 nm). The limiting angle φ by which the molecules may deviate still remains in the undistorted fragment of the layer and is given as arc $\cos(3.3/3.5)$, i.e., $\varphi = 19.5°$. Taking into account that $\alpha = 1.08$, the number of molecules within the undistorted fragment of the layer is determined as $19.5 : 1.08 = 18$ molecules, whereas the protraction of this fragment is $0.6 \times 18 = 10.8$ nm. Hence, despite the fact that the mesogenic groups are twisted with respect to each other, sufficiently protracted areas display explicit layer structure. However, when the layers are twisted, correlation between them becomes conspicuously worse. For instance, the correlation length ξ describing the distance of the between-the-layers correlated ordering in the smectic mesophase exceeds 30 nm (i.e., it expands to more than nine smectic layers), whereas in the cholesteric mesophase this distance is only 20 nm, i.e., roughly equivalent to six layers.

The model suggested to describe the structure of the cholesteric mesophase in polymers seems, at first glance, impossible. Indeed, it is hard to conceive the smectic layers, which are usually considered infinite, being twisted into a helix. However, recently, Renn and Lubensky [104] have predicted theoretically that in the incipient Ch–S$_A$ transition region a cholesteric mesophase displaying certain layer ordering can be formed. A similar layered helical structure was recently discovered for some low molecular weight LC compounds [64]. The specific features of polymer structure in the cholesteric mesophase that have been discussed may be related to the fact that the temperature range spanning this phase is very narrow and close to the transition of the smectic S$_A$ mesophase.

3.7.2 Structure of Cholesterol-Containing Copolymers

As is evident from the data of Table 3.6, the variety of cholesterol-containing copolymers comprises copolymers prepared by copolymerization of cholesterol-containing monomers with either nematogenic or smectogenic monomers. While the former form the cholesteric mesophase only, the latter are liable to display cholesteric as well as various smectic phases. Let us begin the survey of the copolymers structure with mesophases of the smectic type.

Smectic Mesophases

Smectic mesophases are formed in the copolymers of cholesterol-containing monomers with phenylbenzoate monomers bearing a terminal alkyl substituent of 3–4 carbon atoms. Copolymers of series 4 (Table 3.6) are the most scrupulously examined representative of this type of copolymers; it is by way of this example that we will elaborate on the structure of smectic mesophases.

The homopolymer with a phenylbutoxybenzoate mesogenic group was shown to form the S_F phase at room temperature (Table 3.2). At a low content of cholesterol units (17 and 20 mol.%), it is the packing of phenylbenzoate units that determines the mesophase structure in copolymers. The x ray patterns of these copolymers at room temperature display a single sharp reflection at wide scattering angles and three reflections at small angles; all reflections coincide with those characteristic of the homopolymer. Hence, the copolymers retain the tilted arrangement of mesogenic groups. At the same time, the chiral cholesteric units initiate helicoidal twisting rendering the copolymers selectively reflecting light. Thus, the chiral smectic S_F^* phase is realized in these copolymers.

Copolymers containing more than 25 mol.% cholesterol units demonstrate a different structure. In these, the sharp reflection at wide angles is transformed into an amorphous halo, the interplanar distances, calculated from the small-angle reflections, exceeding the corresponding values for the homopolymer. On a uniaxial orientation of the samples, small-angle reflections are shaped to form equatorial arcs, whereas the wide-angle reflection is transformed to a meridional arc, proving the existence of the S_A mesophase. These copolymers "inherit" the bilayer packing characteristic of the phenylbutoxybenzoate homopolymer; in copolymers with a 25–40 mol.% cholesterol units content the bilayer thickness equals the double length of the phenyl benzoate group, whereas on a further increase in the cholesterol units content the values for interplanar distances get closer to the corresponding values for the cholesterol-containing homopolymer.

The transition of the S_F phase characteristic of phenylbenzoate homopolymers to the S_A phase in the copolymers may be accounted for within the framework of the speculations that follow. As suggested by Gray and Goodby [105] the formation of tilted smectic mesophases is associated with the tendency of the dipoles, positioned at an angle to the long axes of mesogenic groups, to acquire an orientation parallel to the layer plane. In the copolymers, bulky cholesterol-containing units disguise the interaction between the dipoles. As a result, the ordering within the layers is disturbed and the mesogenic groups become oriented normal to the layer plane.

Cholesteric Mesophase

The structure of copolymers in the cholesteric mesophase depends on both the copolymer composition and the temperature.

For all copolymers surveyed above the cholesteric mesophase arrives as a result of the fusion of the smectic S_F or S_A phases. X ray patterns of these copolymers in the cholesteric mesophase display the same small-angle reflections as those featuring the smectic phase; however, their intensity decreases substantially when the temperature is increased.

The data accumulated in the x ray studies of the copolymers with nematogenic monomers (series 1 and 3 in Table 3.6) demonstrated that at a low (less than 25 mol.%) cholesterol units content the structure of copolymers retains the packing pattern of mesogenic groups in the corresponding phenylbenzoate homopolymers.

For instance, x ray patterns of the copolymers of series 1 (Table 3.6) and analogous copolymers with other cholesterol-containing monomers with a content of cholesterol units less than 25 mol.% at room temperature are similar to the x ray patterns of the corresponding (IX) [106]. When the content of the cholesterol units is increased the distance between the mesogenic side groups increases and small-angle reflections emerge on the x ray patterns. The latter fact discloses the tendency of cholesterol side groups to a certain layer ordering. When the temperature is elevated the intensity of small-angle reflections gradually decreases while the layer ordering vanishes.

At room temperature, the x ray pattern of the copolymer (3.1) of Table 3.6 reveals only a single sharp reflection corresponding to the interplanar distance of 0.44 nm, whereas no small-angle reflections are observed. At the same time, the copolymer exhibits a planar texture selectively reflecting visible light. These results have led us to assume a cholesteric structure with mesogenic groups arranged in hexagonal order in the direction coinciding with the axis of the cholesteric helix. This mesophase type was called the chiral nematic N_B^* phase. At temperatures above 40 °C, a conventional cholesteric structure, indiscernible from the structure of the copolymers of series 1 (Table 3.6), is formed. Hence, the structure of cholesterol-containing copolymers appears to result from the superposition of the structures of the corresponding homopolymers, containing respective mesogenic side groups. This results in the formation of unusual structures of the S_F^* and N_B^* types, as well as in the specific features arriving in the structure of the cholesteric phase.

References

1. V.P. Shibaev and N.A. Platé, Vysokomolek. Soedin. A **19**, 923 (1977).

2. S.P. Papkov and V.G. Kulichikhin, *Liquid Crystalline State of Polymers* (Chemistry, Moscow, 1977).

3. *Mesomorphic Order in Polymers*, edited by A. Blumstein. ACS Symposium Ser., No. 74, Washington, 1978.

4. *Liquid Crystalline Order in Polymers*, edited by A. Blumstein (Academic Press, New York, 1978).

5. N.A. Platé and V.P. Shibaev, *Comb-Shaped Polymers and Liquid Crystals* (Chemistry, Moscow, 1980 (in Russian); new edition, Pergamon Press, New York, 1987).

6. Ya.B. Amerik and B.A. Krentsel, *Chemistry of Liquid Crystals and Mesomorphic Polymer Systems* (Chemistry, Moscow, 1981).

7. *Liquid Crystal Polymers* I, II, III, *Advances in Polymer Science*, edited by M. Gordon and N.A. Platé (Springer-Verlag, Berlin, 1984).

8. *Polymer Liquid Crystals*, edited by A. Ciferri, W.R. Krigbaum, and R.B. Meyer (Academic Press, New York, 1982).

9. *Recent Advances in Liquid Crystal Polymers*, edited by L. Chapoy (Elsevier, London, 1985).

10. *Polymeric Liquid Crystals*, edited by A. Blumstein (Plenum, New York, 1985).

11. *Liquid Crystalline Polymers*, edited by N.A. Platé (Chemistry, Moscow, 1988). (English translation by Plenum, New York, 1993.)

12. *Side Chain Liquid Crystal Polymers*, edited by C. McArdle (Blackie, London, 1989).

13. V.P. Shibaev, *Chemical Fibres*, No. 3, 4 (1987) (in Russian).

14. V.P. Shibaev and S.V. Belyaev, Vysokomolek. Soedin. A **32**, 2266 (1990).

15. V. Kulichikhin and N.A. Platé, Vysokomolek. Soedin. A **33**, 3 (1991).

16. L. Onsager, Ann. N.Y. Acad. Sci. **5**, 627 (1949).

17. P. Flory, Proc. Roy. Soc. London A **324**, 73 (1956).

18. C. Robinson, Trans. Faraday Soc. **52**, 571 (1955); Tetrahedron **13**, 219 (1961).

19. V. Tzvetkov, *Rigid-Rode Polymer Molecules* (Nauka, Leningrad, 1986).

20. V.P. Shibaev, Dissertation for the degree of Doctor of Sciences, Moscow State University, Moscow, 1974.

21. Ya.S. Freidzon, V.P. Shibaev, and N.A. Platé, Abstracts of papers at the 3rd All Union Conference on Liquid Crystals, Ivanovo (1974), p. 214.

22. V.P. Shibaev, Ya.S. Freidzon, and N.A. Platé, Abstracts of papers at the 11th Mendeleev Congress on General and Applied Chemistry, Vol. 2, p. 164 (Nauka, Moscow, 1975).

23. V.P. Shibaev, Ya.S. Freidzon, and N.A. Platé, USSR Inventor's Certificate No. 525709, Byull. Isobreteneij, No. 31, 1976.

24. O. Hermann-Schonherr, J. Wendorff, H. Ringsdorf, and P. Tschirner, Makromol. Chem., Rapid Commun. **7**, 791 (1987).

25. M. Ebert, O. Herrmann-Schonherr, J. Wendorff, H. Ringsdorf, and P. Tschirner, Makromol. Chem., Rapid Commun. **9**, 445 (1988).

26. A. Wiesemann and R. Zentel, *20 Freiburger Arbeitstagung Flussige Kristallen* (1991).

27. S. Ujiie and K. Iimura, Chem. Lett. 411 (1991); Macromolecules **25**, 3174 (1992).

28. A. Gottschalk and H. Schmidt, Polymer Preprints **30**, 507 (1989).

29. *Applied Liquid Crystal Polymers*, edited by M. Takeda, K. Iimura, N. Koide, and N.A. Platé, Mol. Cryst. Liq. Cryst. **169** 1–192 (1989).

30. V.P. Shibaev, S. Kostromin, and N.A. Platé, European Polymer J. **18**, 651 (1982).

31. S. Berg, V. Krone, and H. Ringsdorf, Makromol. Chem., Rapid Commun. **8**, 389 (1986).

32. V.P. Shibaev and N.A. Platé, Adv. Polymer Sci. **60/61**, 173 (1984).

33. H. Finkelmann and G. Rehage, Adv. Polymer Sci. **60/61**, 99 (1984).

34. V.P. Shibaev, Comb-Shaped Liquid Crystalline Polymers, in *Liquid Crystalline Polymers*, edited by N.A. Platé (Plenum, New York, 1993).

35. Ya.S. Freidzon, N. Boiko, V.P. Shibaev, and N.A. Platé, Dokl. Akad. Nauk USSR **308**, 1419 (1989).

36. S. Kostromin, R. Talroze, V.P. Shibaev, and N.A. Platé, Makromol. Chem., Rapid Commun. **3**, 803 (1982).

37. G. Gray, Synthesis and Properties of Side-Chain Liquid Crystal Polysiloxanes, in *Side-Chain Liquid Crystal Polymers*, edited by C. McArdle (Blackie, London, 1989).

38. S. Kostromin, Ngo Duy Cuong, E. Garina, and V. Shibaev, Mol. Cryst. Liq. Cryst. **193**, 177 (1990).

39. D. Demus, H. Demus, and H. Zaschke, *Flussige Kristallen in Tabellen* (VEB Deutscher Verlag fur Grundstoffindustrie, Leipzig, 1974).

40. Ya.S. Freidzon and V.P. Shibaev, Liquid Crystalline Cholesteric Polymers, in *Liquid Crystalline Polymers*, edited by N.A. Platé (Plenum, New York, 1993).

41. S. Kostromin, V.P. Shibaev, and N.A. Platé, Liq. Cryst. **2**, 195 (1987).

42. H. Finkelmann, D. Naegele, and H. Ringsdorf, Makromol. Chem. **180**, 803 (1979).

43. V.P. Shibaev, S.G. Kostromin, R.V. Talroze, and N.A. Platé, Dokl. Akad. Nauk SSSR **259**, 1147 (1981).

44. H. Finkelmann, H. Ringsdorf, and J.-H. Wendorff, Makromol. Chem. **179**, 273 (1978).

45. P. Zugenmaier. Makromol. Chem. Suppl. **6**, 31 (1984).

46. P. Davidson, A.M. Levelut, M.F. Achard, and F. Hardouin, Liq. Cryst. **4**, 561 (1989).

47. S. Diele, B. Hisgen, B. Reck, and H. Ringsdorf, Makromol. Chem., Rapid Commun. **7**, 267 (1986).

48. S.G. Kostromin, V.V. Sinitzyn, R.V. Talroze, V.P. Shibaev, and N.A. Platé, Makromol. Chem., Rapid Commun. **3**, 809 (1982).

49. R. Zentel and H. Ringsdorf, Makromol. Chem., Rapid Commun. **5**, 393 (1984).

50. G. Decobert, F. Soyer, J.C. Dubois, and P. Davidson, Polymer Bull. **14**, 549 (1985).

51. M. Mauzac, F. Hardouin, H. Richard, M.F. Achard, G. Sigaud, and H. Gasparoux, European Polymer J. **22**, 137 (1986).

52. G. Scherowsky, U. Muller, J. Springer, W. Trapp, A.M. Levelut, and P. Davidson, Liq. Cryst. **5**, 1297 (1989).

53. R.V. Talroze, V.V. Sinitzyn, V.P. Shibaev, and N.A. Platé, Mol. Cryst. Liq. Cryst. **80**, 211 (1982).

54. B. Krücke, H. Zaschke, S.G. Kostromin, and V.P. Shibaev, Acta Polymerica **36**, 639 (1985).

55. V. Tsukruk, V. Shilov, and Yu. Lipatov, Macromolecules **19**, 1308 (1986).

56. P. Davidson and A.M. Levelut, J. Phys. (Paris) **50**, 2415 (1989).

57. Ya.S. Freidzon, N.I. Boiko, V.P. Shibaev, V.V. Tsukruk, V.V. Shilov, and Yu.S. Lipatov, Polymer Commun. **27**, 190 (1986).

58. B. Krücke, M. Schlossarek, and H. Zaschke, Acta Polymerica **39**, 607 (1988).

59. R. Duran, D. Guillon, Ph. Gramain, and A. Skoulios, J. Phys. (Paris) **49**, 1455 (1988).

60. S. Diele, M. Naumann, F. Kuschel, B. Reck, and H. Ringsdorf, Liq. Cryst. **7**, 721 (1990).

61. T.I. Gubina, S.G. Kostromin, R.V. Talroze, V.P. Shibaev, and N.A. Platé, Vysokomolek. Soedin. B **28**, 344 (1986); Liq. Cryst. **4**, 197 (1989).

62. P.Le Barny, J. Dubois, C. Friedrich, and N. Noel, Polymer Bull. **15**, 341 (1986).

63. N. Spassky, N. Lacoudre, A. Le Borqne, J.-P. Vairon, C.L. Jun, C. Friedrich, and C. Noel, Makromol. Chem., Macromol. Symp. **24**, 271 (1989).

64. J.W. Goodby, M.A. Waugh, S.M. Stein, E. Chin, R. Pindak, and J.S. Patel, Nature **337**, 449 (1989).

65. Ye.G. Tropsha, Ya.S. Freidzon, and V.P. Shibaev, Fifth All-Union Scientific Conference on Liquid Crystals, Abstracts, Ivanovo (1985), Vol. 2, p. 117.

66. Ya.S. Fredizon, N.I. Boiko, V.P. Shibaev, and N.A. Platé, Dokl. Akad. Nauk USSR **282**, 934 (1985).

67. D. Demus, S. Diele, S. Grande, and H. Sackmann, in *Advances in Liquid Crystals*, Vol. 6, edited by G.H. Brown (Academic Press, New York, 1983).

68. S. Diele, K. Ziebarth, G. Pelzl, D. Demus, and W. Weissflog, Liq. Cryst. **8**, 211 (1990).

69. S. Diele, G. Pelzl, A. Mädicke, D. Demus, and W. Weissflog, Mol. Cryst. Liq. Cryst. **191**, 37 (1990).

70. C. Casagrande, M. Veyssie, and H. Finkelmann, J. Phys. (Paris) **43**, L671 (1982).

71. H. Benthack-Thoms and H. Finkelmann, Makromol. Chem. **186**, 1895 (1985).

72. G. Sigaud, M.F. Achard, F. Hardouin, M. Mauzac, H. Richard, and H. Gasparoux, Macromolecules **20**, 578 (1987); M.F. Achard, G. Sigaud, P. Keller, and F. Hardouin, Makromol. Chem. **189**, 2159 (1988).

73. S.G. Kostromin and A. Mädicke, Abstracts of VIth All-Union Conference on Liquid Crystals, Tschernigov **3**, 383 (1988).

74. V.V. Tsukruk and V.V. Shilov, *Structure of Polymeric Liquid Crystals* (Naukova Dumka, Kiev, 1990) (in Russian).

75. S.G. Kostromin, V.P. Shibaev, and S. Diele, Makromol. Chem. **191**, 2521 (1990).

76. A.J. Leadbetter, in *Thermotropic Liquid Crystals*, edited by G.W. Gray (Wiley, New York, 1987).

77. P. Davidson and A.M. Levelut, J. Phys. (Paris) **49**, 689 (1988).

78. F. Hardouin A.M. Levelut, and G. Sigaud, J. Phys. (Paris) **42**, 71 (1981).

79. G. Pepy, J.P. Cotton, F. Hardouin, P. Keller, M. Lambert, F. Moussa, A. Lapp, and C. Strazielle, Makromol. Chem. **15**, 251 (1988).

80. Z.X. Fan, S. Buchner, W. Haase, and H.G. Zachmann, J. Chem. Phys. **92**, 5099 (1990).

81. P. Davidson, B. Pansu, A.M. Levelut, and L. Strzelecki, J. Phys. (Paris) II **1**, 61 (1991).

82. Ya.S. Freidzon, R.V. Talroze, N.I. Boiko, S.G. Kostromin, V.P. Shibaev, and N.A. Platé, Liq. Cryst. **3**, 127 (1988).

83. H. Finkelmann, and D. Day, Makromol. Chem. **180**, 2269 (1979).

84. H. Finkelmann, H.J. Kock, W. Gleim, and G. Rehage, Makromol. Chem., Rapid Commun. **55**, 287 (1984).

85. R. Zentel and G.R. Strobl, Makromol. Chem. **185**, 2669 (1985).

86. Yu.S. Lipatov, V.V. Tsukruk, V.V. Shilov, I.I. Konstantinov, and Yu.B. Amerik, Vysokomolek. Soedin. B **25**, 726 (1983).

87. Yu.S. Lipatov, V.V. Tsukruk, V.V. Shilov, S.G. Kostromin, and V.P. Shibaev, Vysokomolek. Soedin. B **29**, 411 (1987).

88. P. Davidson, L. Noirez, J. Cotton, and P. Keller, Liq. Cryst. **10**, 111 (1991).

89. P. Keller, B. Carvalho, J. Cotton, M. Lambert, F. Moussa, and G. Pepy, J. Phys. (Paris) Lett. **46**, L1065 (1985).

90. R.G. Kirste and H.G. Ohm, Makromol. Chem., Rapid Commun. **6**, 179 (1985).

91. J. Kalus, S.G. Kostromin, V.P. Shibaev, A.B. Kunchenko, Yu.M. Ostanevich, and D.A. Svetogorsky, Mol. Cryst. Liq. Cryst. **155**, 347 (1988).

92. L. Noirez, P. Keller, P. Davidson, F. Hardouin, and J.P. Cotton, J. Phys. (Paris) **49**, 1993 (1988).

93. G. Pepy, J.P. Cotton, F. Hardouin, P. Keller, M. Lambert, F. Moussa, L. Noirez, A. Lapp, and C. Strazielle, Makromol. Chem. **15**, 251 (1988).

94. A.B. Kunchenko and D.A. Svetogorsky, J. Phys. (Paris) **47**, 2015 (1986).

95. V.P. Shibaev, Ya.S. Freidzon, and N.A. Platé, Vysokomolek. Soedin. A **20**, 82 (1978).

96. Ya.S. Freidzon, A.V. Kharitonov, V.P. Shibaev, and N.A. Platé, European Polymer. J. **21**, 211 (1985).

97. T. Yamaguchi, T. Asada, H. Hayashi, and N. Nakamura, Macromolecules **22**, 1141 (1989).

98. T. Yamaguchi and T. Asada, Liq. Cryst. **10**, 215 (1991).

99. A. Blumstein, Y. Osada, S.B. Clough, E.C. Hsu, and R.B. Blumstein, in *Mesomorphic Order in Polymers and Polymerization in Liquid Crystalline Media*, edited by A. Blumstein, ACS Symposium Ser., No. 74, Washington, DC, 1978.

100. E.C. Hsu, S.B. Clough, and A. Blumstein, J. Polymer Sci., Polymer. Lett. Ed. **15**, 545 (1977).

101. J.H. Wendorff and F.P. Price, Mol. Cryst. Liq. Cryst. **24**, 129 (1973).

102. P. de Gennes, *The Physics of Liquid Crystals* (Clarendon Press, Oxford, 1974).

103. Ya.S. Freidzon, Ye.G. Tropsha, V.V. Tsukruk, V.V. Shilov, V.P. Shibaev, and Yu.S. Lipatov, Vysokomolek. Soedin. A **29**, 1371 (1987).

104. S.R. Renn and T.C. Lubensky, Phys. Rev. A **38**, 2132 (1988).

105. G.W. Gray and J.W. Goodby, in *Smectic Liquid Crystals* (Leonard Hill, Glasgow, 1984).

106. Ya.S. Freidzon, N.I. Boiko, V.P. Shibaev, and N.A. Platé, European Polymer. J. **22**, 13 (1986).

4

Phase Behavior of High- and Low-Molar-Mass Liquid Crystal Mixtures

F. Hardouin, G. Sigaud, and M.F. Achard

4.1 Introduction

Over the past twenty years a rough estimate of the literature related to the physics and chemistry of liquid crystalline mixtures amounts to 5000 references! Our purpose is thus neither to present an exhaustive review nor to justify the interest in this subject.

Studying mesomorphic mixtures requires primarily the achievement of temperature–composition phase diagrams at atmospheric pressure. From this point of view, binary liquid crystalline systems compare qualitatively to any other solution. In particular, we can differentiate two behaviors:

(i) ideal (or nearly ideal) with no (or little) excess free energy connected to mixing (such systems are of course fully miscible); or
(ii) or real with excess free energy of mixing resulting in nonideal properties [1,2].

The various topologies for phase diagrams of mesomorphic mixtures have been described in the past years in different theoretical works applying standard thermodynamics [3–5] or lattice models [6,7]. In particular, the latter account for the specific influence of the order parameter of the liquid crystalline phase upon the nonideal behavior of the solution through a coupling term in the excess free energy. However, as far as "fluid" mesophases are concerned (noncrystalline [8], i.e., except S_B, S_E, S_G, S_H, S_J, S_K), the fundamental criterions and techniques which apply to the analysis of a liquid crystalline binary mixture are basically the same as that for the usual non-ordered liquids. Some are briefly summarized in the following.

In the ideal case it follows from the Schroeder–Van Laar equations [3] that the boundaries of the phase transition under consideration must lie completely in between T_A and T_B, the transition temperatures for the two pure compounds A and B (Fig. 4.1). Moreover, these curves limiting the spindle, must be monotonous from one limit to the other. All other systems

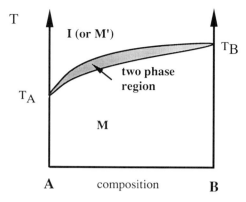

FIGURE 4.1. Example of an ideal behavior for M (mesomorphic) → I (isotropic) phase change similar to liquid → vapor (alt.: M (low T mesomorphic) → M' (high T mesomorphic)).

refer to nonideal behavior (Fig. 4.2). Any typical nonideal character is likely to be expected in mesomorphic solutions: azeotropic features (maximum or minimum, see Fig. 4.2) but also consolute points and gaps of miscibility (Fig. 4.3). These remarks hold for the transition to the isotropic phase ("clearing" or "isotropization") as well as for the mesophase to mesophase changes provided that both are "fluid." Of course, increasing the polymorphism of the pure compounds leads to more complex topologies as described further.

The techniques used for the experimental studies of liquid crystalline mixtures are also standard methods for the study of ordinary solutions. The contact method is derived from the study of solid solutions [9], and has been

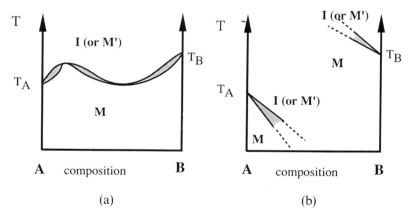

(a)   (b)

FIGURE 4.2. Examples of nonideal behaviors. (a) Curves lie between T_A and T_B but are not monotonous. (b) At least one part of the curve is outside the $[T_A, T_B]$ interval: either T decreases from the low limit T_A or T increases from the high limit T_B.

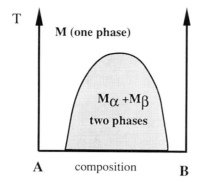

FIGURE 4.3. Nonideal behavior: gap of miscibility. M_a and M_b are the same phase with different composition (a) and (b).

smartly extended to liquid crystalline systems in which diffusion remains limited. In this technique, the characteristic temperatures of the phase diagram are obtained from microscopic observation of an unique sample in which the two compounds are melted separately to merge at a convenient place, and across this zone of diffusion all compositions are present (Fig. 4.4). The shifts with thermal variation of the lines separating two different phases, the temperatures at which they join, vanish, or split indicate the locations of triple points or extrema. The drawback is that there is no way to measure the corresponding compositions and a diagram from the contact method will not show this scale.

The second way to draw the phase diagram is, of course, the classic characterization of mixtures at different compositions with the help of three fundamental techniques: optical polarized microscopy, differential scanning calorimetry, and x ray scattering. These preliminary comments will serve in the following parts of the paper.

The first section illustrates the interest of liquid crystalline solutions through selected works in which mixing two low-molar-mass mesogens (LMMs) has provided significant results for the physics of liquid crystals. In particular, the use of homologous compounds giving ideal (or nearly ideal) behavior has been fruitful in many cases. Alternatively, the use of combinations of unlike mesogens to cause strong nonideal behavior also offered great interest. The second and main section is specifically devoted to the description of the

FIGURE 4.4. Preparation of a contact sample.

behavior of liquid crystalline polymers (LCPs) considered in terms of the usual partner to mix with LMMs.

4.2   Binary Mixtures of Low-Molar-Mass Mesogens

In the early stages of liquid crystalline research, mixing two mesogens had been readily used for mesomorphic assignments [10]. The basic concept is the principle of isomorphism: provided that two compounds are fully miscible, the nature of the phase in the pure materials is the same (reciprocal not true!). This criterion is now classic and comparison of the standard material with a newly synthesized one is one of the first steps in its identification process. Since the availability of new compounds is generally poor, it follows that the contact method has been largely developed at the same time mainly by Halle's group [11]. However, the continuity of a texture is sometimes impossible to establish if another phase occurs owing to entropic or energetic reasons. Therefore, complete miscibility between phases of the same type is made easier by using chemically related components with comparable temperatures of transition. Finally, this miscibility technique is peculiarly helpful where a direct textural identification is doubtful; it is thus currently applied to analyze the polymorphism of LCPs (see Section 4.3.1).

Aside from this essential application, using mixtures is a powerful tool for physical studies. Mixing homologous compounds, i.e., which differ only by the length of their flexible aliphatic chains is, in most cases, a convenient way to produce ideal (or nearly ideal) solutions which allow us to move continuously along the "lines"* of transition. This property is used as a substitute to T (temperature) and P (pressure) diagrams which require experimental setups producing high pressures up to 10 kbar.

An illustrative example is the search for tricritical behavior at the nematic–smectic A (N–S_A) transition. Theoretical models predicted a change from the first to the second order of the transition as the difference between the N–S_A and the N–I (nematic–isotropic) temperatures increases, thus defining a Landau tricritical point [12–14]. The investigations on different homologous series have shown that the ratio T_{NA}/T_{NI} can be easily controlled by the composition of the mixtures (Fig. 4.5) [15]. It has then been possible, using standard DSC equipment, to show a reliable relationship between the progressive vanishing of the heat at the N–S_A transition and the decrease of T_{NA}/T_{NI} in qualitative agreement with the theories [16,17]. Later, high-resolution calorimetric measurements could be performed on such mixtures which have claimed the occurrence of a tricritical point for a very narrow N

---

* The term "line" describes improperly the separation between two phases in a binary diagram since it is actually a two-phase domain with finite width (first-order transition). However, it is convenient and of common use. For the same reasons of convenience, the spindle is not specified in the diagrams when practically unobservable.

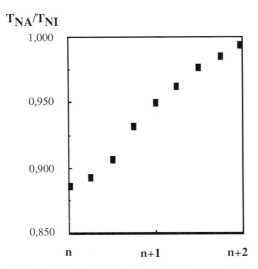

FIGURE 4.5. Evolution of the ratio T_{NA}/T_{NI} with the constitution of mixtures in adjacent binary systems of homologues (n refers to the length of the aliphatic tail).

range [18]. Extending this result, varying P, proved to be more difficult due to the crystallization at high pressure. So studies in P and T diagrams are much less than in mixtures [19,20].

A second example is the case of the N–A–C (for smectic C) point [15]. Three coexisting phases at given P and T is a common feature in a binary system and N–A–C triple points are likely to be observed in mixtures (Fig. 4.6). However, the possible multicritical character of this location was theoretically pointed out by Chen and Lubensky [21], and Chu and McMillan

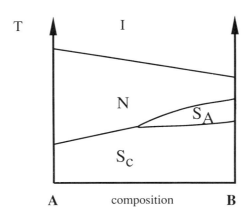

FIGURE 4.6. Schematic representation of a binary system exhibiting an N–A–C point.

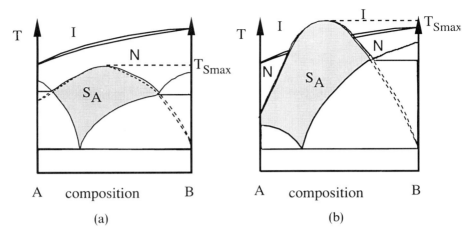

FIGURE 4.7. Schematic representation of binary systems exhibiting induced S_A phases: (a) maximum along the $N-S_A$ line and (b) maximum along the $I-S_A$ line.

[22]. A thermodynamical analysis of binary systems involving the nematic, smectic A, and smectic C phases provided almost simultaneously the experimental support [23,24]. Thus, mixtures of two compounds were found to present a N–A–C multicritical point and could be used for high-resolution experiments [25]. The confirmation in a P, T diagram was given much later [26]. Data on seven different types of mixtures are available so far, and have been used to demonstrate the universal topology of this widely studied multicritical point [15,27]. It is clear that mixtures for such physical studies must be designed to approach as much as possible the behavior of a pure compound. However, the nonideal properties of the smectic A solutions remain one of the main attractive subjects of investigation in liquid crystal research. Among these various observations the most common peculiarity is the frequent "azeotropic" behavior of the smectic A mixtures, i.e., the observations of one or two extrema (mostly a single maximum) along the $N-S_A$ or $I-S_A$ phase lines (Fig. 4.7) [28,29,30]. Molecular complexing is one of the interpretations proposed for these induced smectic phases (ISP in the following) [30]. It must be noticed that we observe an extremum of the layer spacing at these points [28]. In addition, the N–I phase line is usually merely affected by the underlying phenomenon, except when the $N-S_A$ line approaches the isotropic phase in these mixtures: this also suggests that the smectic ordering is the only pertinent parameter, and the Brochard et al. theoretical arguments [7] could have some relevance to this problem. The last remark regards the nature of the components; as stated in the introduction, nonideality is related to unlike characters. Although there exist some exceptions [29–31], the ISPs are mainly observed when *one* component presents a strong dipole, typically, cyano or nitro [30]. This contribution of "polar mesogens" in the

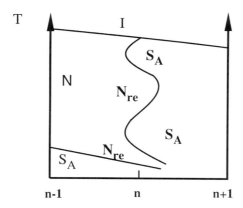

FIGURE 4.8. Evidence for multi-reentrant $N-S_A$ behavior across two contiguous binary diagrams of homologous compounds [37].

unusual properties of smectic A solutions will be constant up to the end of this section.

The most famous example of the mixture of polar compounds is certainly connected to the discovery of nematic reentrance by Cladis [32]. It is worth noting that this behavior has been now observed, commonly in mixtures of homologous polar compounds, and in this context the N–I curve does not show any nonideal feature [33–35]. This remark also applies to the even more surprising multi-reentrance (Fig. 4.8) [36,37]. Here again the nonideal character appears restricted to the smectic A phase and is connected essentially to the variability of the modulation of the layers in the partially bilayer smectic A (S_{Ad}) regardless of the similarity of the components.

Another important result arising from the study of phase diagrams involving polar compounds once more are the observations of the cases of smectic A–smectic A transitions. Most references on the description of the S_A polymorphism can be found in [37,38]. Initially, the evidence of an S_A–S_A phase transition was given by mixing a polar compound with a bilayer structure and mixing a nonpolar compound with a monolayer structure (Fig. 4.9) [39]. It has motivated a systematic investigation of related compounds and their mixtures. These works, and the still ongoing works [40,41], provide a variety of new topologies for physical studies. Closely connected to the S_A solutions, we can distinguish [42–45] the intermediate modulated phases $(S_{\tilde{A}}, S_{\tilde{A}cre}, \ldots)$ between two types of S_A phases; the incommensurate smectic A phase in mixtures of two polar compounds with different molecular lengths [46]; the so-called "nematic bubble" (primarily cholesteric [47]) [48] surrounded with an S_A medium (Fig. 4.10); and the critical point between S_{Ad} (partially bilayer) and S_{A2} (bilayer) phases obtained by lengthening the aliphatic chains in homologous series (i.e., decreasing pressure) [49–52].

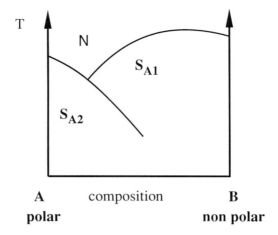

FIGURE 4.9. Schematic representation of a polar–nonpolar system exhibiting a S_{A_1} (monolayer)–S_{A_2} (bilayer) transition line [39].

At this point, it should be noted that no S_A–S_A gap of miscibility is observed for these special critical points. Such a typical phase separation with a consolute point (see Fig. 4.3) was observed only recently, mixing chemically unlike mesogens [53] (Fig. 4.11).

Mixtures of disc-like mesogens also deserve a short comment. Although some original results have been reported, such as the existence of a reentrant isotropic phase [54], phase diagrams are scarce in this domain. We can explain this deficiency by the poor miscibility of these compounds in the columnar phases. Indeed, the little difference in the parameters of the two-

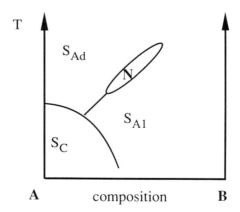

FIGURE 4.10. Schematic representation of a nematic bubble [47,48].

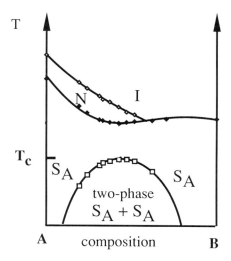

FIGURE 4.11. Consolute point and phase separation in a smectic A solution [52].

dimensional lattice, created in an homologous series by the lengthening of the chains of the paraffinic crown, results rapidly in the vanishing of the mesomorphic properties in the binary mixtures [55].

4.3 Mixtures of a Liquid Crystalline Polymer with a Low-Molar-Mass Liquid Crystal

It is the purpose of this introduction to outline at first some peculiarities of the study of LCP–LMM solutions.

Unlike pure LMM, a two-phase domain, a few degrees large at least, is always observed at the isotropization for a neat LCP. Whether this is the result of weight distribution or of kinetic effects is not clear. However, for convenience, it must be noted that it is usually skipped in the phase diagrams.

Then the viscosity of a LCP is high. In mixtures rich in polymer this fact has two consequences:

(i) the optical textures take a long time to form; and
(ii) large differences in composition can be created in the samples across a biphasic separation (especially at isotropization); recovering a homogeneous sample after a phase change appears impossible in a binary system without stirring, owing to the slow diffusion in such a viscous medium.

On the other hand, the contact method remains a rapid and reliable technique to obtain qualitative information on many aspects of the behavior of

LCP–LMM solutions* [75]. It has been successfully applied for phase characterization, either with main-chain LCPs or with side-chain LCPs, in systems for which a nearly ideal behavior allows us to observe an uninterrupted monophasic domain throughout the phase diagram [76–85]. Recently, it has provided strong support to the difficult identification of undulated phases (S-type) in side-chain LCPs [58,69]. However, nonideality is inherent to these LCP–LMM solutions as a consequence of the size and form of the macromolecular component.

For example, gaps of miscibility which appear unusual in mixtures of two LMM (see the previous section) are commonly observed when the second component is a LCP, since the term of entropy of mixing is dramatically reduced as the degree of polymerization (i.e., the size) of the macromolecule increases. Widely studied, the liquid–liquid phase separation in solutions of nonliquid crystalline polymers has thus been generalized in this last decade to nematic solutions combining a LCP and a LMM [56–62]. Although such a nematic–nematic phase separation sometimes makes it difficult to base on an isomorphism for phase assignment, they definitely are interesting from both the academic and application points of view. We must note that since the macromolecular character of one of the species is the driving factor for this phase separation, a large difference in chemical constitution, between the mesogenic moieties of the polymer and the LMM solvent, is not necessarily required. On the other hand, phase separation need not occur in every binary system involving a LCP. So, the way in which the solubility of a LCP can be controlled from phase separation to complete miscibility, through modifications of the constitutions of the components in the mixture, is the subject of the first part in this section.

Also of interest are other nonideal behaviors such as the reentrance [69,74] enhanced nematic phase [73] or the induced smectic phase (ISP, see the previous section). This latter case will be discussed in the last part. It is worth noting that ISP links to phase separation from several points of view. Since it occurs at richer polymer concentration than gaps of miscibility, the ISP can be observed in the same binary diagram, and it appears that in most cases the ISP is sensitive to the same chemical modifications as is phase separation. Moreover, once a LCP is involved, a strong chemical difference between the components is not required: unlike LMM mixtures, the ISP are common in the absence of polar groups.

4.3.1 Miscibility Properties

The behavior of a macromolecular solution can be described by the interaction parameter χ introduced by the Flory–Huggins theory and charac-

---

* Although the application to a LC polydispersed polymer could be less trivial. Nevertheless, the time scale of the mass diffusivity is long enough compared to the rate of heating/cooling to avoid segregation starting from a homogeneous sample.

terizing the compatibility of the polymer solvent couple [63]

$$\chi(T) = \tfrac{1}{2} - \psi\left(1 - \frac{\theta}{T}\right).$$

In this theory, the critical temperature T_c at the consolute point directly connects to the above parameters θ and ψ and to the degree of polymerization x

$$\frac{1}{T_c} = \frac{1}{\theta}\left[1 + \frac{1}{\psi}\left(\frac{1}{\sqrt{x}} + \frac{1}{2x}\right)\right].$$

Given x, T_c is thus a relevant experimental parameter to describe the interaction parameter for a polymer-solvent system and, in our particular interest, to characterize the miscibility of a LC polymer in a LMM solvent: the lower T_c the better the solubility.

In these mixtures however, the phase separation may occur in the nematic state ($T_{c,N}$) or in the isotropic melt ($T_{c,I}$) as classically observed in non-mesogenic macromolecular solutions. At any rate, the critical concentration is located at approximately 10% or less in polymer, since $\phi_c \approx 1/(1 + \sqrt{x})$ with $x \approx 100$ or more. The diagrams are thus highly disymmetric (Fig. 4.12). Experimentally, the contact method is well suited to the determination of the critical temperature T_c in numbers of binary systems.

These sets of observations provide a provisional criterion for the miscibility of a mesogenic polymer in LMM solvent. We can emphasize the influence on the miscibility of the molecular parameter of each component:

(i) the lengths of the flexible parts, the size and nature of the rigid core in conventional rod-like LMM solvents;

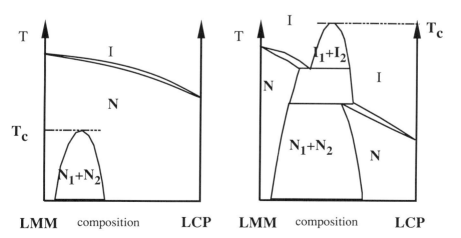

FIGURE 4.12. Schematic representation of phase separation occurring either in the nematic phase or in the isotropic state.

(ii) the chemical architecture of the polymer, the nature of the backbone, the nature of the mesogenic groups, the proportion of mesogenic groups, and the degree of polymerization in the LCP solutes.

Evidence is also given for a significant effect of the nature of the medium, isotropic or nematic, on the Flory interaction parameter [64].

Influence of the Molecular Parameters of Low-Molar-Mass Solvent on the Miscibility

Two examples summarize well the typical evolutions of T_c, observed at constant mesogenic polymer, according to the variations of the molecular parameters of the LMM. In the first one, we select the following series of homologous mesogenic solvents:

$$C_{n'}H_{2n'+1}O-\bigcirc-CO_2-\bigcirc\bigcirc-OC_{m'}H_{2m'+1} \qquad \underline{1}$$

with variables m' and n' in order to evaluate their miscibility of a given side-chain polysiloxane

$$PMS-(CH_2)_nO-\bigcirc-OCO-\bigcirc-OC_mH_{2m+1} \qquad \underline{A}$$

with $n = 4$ and $m = 1$.

Figure 4.13 shows T_c versus the total number $(m' + n')$ of carbon atoms in the two aliphatic tails of the LMM. For short chains ($m' + n' = 4$–5), complete miscibility is observed down to room temperature. As the number of C in the aliphatic chains increases, a phase separation first arises in the nematic phase and then in the isotropic state, i.e., T_c increases with $(m' + n')$. Thus, the interactions between the solvent and the solute are not favored by an increase in the aliphatic content of the LMM solvent. In other words, the miscibility of the polymer increases as the overall flexibility of the solvent decreases.

In the second example, the influence of the nature of the core of the solvent is evidenced [60]. Three solvents are used with congruent chains (13 carbon atoms) but different cores

$$-\bigcirc-CH=N-\bigcirc- \qquad 2$$

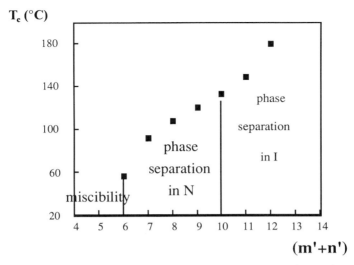

FIGURE 4.13. Evolution of the critical temperature T_c versus the total number of carbons $(m' + n')$ in the flexible chains of the LMM solvent $\underline{1}$ for constant polymer $\underline{A}$.

The length and the stiffness of these molecules are noticeably different. From the first to the third one, the flexibility decreases (with decreasing length of the core and with the replacement of a flexible CH_2—CH_2 group by a rigid naphtalene part) and at the same time T_c undergoes a drastic decrease (Table 4.1). To conclude, every molecular parameter that decreases the flexibility of the LMM solvent (tail or core) promotes the miscibility of a given LCP. Note that this result is general whatever the considered LCP.

Influence of the Molecular Parameters of Liquid Crystalline Polymers on the Miscibility

In this section the evolution of the behavior of solvent–solute couples consisting of a constant LMM solvent with different LCPs is described. We can remark that the nematic–nematic immiscibility was first proved [56,57] in binary systems of side-chain polymers (Fig. 4.14(A1)) with LMM, and that only examples with classical side-chain polymers have been reported so far [59–62]. Is this phenomenon restricted to this kind of mesogenic polymer?

TABLE 4.1. Evolution of T_c for constant polymer $\underline{A}$ in three solvents with the same aliphatic content but different rigid cores.

Solvent	T_c (°C)
$\underline{2}$	152
$\underline{3}$	108
$\underline{4}$	57

Recent results show that it can be extended to other polymer architectures as well. In particular, we observed a gap of miscibility with "side-on fixed" side-chain polymers (Fig. 4.14(A2)) to compare with the classical "side-end fixed" (Fig. 4.14(A1)), with main-chain polymers (Fig. 4.14(B)), and with combined polymers (Fig. 4.14(C)). Of course, this list is not exhaustive and other architectures more complex should be tested in the future.

Let us detail now the evolution of T_c, with constant mesogenic solvent of the series $\underline{1}$, as a function of modifications connected to the backbone or to the mesogenic groups of the LCP.

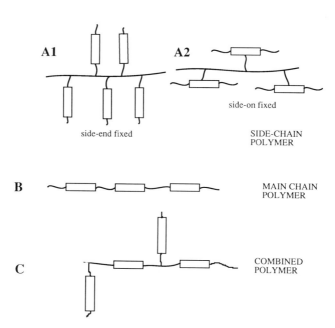

FIGURE 4.14. Scheme of some chemical architectures of LC polymers (rectangles represent mesogenic groups).

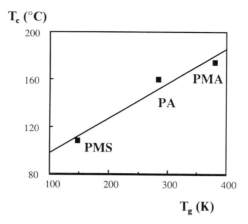

FIGURE 4.15. Evolution of T_c for three side-chain LCPs as a function of the glass transition temperature T_g of the corresponding backbone.

Parameters Related to the Backbone [62]

The influence of the nature of the backbone on the miscibility properties is studied through the measurement of T_c in solutions between different types of side-chain LCPs bearing identical mesogenic groups and the same LMM solvent.

Three mesogenic polymers of a comparable degree of polymerization are considered, a polymethylsiloxane (PMS), a polyacrylate (PA), and a polymethacrylate (PMA) in combination with a solvent of series $\underline{1}$ ($m' + n' = 15$). Their mesogenic moiety is similar to those of $\underline{A}$ with $n = 6$ and $m = 1$. The glass transition temperatures T_g of the bare backbone provide a reliable comparison of their flexibility. As shown in Fig. 4.15, there is a straightforward relation between T_c and T_g: the lower the T_g, the better the miscibility. The following question is: Does the flexibility of the mesogenic moieties play the same role as the flexibility of the backbone does?

Parameters Related to the Mesogenic Groups

Depending on the architecture of the polymer (see Fig. 4.14), the relevant parts can be defined as follows:

(i)   the spacer which decorrelates the mesogenic core from the backbone, and the aliphatic tail of the mesogenic moiety for side-chain polymers;
(ii) the flexible segments separating two mesogenic groups for main-chain polymers; and
(iii) these combined parameters for combined polymers.

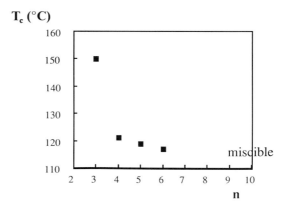

FIGURE 4.16. Evolution of T_c for side-chain polysiloxanes as a function of the spacer length n (data nonavailable for $n = 7, 8, 9$) in the solvent $\underline{1}$.

Side-Chain Liquid Crystalline Polymers

(i) Spacer length [60]

Polymers of type $\underline{A}$ are chosen with a variable number n of methylene groups in the spacer (3–6 and 10) in a given solvent of series $\underline{1}$ ($m' + n' = 17$). The evolution of T_c as a function of n is reported in Fig. 4.16. T_c clearly decreases as the spacer length increases. For the longest ($n = 10$), complete miscibility is observed in the nematic domain. Thus the miscibility is improved by the uncoupling of the mesogenic part from the backbone.

(ii) Aliphatic tail

In Fig. 4.17 are reported two evolutions of T_c for a solvent of series $\underline{1}$ ($m' + n' = 15$) as a function of the length m of the aliphatic chains, either

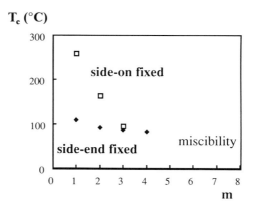

FIGURE 4.17. Evolution of T_c as a function of the length m of the aliphatic chains of a "side-end" and a "side-on" polysiloxane in the solvent $\underline{1}$.

of a polymer $\underline{A}$ $(n = 4)$ [60] or of a "side-on fixed" polysiloxane (see Fig. 4.14) with a similar spacer [66]:

$$
\begin{array}{c}
-\text{PMS}- \\
| \\
(\text{CH}_2)_4 \\
| \\
\text{O} \\
| \\
\text{C}{=}\text{O}
\end{array}
$$

$$C_mH_{2m+1}O-\!\!\!\bigcirc\!\!\!-CO_2-\!\!\!\bigcirc\!\!\!-O_2C-\!\!\!\bigcirc\!\!\!-OC_mH_{2m+1}$$

In both cases, the miscibility increases as the length of the flexible extremities increases. In addition, we note that miscibility seems lesser for "side-on" than for "side-end" polymers (Fig. 4.17). This latter result may originate from the differences in the mesogenic interactions of the two classes of polymers, since "side-end" are essentially smectogenic [65] while "side-on" are only nematogenic [66].

Main-Chain Liquid Crystalline Polymers

In semiflexible main-chain polymers, the flexibility depends largely on the length of the aliphatic parts bridging two mesogenic cores. The miscibility tests have been carried out with a family of mesogenic polyesters [67]

$$\left[-O-\!\!\!\bigcirc\!\!\!-\underset{CH_3}{N}{=}\overset{O}{\underset{}{N}}-\!\!\!\underset{CH_3}{\bigcirc}\!\!\!-OCO-(CH_2)_n-CO-\right]_x \quad \underline{B}$$

In contrast to side-chain LCPs, cases of phase separation seldom appear with main-chain polymers. Only solvents with saturated rings give rise to a gap of miscibility. The solvent used is the following:

$$C_3H_7-\!\!\!\bigg\langle H\bigg\rangle\!\!\!-\!\!\!\bigcirc\!\!\!-COO-\!\!\!\bigg\langle H\bigg\rangle\!\!\!-C_3H_7 \qquad \underline{5}$$

We can observe in Table 4-2 a regular decrease of T_c as n increases. Thus, as for side-chain polymers, miscibility is favored increasing the length of the flexible parts.

Combined Liquid Crystalline Polymers

It is suggested that this family combines the properties of a side-chain polymer and of a main-chain polymer. This ambivalence is confirmed by the miscibility properties, since the general rules stated above for flexible parts

TABLE 4.2. Evolution of T_c for a series of main-chain polymers $\underline{B}$ as a function of the length of the flexible parts n in solvent $\underline{5}$.

n	5	6	7	10	11	13	14
T_c (°C)	280	230	208	141	120	95	82

located either in the side chain or in the main chain are valid. As an example of this regular behavior, Table 4.3 shows the evolution of T_c with respect to the length of the flexible segment in the main chain of the polymers having the following formula [68]:

$$\left[\begin{array}{l} -OCO-CH-COO(CH_2)_nO-\bigcirc\!\!-\!\!\bigcirc-O(CH_2)_nO- \\ \overset{|}{(CH_2)_6O}-\bigcirc\!\!-N{=}N\!\!-\!\!\bigcirc-OCH_3 \end{array}\right] \qquad \underline{C}$$

Proportion of Mesogenic Groups X

The parameter X first introduced in this section represents the average number of mesogenic groups per repeat unit: only homopolymers corresponding to $X = 1$ have been considered so far, but the amount of mesogenic groups can be reduced progressively in the following type of copolymers:

$$\cdots\left[\begin{array}{c}CH_3 \\ | \\ Si-O \\ | \\ \Box \\ | \end{array}\right]_a\cdots\left[\begin{array}{c}CH \\ | \\ Si-O \\ | \\ CH_3 \end{array}\right]_b\cdots$$

$$X = a/(a + b)$$

TABLE 4.3. Evolution of T_c for polymer $\underline{C}$ in solvent $\underline{1}$ ($m' + n' = 18$) as a function of the length n of the flexible segments in the main chain.

n	2	6
T_c (°C)	205	115

TABLE 4.4. Evolution of T_c as a function of the proportion of mesogenic groups X for polymer s<u>A</u> in a solvent <u>1</u>.

X	1	0.95	0.8	0.7	0.6	0.5	0.3
T_c (°C)	220	210	191	144	137	90	Miscible

An example of the evolution of T_c as a function of X for a polymer <u>A</u> ($n = 4$, $m = 1$) in a solvent of series <u>1</u> ($m' + n' = 17$) is shown in Table 4.4. T_c drops as X decreases, i.e., the solubility of the copolymer in a LMM improves for lower mesogenic content, and despite a gradual loss of the liquid crystalline properties of the neat copolymer. Complete miscibility is reached at about $X \approx 0.5$. However, since the bare dimethylsiloxane backbone of A ($X = 0$) is quite unsoluble in ay LMM, the compatibility of the components should break anew at some lower limit in X. This value could be evaluated for copolymers with long polar groups: it lies between $X = 0.3$ and $X = 0.09$, since the former compound is still fully miscible while the second is no longer compatible with the same solvent. Thus the influence of X can be summarized in Table 4.5.

Degree of Polymerization x

All the comparative results previously described have been obtained using polymers of the same degree of polymerization x. Its influence on the miscibility of polymer <u>A</u> ($n = 4$, $m = 1$, variable x) in solvent <u>1</u> ($m' + n' = 19$) is illustrated in Fig. 4.18, which represents the evolution of T_c as a function of the degree of polymerization. Miscibility is clearly improved lowering x as usually observed in nonliquid crystalline macromolecular solutions. It is worth noting that these measurements of T_c, as a function of the degree of polymerization x, are a convenient method to determine the Flory interaction parameter χ ranking the compatibility of the polymer–solvent systems [64].

This quantitative approach to the solubility of a LCP shows an important decrease in the miscibility from the isotropic to the anisotropic phase, since the interaction parameter χ is largely smaller in the isotropic state (0.70) than

TABLE 4.5. Influence of the proportion of mesogenic groups X on the liquid crystalline properties of a polysiloxane and on its miscibility properties with a given LMM solvent.

	X			
	1	≈ 0.5	≈ 0.1	0
LC properties for the neat polymer	Yes	Yes	Yes	No
Compatibility with a LMM	No	Yes	No	No

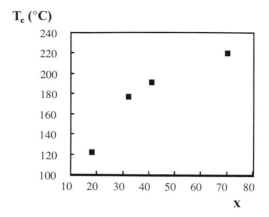

FIGURE 4.18. Influence of the degree of polymerization x on the critical temperature T_c.

in the nematic phase (1.11). A coupling with the nematic order parameter, as proposed by Brochard et al. [7], is in good agreement with this result.

Concluding Remarks

The results summarized in the above section establish a set of empirical rules accounting for the miscibility of a LCP in a LMM. Depending on the need, either to get rid of the difficulties introduced by a gap of miscibility or to achieve inhomogeneous liquid crystalline mixtures for applications, phase separation can be avoided or created according to modifications in the molecular structure of each component. Some parameters appear peculiarly efficient: the flexibility of either the LCP or the LMM and the proportion of mesogenic groups. More emphasis is put in the next part on their role on other nonideal behaviors.

4.3.2   Other Nonideal Behaviors

At the beginning of this section, we have noted that an enhanced or induced S–A phase (ISP) is the other usual nonideal feature in LCP–LMM binary diagrams. However, they are only observed in solutions of "side-end fixed" side-chain polymers, owing to their high smectogenic character. In this case, this phenomenon is common to different backbones [62]. By comparison with the characterization of the gap of miscibility through its temperature at the consolute point T_c, the ISP can be depicted by the maximum temperature T_{S_A} observed for the S_A in the phase diagram (Fig. 4.19) [61,62]. T_{S_A} usually shows regular evolution according to the chemical variations of the constituents as T_c does.

In Table 4.6 are reported both T_c and T_{S_A} for a polymer $\underline{A}$ ($n = 4, m = 1$) in

T

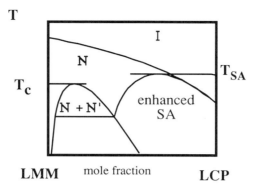

LMM   mole fraction   LCP

FIGURE 4.19. Schematic representation of the nematic–nematic phase separation and of the enhanced (or induced) S_A phase.

solution with the LMM mesogens of series 1 (see Section 4.3.1). A given length of the aliphatic chain of the solvent is required to observe an ISP. Then T_{S_A} barely increases as the chain length of solvent increases. Thus T_{S_A} is influenced in the same way but not to the same extent as T_c by this modification of the solvent parameter. At longer LMM the ISP is concealed either by the occurrence of the isotropic phase or by the widening of the two-phase domains of noncomplete miscibility. The molecular parameters of the polymer also play a role with respect to the ISP. The evolutions of T_c and T_{S_A} as a function of the spacer length n of polymer $\underline{A}$ ($m = 1$) in solvent 1 ($n' + m' = 16$) are compared in Table 4.7.

We can remark that the occurrence of ISP and better miscibility proceed together from greater flexibility of the polymer. This result is corroborated by the proportion of mesogenic groups [71]: in Table 4.8, the temperatures T_{S_A} are listed as a function of X for polymer $\underline{A}$ ($n = 4$, $m = 1$) in solvent 1 ($m' + n' = 16$). The "absolute" value of T_{S_A} slightly increases as X is lowered. But we note that the difference between T_{S_A} and the transition temperature to the S_A phase in the pure polymer (T_{NS_A} or T_{IS_A}) increases dramatically as X becomes smaller. Thus, with respect to this temperature, the ISP presents a considerable stabilization in solutions of "diluted LCPs" in a LMM. These

TABLE 4.6. Evolution of T_c and T_{S_A} for a polymer $\underline{A}$ as a function of the aliphatic part ($n' + m'$) of a solvent 1.

	$n' + m'$									
	9	10	11	12	13	14	15	16	17	18
T_c (°C)	M	M	M	M	57	91	108	121	133	149
T_{S_A} (°C)	—	—	86	88	90	91	92	93	—	—

TABLE 4.7. Evolutions of T_c and T_{S_A} as a function of the spacer length n of polymer $\underline{A}$ in a solvent $\underline{1}$.

	n	
	4	6
T_c (°C)	121	117
T_{S_A}	93	115

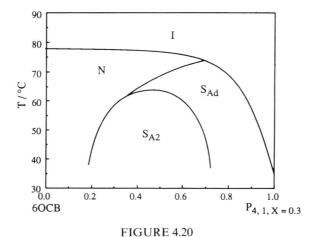

FIGURE 4.20

ISPs also help demonstrate the complex role of partial fixation and its practical interest. Lower X means higher miscibility (Section 4.3.1 and Table 4.8) and lower viscosity which make easier the investigations on mixtures [58,72]. The assignment of phases can be completed through x ray analysis. In this regard the phase diagrams presented in Fig. 4.20, which correspond to polymer A ($n = 4$, $m = 1$) with two different proportion of mesogens ($X = 1$, $X = 0.3$) in solution in a polar LMM solvent (hexyloxycyanobiphenyl 6OCB), underline the interest of partial fixation since evidence is given here for a S_{Ad}–S_{A_2} phase transition [58].

4.4   Conclusions

The overview of studies in nematic and smectic A binary solutions highlights the various interests of such systems.

TABLE 4.8. Evolutions of T_c and T_{S_A} as a function of the proportion of mesogenic groups X for polymer $\underline{A}$ in a solvent $\underline{1}$.

	X			
	1	0.5	0.3	0.09
Transition temperatures (pure polymer) (°C)	$T_{I_N} = 104$ $T_{NS_A} = 78$	$T_{I_N} = 56$ $T_{NS_A} = 44$	$T_{IS_A} = 27$	no mesophase
T_c (solution) (°C)	121		miscible	
T_{S_A} (solution) (°C)	93	95	102	107

(i) In ideal (or nearly ideal) solutions the parameter composition offers a convenient way to modify regularly the liquid crystalline properties. It often provides physical experiments with systems easier to analyze than pure compounds (lower temperatures, atmosphere pressure, . . .). Unfortunately LCP–LMM mixtures and blends of two LCPs most often show poor or no compatibility. Averaging the liquid crystalline properties is nevertheless possible in a polymer system through the chemical mixing of different mesogenic moieties via a copolymer [86,87]: the pseudobinary diagrams obtained as a function of the copolymer composition present large similarities with the usual diagrams of mixtures of two LMMs [86].

(ii) Nonideal solutions are also interesting for their multiple original topologies (ISP, reentrance) as well as for composition driven phase separation which occurs in an anisotropic medium. In these cases the role of the coupling between the composition and order parameter is largely unknown. This is especially true in LCP–LMM systems in which the anisotropy of the shape of the macromolecules and associations among their mesogenic groups depend strongly on the environment.

We would like to stress the potential interest of plural or pseudoplural liquid crystalline systems, i.e., which combine more than two components. Multicomponent lyotropic systems have been extensively and fruitfully studied for a long time. But the study of the ternary mixtures of thermotropic liquid crystals or of mixtures combining lyotropic and thermotropic liquid crystals should deserve development. We name as pseudoplural these binary systems in which an intramolecular parameter of composition between two unlike parts can be defined and controlled in one component. Examples in LMMs could be the analysis of phase diagrams of polycatenar mesogens in which the ratio aliphatic part/aromatic part can be regularly and largely varied or the study of mixtures with perfluorinated or semiperfluorinated liquid crystals. In LCPs, copolymers are typical of such intramolecular variable composition. Of peculiar interest in this regard are copolymers with variable content in mesogenic groups, which form smectic phases with micro-segregated structures and which can be easily mixed with a LMM owing to their lower viscosity and good miscibility. We can expect original advances both for academic research and application from the investigations on these complex mesomorphic fluids.

The latest developments of interest in the study of phase diagrams of molecular mesogens regards special systems involving chiral compounds. Either these diagrams are the combination of two enantiomeric mesogens thus going through the racemic mixture at a 1:1 composition, or they mix the racemic material and one of the enantiomers. The use of such phase diagrams is of great help in solving the highly complex polymorphism among newly discovered phases in chiral mesogens such as $TGB_A^\dagger$, TGB_C, SC^*_{ferro}, SC^*_α, SC^*_{ferri}, $SC^*_{antiferro}$ [88–91]. For example, the contact method between two enantiomers has been nicely applied as a convenient way to definitely distinguish the TGB_A phase, for which the textural identification is uncertain, from the cholesteric one [92]. Also the variation in the composition of this type of mixture can be considered as a change in the chirality field. This approach has been used to analyze how near critical mixtures of polar compounds are susceptible to this field [93].

As for LCP–LMW systems, the uncompatibility of side-chain polymers is now described in the smectic A phase by the use of fluorinated LMW solvents [94]. At last the study of phase diagrams between LMW liquid crystals and nonliquid crystalline precursors of cross-linked matrices also appears of current interest in view of optimization of PDLC materials for display applications [95].

References

1. F.H. Hildebrand and R.L. Scott, *Regular Solutions* (Prentice-Hall, Englewood Cliffs, 1962).

2. J.S. Rowlinson and F.L. Swinton, *Liquid and Liquid Mixtures* (Butterworth, London, 1982).

3. M. Domon and J. Billard, Pramana, Suppl. **1**, 131 (1975);
   M. Domon, Thesis, Université de Lille, 1973.

4. G.R. Van Hecke, J. Phys. Chem. **83**, 2344 (1979).

5. G.R. Van Hecke, T.S. Cantu, M. Domon, and J. Billard, J. Phys. Chem. **84**, 363 (1980).

6. J. Sivardiere, J. Phys. (Paris) **41**, 1081 (1980).

7. F. Brochard, J. Jouffroy, and P. Levinson, J. Phys. (Paris) **45**, 1125 (1984), and references therein.

---

† TGB for twist grain boundary.

8. P.S. Pershan, *Structure of Liquid Crystal Phases* (World Scientific, Singapore, 1988).

9. L. Kofler and A. Kofler, *Thermomicromethoden zur kennzeichnung Organisher Stoffe und Stoffgemishe* (Verlag Chemie, Weinheim, 1954).

10. H. Arnold and H.Z. Sackmann, Z. Phys. Chem. (Leipzig) **213**, 137 (1960); **213**, 145 (1960).

11. D. Demus and L. Richter, *Textures of Liquid Crystals* (Verlag Chemie, Weinheim, 1978).

12. K.K. Kobayashi, Phys. Lett. A **31**, 125 (1970).

13. W.L. McMillan, Phys. Rev. A **4**, 1238 (1971).

14. P.G. de Gennes, Solid State Commun. **10**, 753 (1972).

15. Most references on the use of mixtures in the investigation of critical phenomenon in liquid crystals are to be found in the following reviews:
D.L. Johnson, J. Chim. Physique **80**, 45 (1983);
M.A. Anisimov, Mol. Cryst. Liq. Cryst. **162** A, 1 (1988).

16. M.F. Achard, H. Gasparoux, F. Hardouin, and G. Sigaud, J. Phys. (Paris) Coll. C3, **37**, C3-107 (1976).

17. D.L. Johnson, C. Maze, E. Oppenheim, and R. Reynolds, Phys. Rev. Lett. **34**, 1143 (1975).

18. J. Thoen, H. Marynissen, and W. Van Dael, Phys. Rev. Lett. **52**, 204 (1984).

19. P.H. Keyes, H.T. Weston, and W.B. Daniels, Phys. Rev. Lett. **31**, 628 (1973).

20. G.B. Kastings, K.J. Lushington, and C.W. Garland, Phys. Rev. B **22**, 321 (1980).

21. J. Chen and T. Lubensky, Phys. Rev. A **14**, 1202 (1976).

22. K. Chu and W. McMillan, Phys. Rev. A **15**, 1181 (1977).

23. D.L. Johnson, D. Allender, R. de Hoff, C. Maze, E. Oppenheim, and R. Reynolds, Phys. Rev. B **16**, 470 (1977).

24. G. Sigaud, F. Hardouin, and M.F. Achard, Solid State Commun. **23**, 35 (1977).

25. M.A. Anisimov, V.P. Voronov, A.O. Kulkov, V.N. Petukhov, and F. Kholmurodov, Mol. Cryst. Liq. Cryst. **150**, 399 (1987).

26. R. Shashidar, B.R. Ratna, and S. Krishna Prasad, Phys. Rev. Lett. **53**, 2141 (1984).

27. D. Brisbin, D.L. Johnson, H. Fellner, and M.E. Neubert, Phys. Rev. Lett. **50**, 178 (1983).

28. B. Engelen, G. Heppke, R. Hopf, and F. Schneider, Mol. Cryst. Liq. Cryst. Lett. **49**, 193 (1979).

29. M. Domon and J. Billard, J. Phys. (Paris) Coll. C3, **40**, C3–413 (1979).

30. W.H. De Jeu, L. Longa and D. Demus, J. Chem. Phys. **84**, 6410 (1986).

31. G. Pelzl, J. Szabon, A. Wiegeleben, S. Diele, W. Weissflog and D. Demus, Cryst. Res. Technol. **25**, 223 (1990).

32. P.E. Cladis, Phys. Rev. Lett. **35**, 48 (1975).

33. D. Guillon, P.E. Cladis, and J. Stamatoff, Phys. Rev. Lett. **41**, 1598 (1978).

34. H.T. Nguyen, G. Sigaud, M.F. Achard, H. Gasparoux, and F. Hardouin, in *Advances in Liquid Crystal Research and Applications* (Pergamon Press, Oxford, 1980).

35. G. Sigaud, H.T. Nguyen, F. Hardouin, and H. Gasparoux, Mol. Cryst. Liq. Cryst. **69**, 81 (1981).

36. H.T. Nguyen, F. Hardouin, and C. Destrade, J. Phys. (Paris) **43**, 1127 (1982).

37. F. Hardouin, A.M. Levelut, M.F. Achard, and G. Sigaud, J. Chim. Physique **80**, 53 (1938).

38. F. Hardouin, Physica **140A**, 359, (1986).

39. G. Sigaud, F. Hardouin, M.F. Achard, and H. Gasparoux, J. Phys. (Paris) Coll. C3, **40**, C3-356 (1979).

40. L.K.M. Chan, G.W. Gray, D. Lacey, T. Srithanratana, and K.J. Toyne, Mol. Cryst. Liq. Cryst. **150B**, 335 (1987).

41. G. Pelzl, C. Scholz, D. Demus, and H. Sackmann, Mol. Cryst. Liq. Cryst. **168**, 147 (1989).

42. G. Sigaud, F. Hardouin, M.F. Achard, and A.M. Levelut, J. Phys. (Paris) **42**, 107 (1981).

43. A.M. Levelut, R.J. Tarento, F. Hardouin, M.F. Achard, and G. Sigaud, Phys. Rev. A **24**, 2180 (1981).

44. A.M. Levelut, J. Phys. Lett. (Paris) **45**, L-603 (1984).

45. G. Sigaud, F. Hardouin, and M.F. Achard, Phys. Rev. A **31**, 547 (1985).

46. B.R. Ratna, R. Shashidar, and V.N. Raja, Phys. Rev. Lett. **55**, 1476 (1985).

47. P.E. Cladis and H.R. Brand, Phys. Rev. Lett. **52**, 2261 (1984).

48. F. Hardouin, M.F. Achard, H.T. Nguyen, and G. Sigaud, Mol. Cryst. Liq. Cryst. Lett. **3**, 7 (1986).

49. F. Hardouin, M.F. Achard, C. Destrade, and H.T. Nguyen, J. Phys. (Paris) **45**, 765 (1984).

50. F. Hardouin, M.F. Achard, H.T. Nguyen, and G. Sigaud, J. Phys. (Paris) Lett. **46**, L-123 (1985).

51. R. Shashidhar, B.R. Ratna, S. Krishna Prasad, S. Somasekhara, and G. Heppke, Phys. Rev. Lett. **59**, 1209 (1987).

52. Y.H. Jeong, G. Nounesis, C.W. Garland, and R. Shashidhar, Phys. Rev. A **40**, 4022 (1989).

53. G. Sigaud, M.F. Achard, H.T. Nguyen, and R.J. Twieg, Phys. Rev. Lett. **65**, 2596 (1990).

54. C. Destrade, P. Foucher, J. Malthete, and H.T. Nguyen, Phys. Lett. **88A**, 187 (1982).

55. C. Destrade, M.C. Mondon, and J. Malthete, J. Phys. (Paris) Coll. C3, **40**, C3-17 (1979).

56. H. Ringsdorf, H.W. Schmidt, and A. Schneller, Makromol. Chem., Rapid Commun. **3**, 745 (1982).

57. C. Casagrande, M. Veyssie, and H. Finkelmann, J. Phys. (Paris) Lett. **43**, L-371 (1982).

58. F. Hardouin, G. Sigaud, P. Keller, H. Richard, H.T. Nguyen, M. Mauzac, and M.F. Achard, Liq. Cryst. **5**, 2 (1989).

59. H. Benthack-Thoms and H. Finkelmann, Makromol. Chem. **186**, 1895 (1985).

60. G. Sigaud, M.F. Achard, F. Hardouin, M. Mauzac, H. Richard, and H. Gasparoux, Macromolecules **20**, 578 (1987).

61. G. Sigaud, M.F. Achard, F. Hardouin, and H. Gasparoux, Mol. Cryst. Liq. Cryst. **155**, 443 (1988).

62. M.F. Achard, G. Sigaud, P. Keller, and F. Hardouin, Makromol. Chem. **189**, 2159 (1988).

63. P.J. Flory, *Principles of Polymer Chemistry* (Cornell University Press, Ithaca, 1953).

64. G. Sigaud, M.F. Achard, F. Hardouin, C. Coulon, H. Richard, and M. Mauzac, Macromolecules, **23**, 5020 (1990).

65. M. Mauzac, F. Hardouin, H. Richard, M.F. Achard, G. Sigaud, and H. Gasparoux, European Polymer. J. **22**, 137 (1986).

66. P. Keller, F. Hardouin, M. Mauzac, and M.F. Achard, Mol. Cryst. Liq. Cryst. **155**, 171 (1988).

67. A. Blumstein and O. Thomas, Macromolecules **15**, 1264 (1982).

68. B. Reck and H. Ringsdorf, Makromol. Chem., Rapid Commun. **6**, 291 (1985); **7**, 389 (1986).

69. H.T. Nguyen, M.F. Achard, F. Hardouin, M. Mauzac, H. Richard and G. Sigaud, Liq. Cryst. **7**, 385 (1990).

70. J.W. Doane, A. Golemme, J.L. West, J.B. Whitehead, Jr, and B.G. Wu, Mol. Cryst. Liq. Cryst. **165**, 511 (1988).

71. H. Richard, M. Mauzac, M.F. Achard, G. Sigaud, and F. Hardouin, Liq. Cryst. **9**, 679 (1991).

72. S. Westphal, S. Diele, A. Mädicke, F. Kuschel, U. Scheim, K. Rühlman, B. Hisgen, and H. Ringsdorf, Makromol. Chem., Rapid Commun. **9**, 489 (1988).

73. T. Kato and J.M.J. Fréchet, Macromolecules **22**, 3818 (1989).

74. G. Sigaud, F. Hardouin, M. Mauzac, and H.T. Nguyen, Phys. Rev. A **33**, 789 (1986).

75. C. Noël, in *Recent Advances in Liquid Crystalline Polymers*, edited by L.L. Chapoy (Elsevier, London, 1985);
in *Polymeric Liquid Crystals*, edited by A. Blumstein (Plenum, New York, 1985);

in *Side Chain Liquid Crystal Polymer*, edited by C.B. McArdle (Blackie, Glasgow, 1989).

76. K. Nyitrai, F. Cser, M. Lengyel, E. Seyfried, and G. Hardy, European Polymer. J. **13**, 673 (1977).

77. F. Cser, K. Nyitrai, G. Hardy, J. Menczel, and J. Varga, J. Polymer. Sci., Polymer. Symp. **69**, 91 (1981).

78. H. Finkelmann, H. J. Kock, and G. Rehage, Mol. Cryst. Liq. Cryst. **89**, 23 (1982).

79. B. Millaud, A. Thierry, and A. Skoulios, Mol. Cryst. Liq. Cryst. Lett. **41**, 263 (1978).

80. C. Noël and J. Billard, Mol. Cryst. Liq. Cryst. Lett. **41**, 269 (1978).

81. B. Fayolle, C. Noël, and J. Billard, J. Phys. (Paris) Coll. C3, **40**, C3-485 (1979).

82. A.C. Griffin and S.J. Havens, J. Polymer. Sci., Polymer. Lett. **18**, 259 (1980).

83. A.C. Griffin and S.J. Havens, Mol. Cryst. Liq. Cryst. Lett. **49**, 239 (1979).

84. C. Noël, J. Billard, L. Bosio, C. Friedrich, F. Lauprêtre, and C. Strazielle, Polymer **25**, 263 (1984).

85. J. Billard, A. Blumstein, and S. Vilasagar, Mol. Cryst. Liq. Cryst. Lett. **72**, 163 (1982).

86. See, for example, *Comb-Shaped Polymers and Liquid Crystals*, edited by N.A. Platé and V.P. Shibaev (Plenum, New York, 1987), Chapter 4.

87. M.F. Achard, M. Mauzac, H. Richard, G. Sigaud, and F. Hardouin, European Polymer. J. **25**, 593 (1989).

88. J.W. Goodby, M.A. Waugh, S. Stein, E. Chin, R. Pindak, and J.S. Patel, J. Am. Chem. Soc. **111**, 8119 (1989);
Y.S. Freidzon, Y.G. Tropsha, V.V. Tsuruk, V.V. Shibaev, V.P. Shibaev, and Y.S. Lipatov, J. Polym. Chem. **29**, 1371 (1987).

89. A.D.L. Chandani, Y. Ouchi, H. Takezoe, A. Fukuda, K. Terashima, K. Furukawa and A. Kishi, Jpn. J. Appl. Phys. **28**, L 1261 (1989).

90. A.J. Slaney and J.W. Goodby, Liq. Cryst. **9**, 849 (1991).

91. H.T. Nguyen, A. Bouchta, L. Navailles, P. Barois, N. Isaert, R.J. Twieg, A. Maaroufi, and C. Destrade, J. Phys. II (Paris) **2**, 1889 (1992).

92. A. Bouchta, H.T. Nguyen, M.F. Achard, F. Hardouin, C. Destrade, R.J. Twieg, A. Maaroufi, and N. Isaert, Liq. Cryst. **12**, 575 (1992).

93. Gorecka E. Pyzuk and J. Mieczowski, Europhys. Lett. **22**, 71 (1993).

94. J.D. Laffitte, G. Sigaud, M.F. Achard, F. Hardouin, and H.T. Nguyen, to be published.

95. G.W. Smith, Phys. Rev. Lett. **70**, 198 (1993).

5

Polymer-Dispersed Liquid Crystal Films

G.P. Montgomery, Jr., G.W. Smith, and N.A. Vaz

5.1 Introduction

Polymer films containing microdroplets of liquid crystal (LC) are a recently discovered class of materials with considerable potential for a variety of light control applications [1–10]. These films can be switched from a light-scattering to a transparent state by the application of a modest voltage. This transmittance change is achieved by a field-induced reorientation of the LC molecules which improves the match between the refractive indices of the droplets and the polymer matrix. Electrooptic devices based on these films have many attractive features: they require no polarizers, operate at very low power levels, have rapid response times, are easily fabricated, and can be made in a large, mechanically flexible format. Potential applications of these films include electronic displays, signs, windows for buildings and vehicles, automobile sunroofs, room dividers, light valves, and temperature sensors/indicators.

Two basic types of these films are presently under study. The "NCAP" film, the first to be discovered [1–3], is an emulsion-based system in which LC droplets are dispersed in a water-borne polymer which is subsequently dried and hardened. The second major type is the polymer-dispersed liquid crystal (PDLC) film [4–8], which is produced by phase separation (PS). Several different PS formation techniques have been demonstrated. All involve the hardening of a LC/polymer solution, during which LC microdroplets separate from the matrix. The formation processes of NCAP and PDLC films are quite different, although aspects of their electrooptic behavior can be similar. In this chapter we will focus on PDLC films. We will examine their operating principle, formation mechanisms, component materials, microstructure, electrooptic behavior, and light scattering characteristics. For the most part, data acquired in our laboratory will be used to illustrate the discussion.

5.2   Film Structure and Operation

Figure 5.1 is a scanning electron micrograph of a typical PDLC film comprised of roughly spherical LC droplets, approximately 1.5 μm in diameter, surrounded by polymer matrix material. The fact that the droplets are never perfectly spherical significantly affects the electrooptic behavior of the films (Section 5.7). Although smectic or cholesteric LC materials have been used in a few films [4,11–13], most have been formed from nematic LC materials; therefore, in this chapter we will restrict our discussion to PDLC films containing nematic droplets. With one exception (Section 5.7.1), the polymer matrix materials used in PDLC films have been optically isotropic; therefore, we will assume optically isotropic matrix materials in our discussions. In practice, the thin PDLC film is sandwiched between two electrode-coated transparent substrates which may be either rigid or flexible.

In the absence of external fields, the LC molecules within each droplet will adopt the configuration that minimizes the droplet free energy for the boundary conditions determined by interactions at the droplet–polymer interface. Figure 5.2 schematically illustrates those configurations most often discussed in the literature; other configurations are possible [14–19]. While the radial configuration is spherically symmetric, the bipolar and axial configurations are cylindrically symmetric and the axis of symmetry (droplet director) is an

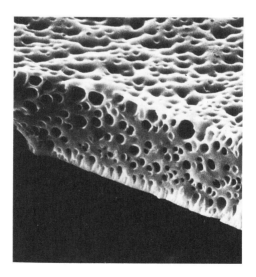

FIGURE 5.1. Scanning electron micrograph showing typical PDLC film structure. The film thickness is about 17.5μm and the droplets are approximately 1.5 μm diameter.

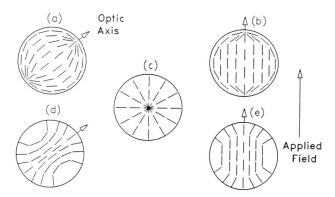

FIGURE 5.2. Common PDLC droplet configurations: (a) randomly oriented bipolar—zero field; (b) aligned bipolar—strong electric field; (c) randomly oriented axial—zero field; (d) aligned axial—strong electric field; and (e) radial—zero field. In a strong electric field a radially aligned droplet assumes the axial configuration shown in (d).

optic axis for the droplet. The bipolar configuration is most commonly found in PDLC films of practical interest and will be used to illustrate the operating principle of these films.

For light polarized along the optic axis, the droplet refractive index is close to n_e, the extraordinary refractive index of the LC; for light polarized normal to this axis, the droplet refractive index is close to n_o, the ordinary LC refractive index. Since the orientation of the optic axis varies randomly from droplet to droplet in the absence of external fields or applied stresses (Fig. 5.3(a)), incoming light propagating normal to the film will probe a range of refractive

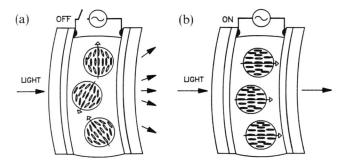

FIGURE 5.3. (a) Schematic diagram of a PDLC film in its off-state illustrating scattering of light by randomly oriented droplet directors. (b) Schematic diagram of a PDLC film in its on-state illustrating high transparency when droplet directors are aligned by an applied electric field. The substrates are shown curved to illustrate the flexibility achievable in PDLC devices.

indices between n_e and n_o. Since these cannot all be equal to the refractive index n_p of the polymer matrix, light will be scattered by the droplets. For typical droplet diameters (0.1–10 μm), droplet volume fractions (up to 50%) and film thicknesses (10–60 μm), light will be scattered by many droplets before emerging from the film. This multiple scattering causes the films to be cloudy in their scattering state (off-state). To maximize off-state scattering, the birefringence of the LC material should be as large as possible over the operating temperature range of the film.

An electric field applied across the film alters the orientation of the droplet directors, the molecular configuration within each droplet, or both. If the LC material has positive dielectric anisotropy, the molecules will attempt to align with their long axes parallel to the applied field (Fig. 5.3(b)). For sufficiently strong fields, nearly all the molecular axes will point along the field and each droplet will present to the incoming light a refractive index close to n_o. If the polymer matrix has been chosen so that its refractive index n_p, is close to n_o (the usual case in PDLC systems studied to date), there will be very little light scattering and the film will be highly transparent (on-state). Effects of index matching on film transparency will be discussed in Section 5.7.1. When the electric field is removed, the LC molecules revert to their initial off-state configuration.

In the remainder of this chapter we will discuss only PDLC films which have positive dielectric anisotropy and operate as we have just described. We note, however, that by using a LC with negative dielectric anisotropy in shaped droplets with carefully controlled molecular alignment at the droplet/polymer interface, it is possible to make PDLC films which are transparent in the absence of an applied electric field and scattering when a field is applied [20]. Such "reverse-mode" PDLC films are very new and have not been extensively studied.

It is important to note that the electric-field-induced reorientation will be opposed by elastic torques within the droplet; at each field value, there will be an equilibrium configuration that minimizes the droplet free energy, which consists of elastic, surface, and field contributions. Since the droplet director orientation and/or the molecular configuration within the droplets can be continuously varied by adjusting the magnitude of the applied field, the degree of light scattering and, hence, film transparency can also be varied continuously. The equilibrium configuration at each applied field value will depend on temperature.

In practice, PDLC films are driven by a.c. (sinusoidal or square wave) voltages to minimize electrochemical degradation which occurs in LC devices exposed to d.c. fields [21]. The molecular reorientation within the nematic droplets depends on the rms value of the applied electric field. The precise frequency range over which a PDLC film operates depends on the materials from which it is formed; typically, good electrooptic performance is observed for frequencies ranging from ~ 30 Hz to ~ 10 kHz.

5.3   Film Formation: General Aspects

Although film fabrication will be described in detail below, we present here an overview of the techniques for producing films containing microdroplet dispersions as a basis for our discussion of the energetics and kinetics of film formation. We consider two quite different methods for producing these films: emulsification and phase separation.

5.3.1   Emulsification—NCAP Films

In the simplest NCAP fabrication method, LC droplets are dispersed in a waterborne polymer to form an emulsion. The aqueous phase either can contain a water soluble polymer (such as polyvinyl alcohol) or can consist of a colloidal suspension of a water insoluble polymer (e.g., a latex emulsion) [3]. The emulsion is then coated onto a transparent conductive substrate, dried, and laminated to a second conductive substrate [3]. Another means of preparing a NCAP film is to microencapsulate the emulsified LC droplets with a polymer shell immediately after forming the emulsion. The encapsulated droplets are then suspended in a binder to form a film [3]. Generally, droplets having diameters in the range 1 μm–30 μm are produced by the NCAP process [1,2,22–25].

5.3.2   Phase Separation—Polymer-Dispersed Liquid Crystal Films

The first step in all PS methods is to mix a low molecular weight LC with a liquid polymer or polymer precursors to form a solution. Subsequently, when the polymer matrix is caused to harden, LC microdroplets separate from the matrix. Three major hardening processes have been investigated:

  (i) polymerization (curing or cross-linking) of a thermoset matrix precursor [4–8,26];
 (ii) cooling of a molten thermoplastic matrix [6,26,27]; and
(iii) evaporation of a common solvent in which the thermoplastic and the LC have both been dissolved [6,26,27].

Acronyms for the three processes have been devised: polymerization-induced phase separation (PIPS), thermally-induced phase separation (TIPS), and solvent-induced phase separation (SIPS) [6,26,28].

Polymerization-Induced Phase Separation (Thermoset Matrices)

Both thermal and radiation-induced cures have been used to form films by PIPS. Some precursors (e.g., multifunctional epoxies, polyurethanes) can be thermally polymerized: the components are mixed and their rate of cross linking is adjusted by controlling the temperature [4–7,26,29,30]. Use of accelerators [29] provides an additional method for controlling the cure rate.

Other precursors (e.g., certain epoxies and thiol-enes) can be cured by exposure to ultraviolet (UV) or visible light [8,31–37]. For the UV-induced cure to take place, the precursors must contain a small amount of a photo-initiator. Particle (electron-beam) radiation can also be used to induce cure [38,39].

Generally speaking, PIPS produces a wider range of droplet sizes (from below 0.02 μm to above 20 μm) than any of the other techniques with fast cure rates yielding small droplet sizes [5,6,26,29,32,40].

Thermally-Induced Phase Separation (Thermoplastic Matrices)

Studies of TIPS and SIPS have been pioneered at Kent State University [6,26,27,41–44]. PDLC films formed by TIPS are produced by dissolving a LC in a thermoplastic polymer (e.g., polyvinylformal, polymethylmetha-crylate, or a non-cross-linking epoxy) which has been heated above its soft-ening temperature, usually between 60 °C and 200 °C [27]. The solution is sandwiched between conducting substrates and then cooled into the misci-bility gap. The resulting decrease in solubility leads to phase separation of the LC in the form of droplets whose diameter depends on the cooling rate, LC concentration, and certain material parameters [6,9,10,45]. Droplets formed by TIPS can have diameters as small as 0.2 μm [45].

TIPS techniques give good control of droplet size. Furthermore, an un-satisfactory film can be reformed by heating to dissolve the LC and then recooling under proper control. Unfortunately, the operating temperature of a TIPS (or SIPS) film is limited by the softening point of the thermoplastic since the LC components may redissolve in the matrix at elevated tempera-tures. In addition (see Section 5.5.1), the refractive indices of thermoplastics are much more temperature sensitive than those of thermosets, an undesir-able attribute for films which must operate over wide temperature ranges. However, these problems may be avoided by choosing matrix materials which cross link after film formation [46].

Solvent-Induced Phase Separation (Thermoplastic Matrices)

SIPS is particularly useful with thermoplastics which melt above the decom-position temperature of the polymer or LC or when solvent coating methods are used [26]. SIPS films are formed by dissolving both the LC and the thermoplastic matrix in a common solvent and coating the solution onto a conducting substrate. After the solvent is rapidly removed, a cover substrate is laminated in place. The rate of solvent removal determines the initial drop-let size; diameters less than 1 μm have been produced for fastest evaporation rates [6,45]. If the initial droplet morphology is undesirable, the film may be heated to redissolve the LC and subsequently cooled at a rate tailored to give the desired droplet size and density [26]. The SIPS process can be slow, requiring many minutes for film formation [6,45].

In all PDLC formation methods, some residual LC is retained by the polymer matrix and some residual precursor matrix material by the LC. One goal of PDLC formation studies is to maximize the degree of phase separation of these components. In the next section we examine the energetics and kinetics of the PIPS formation process in order to assess their influence on droplet microstructure.

5.4   Polymer-Dispersed Liquid Crystal Formation: Polymerization-Induced Phase Separation Systems

5.4.1   Formation Model

During recent years the following qualitative model for PIPS has been developed [6,41,47,48]:

(i) A homogenous solution of LC and polymer precursor is formed and cure is initiated.

(ii) During cure, the molecular weight of the polymer increases, leading to a decrease in the mutual solubility of the LC and matrix.

(iii) Eventually the solubility becomes low enough that phase separation commences. If the initial LC concentration is sufficiently low, LC droplets form in a continuous matrix phase ("Swiss-cheese" morphology [34], Fig. 5.4(a)). On the other hand, if the LC concentration is too high, spheroids of matrix material will separate from a continuous liquid crystalline phase to produce a "reversed-phase" or "polymer-ball" morphology [34,49] (Fig. 5.4(b)). Since the "Swiss-cheese" morphology is the desired microstructure, the initial LC content must be suitably low.

(iv) As phase separation continues, the LC droplets grow by two simultaneous processes: accretion of additional molecules as they come out of solution and diffusion and coalescence of the droplets themselves. The microdroplet diameter continues to increase until the matrix gels and solidifies, essentially "freezing in" the droplet size and number density. A fast cure shortens the interval between initial phase separation and gelation so that droplets have little time to grow and coalesce. Thus, a rapid formation process should yield smaller droplet sizes and higher number densities than would a slow one [30]. This simple model predicts the influence of cure kinetics on droplet morphology. Cure kinetics can be influenced by several factors, including temperature and UV intensity.

We expect the cure kinetics of a thermally cross-linked system to obey the Arrhenius relation

$$\tau_{\text{cure}} = \text{const.} \times e^{E/RT_{\text{cure}}}, \tag{5.1}$$

(a)

(b)

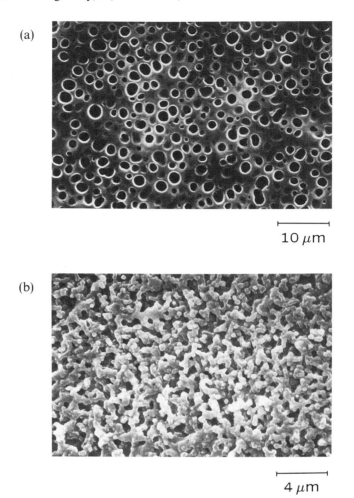

FIGURE 5.4. Scanning electron micrographs of cross sections of PDLC films prepared from two different samples. (a) PDLC film made with Devcon 5 min epoxy and a liquid crystalline mixture containing cyanobiphenyl and Schiff base liquid crystals illustrating the microdroplet or "Swiss-cheese" morphology. (b) PDLC film made with UV-cured NOA65 and EK11650, illustrating the reverse or "polymer-ball" morphology.

where τ_{cure} is the time constant for cure, E is an activation energy, R is the gas constant, and T_{cure} is the cure temperature in kelvins. Thus τ_{cure} should decrease with increasing cure temperature, leading to a decrease in droplet size.

The reaction kinetics of a UV-cured system are determined by the UV intensity, I_{UV} and by the diffusion and lifetime of the reactive species (radicals) generated by the interaction of the UV radiation with the photoinitiator component in the polymer precursor [50]. For a UV-curable photomer, the dependence of τ_{cure} on I_{UV} is given [51] by

$$\tau_{\text{cure}} = f(T_{\text{cure}}) \times I_{\text{UV}}^{-1/2}, \tag{5.2}$$

where $f(T_{\text{cure}})$ contains the temperature dependence of τ_{cure}. Since radical diffusion is slow at low temperatures and lifetime is short at high temperatures, the cure rate should be fastest at intermediate temperatures. Therefore, for a UV-cured PDLC, the temperature dependence of τ_{cure} could depart appreciably from Arrhenius behavior.

The model also suggests that the mutual solubility of the components is lowest when the degree of cure, D_{cure}, of the polymer is greatest (i.e., its final molecular weight is highest) [30]. Hence, a more fully cured PDLC film should have a greater degree of phase separation of the LC and matrix [30].

T_{cure} can influence both D_{cure} and solubility. D_{cure}, like cure rate, is greatest at intermediate temperatures [30,51,52]. As discussed above, a maximum in D_{cure} should lead to the greatest degree of phase separation and, hence, to the largest fractional amount, α, of LC in the PDLC microdroplets. If T_{cure} is too high, components of the LC mixture may remain preferentially dissolved in the matrix, thus complicating the cure process.

By extending the model, we can draw conclusions about the effect of LC concentration on PDLC microstructure. For instance, we expect that droplet size will increase with LC content due to higher supersaturation during cure. However, the limit of solubility of the LC in the cured matrix must be exceeded for this effect to occur. (Indeed, no droplets should be seen for LC concentrations below the solubility limit.) The diluting effect of the LC should also slow the cure kinetics, leading to a further enhancement of droplet diameter. Such dilution could also decrease the degree of cure, thus reducing α.

Let us now describe experimental techniques used to investigate the formation mechanisms of PIPS systems. The methods will be illustrated by specific examples.

5.4.2   Experimental Studies of Polymer-Dispersed Liquid Crystal Formation

Experimental Techniques

Calorimetry and scanning electron microscopy (SEM), the experimental techniques we use to study PDLC formation processes and their effect on film morphology and phase behavior, are simple but effective. The calorimeter is a DSC-2 differential scanning calorimeter [53] modified to permit UV irradiation of the sample, making it possible to study either thermal or radiation-induced cure [7,8,29,30,32]. Isothermal operation of the calorimeter allows us to record the exothermic power released during cross-linking (Fig. 5.5(a)): ΔQ_{tot}, the total heat evolved during cure, is obtained by integration of the exotherm; τ_{cure}, the time constant for the cure process (and hence

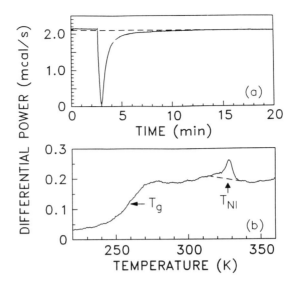

FIGURE 5.5. (a) DSC isothermal heat release curve for UV-cure of a PDLC. The heat of cure is obtained by integration of the curve, the time constant by fit of exponentials to the decay portion. By convention, DSC exotherms are downward-going. (b) DSC temperature scan of a PDLC. Glass transition and nematic–isotropic peak are indicated [32]. The nematic–isotropic transition enthalpy is determined by integration of the nematic–isotropic peak.

for PDLC formation), is determined by exponential fits to the decay portion of the curve [29,30,32,52]. From ΔQ_{tot} we can estimate D_{cure} of the PDLC film [30].

Operated in the temperature scanning mode, the calorimeter reveals the phase behavior of the cured PDLC film (Fig. 5.5(b)), yielding values of T_g (the glass transition temperature of the polymer matrix), T_{NI} (the nematic–isotropic transition temperature of the LC in the microdroplets), and ΔH_{NI} (the nematic–isotropic transition enthalpy, which is determined by integration of the nematic–isotropic peak). Since the nematic–isotropic peak is, in general, due solely to LC contained in the microdroplets, α is directly proportional to ΔH_{NI} [29,30].

Finally, the PDLC microstructure is determined (Section 5.6.2) from SEM [29,30]. Let us now describe results obtained for three PDLC systems produced by the PIPS process—two cured thermally, one by UV irradiation.

Thermally-Cured Epoxy-Based Polymer-Dispersed Liquid Crystals

PDLCs based on thermally cured epoxies are probably the most extensively studied PIPS systems [6,7,29,40,41]. Smith and Vaz [29] have shown that

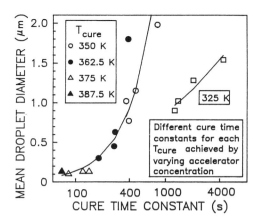

FIGURE 5.6. Droplet diameter versus cure time constant for a thermally cured epoxy-based PDLC [29]. The different dependence for $T_{cure} = 325$ K (52 °C) is thought to be associated with a change in the cure process at the lower temperature.

an increase in either T_{cure} or accelerator concentration shortens τ_{cure} and reduces droplet diameter (Fig. 5.6). They also found that the dependence of τ_{cure} on T_{cure} follows the Arrhenius relation and that the droplet number density decreases with increasing droplet diameter [30]. All these results are consistent with the simple model (Section 5.4.1). Cure temperature also affects α, with lower values occurring at higher T_{cure} (perhaps due to greater solubility or a change in the character of the cure process with increasing temperature) [29].

For LC concentrations above a limiting value, which increases with T_{cure}, droplet size generally increases with LC content [41]. Moreover, for sufficiently large LC concentrations, droplet size decreases with T_{cure} [29,41].

Thermally-Cured Polyurethane-Based Polymer-Dispersed Liquid Crystals

The effect of LC concentration and temperature on cure parameters, phase behavior, and microstructure of a polyurethane-based PDLC has been extensively studied [30]. Due to dilution, τ_{cure} increases moderately with X, the volume percent of LC; τ_{cure} is much more sensitive to temperature and shows the expected Arrhenius behavior. Dilution also leads to a linear decrease of ΔQ_{tot} with increasing X. More importantly, ΔQ_{tot} goes through a maximum at about 375 K (102 °C), indicating that D_{cure} is greatest in this temperature range. This result is consistent with the observation that droplet formation for this system is optimum at 375 K (102 °C) as shown below.

The phase behavior of the PDLC is strongly affected by LC concentration (Fig. 5.7). As X becomes larger, T_g decreases while T_{NI} increases. The decrease in T_g is due to the plasticizing effect of the liquid crystal. The increase in T_{NI}

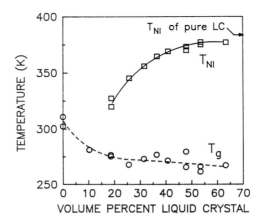

FIGURE 5.7. Glass transition temperature, T_g, and nematic–isotropic temperature, T_{NI}, versus LC concentration for a thermally cured polyurethane-based PDLC [30].

is probably caused by an enhancement in LC purity for LC concentrations far enough above the solubility limit. The greater LC purity for larger X-values is accompanied by an increase in α, as seen in Fig. 5.8: ΔH_{NI} (proportional to α) increases linearly with X for concentrations above the solubility limit.

SEM studies [30] reveal that droplet formation is best at a cure temperature of 375 K (102 °C). They also show that droplet size increases linearly with LC concentration (Fig. 5.9). As expected, no droplets are formed below the solubility limit. Above $X \approx 60\%$, "polymer-ball" rather than "Swiss-

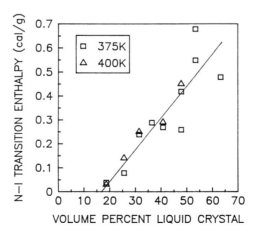

FIGURE 5.8. Nematic–isotropic transition enthalpy versus LC concentration for the polyurethane-based PDLC [30].

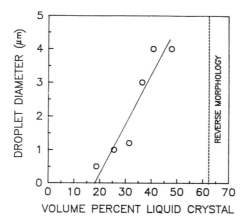

FIGURE 5.9. Droplet diameter versus LC concentration for the polyurethane-based PDLC [30]. The cure temperature was 375 K (102 °C).

cheese" morphology is observed. Finally, as for the epoxy-based system, droplet number density decreases with increasing droplet size. Once again these results are consistent with the simple model (Section 5.4.1).

UV-Cured Thiol-ene-Based Polymer-Dispersed Liquid Crystals

The UV-cure technique possesses an important advantage over other PDLC formation methods: both the onset and kinetics of the cure process can be precisely controlled by the choice of the turn-on time and I_{UV}. T_{cure} can be independently adjusted to obtain a desired solubility and/or D_{cure} [30,31,52]. UV-cured PDLC systems have been described in several publications [8,32,34–36,49,54]. In this section we will discuss recent studies [52] of PDLC formation in 1:1 mixtures of a thiol-ene matrix (Norland Optical Adhesive 65) with a multicomponent cyanobiphenyl/cyanocyclohexyl LC.

As seen in Fig. 5.10, T_{cure} plays a major role in determining the energetics of cure for both the PDLC and the pure matrix material. As expected, ΔQ_{tot} of each goes through a maximum, with that for the PDLC occurring at about 320 K (47 °C), some 50 degrees lower than for the matrix. The greatest D_{cure} for the PDLC (derived from ΔQ_{tot}) is only about 93% of the maximum value for the pure matrix, probably as a result of dilution. Neither ΔQ_{tot} nor D_{cure} are affected by I_{UV}.

The cure time constants of both matrix and PDLC (Fig. 5.11) exhibit the behavior anticipated in the simple model: a minimum at intermediate temperatures. Furthermore, the temperature range for minimum τ_{cure} is close to that for maximum D_{cure}. In this system, τ_{cure} for the PDLC is larger due to dilution by the LC and shows the expected $I_{UV}^{-1/2}$ dependence [52].

As for the thermally cured systems, T_g, T_{NI}, and ΔH_{NI} are sensitive to T_{cure} (and hence to D_{cure}). However, they are unaffected by I_{UV}. Both PDLC and

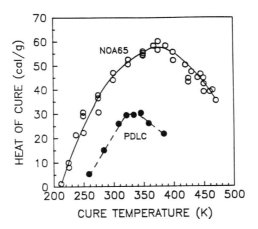

FIGURE 5.10. Total heat of cure versus cure temperature for a UV-cured PDLC and for the pure polymer matrix (NOA65) [52]. The maximum in each curve is associated with the greatest degree of cure of the polymer matrix.

matrix have maximum T_g values at the temperature where D_{cure} is greatest; this behavior is expected since enhanced cure increases the matrix rigidity as well as the degree of phase separation. T_{NI} also increases with D_{cure} [52]. Of most interest is the dependence of ΔH_{NI} on T_{cure} (Fig. 5.12). ΔH_{NI} (and, hence, α) goes through a maximum at the temperature for greatest D_{cure}) This conclusion is consistent with electron micrographs which show that best droplet formation occurs in the same temperature range [52].

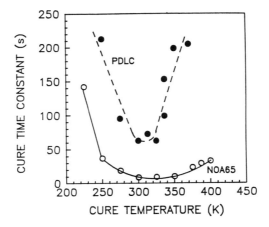

FIGURE 5.11. Cure time constant for the UV-cured PDLC and the pure polymer matrix (NOA65) as a function of cure temperature [52]. The UV intensity was about 3 mW/cm^2.

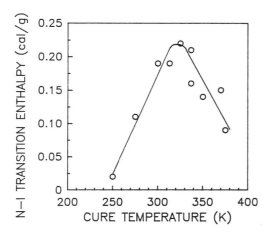

FIGURE 5.12. Nematic–isotropic transition enthalpy versus cure temperature for the UV-cured PDLC [52].

5.4.3 Concluding Remarks Regarding Polymer-Dispersed Liquid Crystal Formation

The results of these three studies are, in general, consistent with the simple model. It appears that calorimetry, coupled with electron microscopy, can provide useful information regarding the details of the PDLC formation process. What is still required is a quantitative theory.

5.5 Materials Selection

In this section we will discuss several material parameters which are important in selecting liquid crystals and polymers for PDLC films:

  (i) refractive indices;
 (ii) miscibility of the LC and polymer matrix;
(iii) temperature range; and
(iv) environmental stability.

Other material parameters (e.g., dielectric constants and electrical resistivities) will be discussed later in connection with their influence on electrooptic performance.

5.5.1 Refractive Indices

The refractive indices of the LC and the polymer matrix have been extensively studied because they strongly affect both the off-state opacity and the

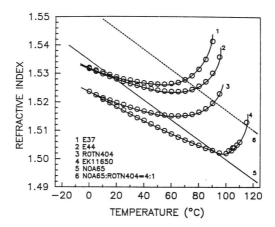

FIGURE 5.13. Temperature dependence of: the ordinary refractive indices n_o of several liquid crystalline materials (circles; lines 1–4); the refractive index n_p of UV-cured polymer NOA65 (solid straight line 5); and UV-cured mixture NOA65:ROTN404 = 4:1 (dashed straight line 6).

on-state transparency of the PDLC film [5,49,55–58]. To obtain refractive index matching we may adjust the composition of either the LC or the polymer resin.

Liquid Crystal

The refractive index of the LC is usually adjusted by varying the composition of a mixture of liquid crystalline materials. Addition of isotropic liquids to adjust the index matching (or other properties like viscosity) is usually not recommended because it depresses T_{NI} and reduces birefringence and, thus, the scattering efficiency of the PDLC off-state. Figure 5.13 shows the temperature dependence of n_o of several liquid crystals identified in Table 5.1 [57,58]. The greatest variation in n_o from material to material is only about 0.02. For most liquid crystals, n_o goes through a minimum at some temperature below T_{NI}. Since n_p usually decreases monotonically with increasing temperature, an index match ($n_p \approx n_o$) is possible at only one or two temperatures. This is illustrated in Fig. 5.13 by plotting n_p data for two UV-cured polymers along with the LC data.

Ultraviolet-Curable Thermoset Resins

An important advantage of the UV-cure method is the relative ease with which we can vary the composition of the formulation to produce polymers with a variety of properties [8,31,37,54,57,59–63]. A typical formulation includes the photoinitiator, oligomers (which give the basic properties to the polymer), and one or more reactive monomers to adjust other properties like flow viscosity, flexibility, glass transition temperature and, of course, n_p.

TABLE 5.1. Selected materials and material properties.

Material	Description	T_{NI} (°C)	Δn	dn/dT (°C^{-1})	Source
Liquid Crystals					
E37	CBs, CT, and E LCs	85	0.247		a
E44	CBs, CT, and CBEB LCs	100	0.263		a
ROTN404	CBs, CT, and PNs LCs	105	0.26		b
EK11650	E LC	116	0.165		c
Polymer Precursors and Polymers					
NOA65	UV-Cured thiol-ene			-3.444×10^{-4}	d
Norcote 02-049	UV-Cured adhesive			-3.136×10^{-4}	e
Dymax 20141	UV-Cured methacrylate			-3.938×10^{-4}	f
Amicon XUV-919	UV-Cured acrylate			-3.477×10^{-4}	g
DGEBA	Aromatic epoxy resin			-3.756×10^{-4}	h
Epon 812	Aliphatic epoxy resin			-3.416×10^{-4}	h
Devcon 5B	5-minute epoxy hardener			-3.474×10^{-4}	i
Capcure 3-800	Mercaptan epoxy hardener			-3.346×10^{-4}	j
Capcure EH-30	Amine epoxy hardener			-4.052×10^{-4}	j
DGEBA:5B = 1:1	DGEBA cured with Devcon 5B			-3.559×10^{-4}	
DGEBA:5B = 1:2	DGEBA cured with Devcon 5B			-3.099×10^{-4}	
Epon:5B = 1:1	Epon cured with Devcon 5B			-2.414×10^{-4}	
Epon:5B = 1:2	Epon cured with Devcon 5B			-3.014×10^{-4}	
Bostik 219	Thermoplastic			-5.984×10^{-4}	k
Hysol 1942	Thermoplastic polyolefin			-6.687×10^{-4}	l
JetMelt 3764	Thermoplastic polyolefin			-7.029×10^{-4}	m

C: cyano; B: biphenyl; T: terphenyl; E: ester; P: phenyl; N: pyrimidine; LC: liquid crystal.

[a] EM Chemicals, Hawthorne, NY.
[b] Hoffmann–La Roche, Nutley, NJ.
[c] Eastman Kodak, Rochester, NY.
[d] Norland Products, New Brunswick, NJ.
[e] Nor-Cote Chem. Co., Crawfordsville, IN.
[f] Am. Chem. & Eng. Co., Torrington, CT.
[g] Amicon Corp., Lexington, MA.
[h] Shell Oil Co., Houston, TX.
[i] Devcon Corp., Danvers, MA.
[j] Diamond Shamrock Co., Morristown, NJ.
[k] Bostik Corp., Reading, Pa.
[l] The Dexter Corp., Industry, CA.
[m] 3M Chem. Corp., St. Paul, MN.

Figures 5.14 a and b show, respectively, the temperature dependence of the refractive index for several polymer precursors and the corresponding cured materials (see Table 5.1). As expected, the indices of the cured materials are larger than those of the precursors, probably due to an increased density of the materials after cure. At each temperature, the large variation in n_p from material to material (about 0.04) will be sufficient to index match a large number of liquid crystalline materials.

Residual LC dissolved in the cured polymer (Section 5.4) raises the effective refractive index of the polymer matrix (Fig. 5.13, dashed line). Therefore, the

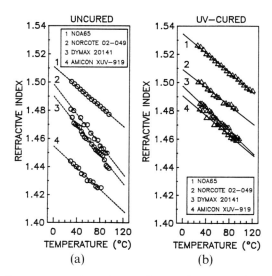

FIGURE 5.14. Temperature dependence of the refractive indices n_p of (a) several UV-curable polymer precursor materials and (b) the corresponding UV-cured polymers. The dashed (solid) straight lines are least squares fits to the data, circles (triangles).

formulation of the PDLC should take this effect into account in order to produce index matching.

Electron-Beam-Curable Thermosets

Electron-beam-curable resins are often similar to UV-curable resins but require no photoinitiators [38,39]. Since the amount of photoinitiator in UV-curable resins is small (about 1–5%), the index matching requirements for electron-beam-curable resins are similar to those for the UV-curable resins.

Epoxy Thermosets

Epoxides consist of two basic components, the epoxy resin and the curing agent. Either can be varied to adjust the refractive index of the cured material. Further index adjustments can be achieved by using mixtures of resins, mixtures of hardeners, or both. Refractive indices of binary mixtures of hardeners generally span a small range (e.g., 1.50–1.515 for binary mixtures of Capcure 3-800 and Capcure EH-30) [58]. However, mixtures of low-index aliphatic resins and high-index aromatic resins can span a very large range (e.g., 1.483–1.57 for binary mixtures of DGEBA and EPON 812) [58]. Thus, compositional changes in epoxides may also lead to matrix materials that can index match a variety of liquid crystals.

The temperature dependence of the refractive indices of epoxy resins, har-

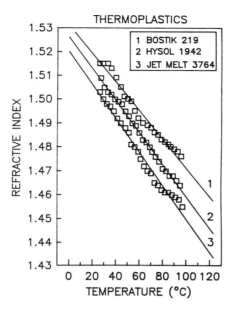

FIGURE 5.15. Temperature dependence of the refractive indices of some thermoplastics (squares) [49]. The straight lines are least squares fits to the data.

deners, and cured epoxides is similar to that of UV-cured polymers and their precursors [58].

Thermoplastics

The refractive indices of several thermoplastic polymers (Table 5.1 and Fig. 5.15) vary little (about 0.01) from sample to sample [49,57]. This small index range may not be sufficient to index match many liquid crystalline materials. The most adverse feature of the refractive indices of thermoplastics, however, is their large temperature variation [49]. From inspection of Table 5.1, which lists values of dn/dT, the slope of the temperature dependence of the refractive index for several polymer materials, it is clear that dn/dT values of thermoplastics are about twice as large as those of thermosets. Thus, index matching over wide temperature ranges may be more difficult with thermoplastics than with thermosets.

5.5.2   Solubility of Liquid Crystal in Matrix

Ideally, the solubility of the LC in the matrix should be large at the onset of cure and rapidly decrease towards zero during cure. If this were the case, the LC would initially dissolve in the prepolymers and completely phase separate from the polymer matrix after cure. In practice this ideal case is not realized. Many LC mixtures partially separate from the polymer precursors at low temperatures; furthermore, after cure, some LC remains dissolved in the matrix (Section 5.4).

In a study of the relative solubility of a LC in polymer precursors, Hirai et al. [54] determined a phase diagram for mixtures of LC and a UV-curable resin consisting of monomers and oligomers. The diagram exhibited single and biphasic regions and features of a spinodal decomposition process.

In related work, Yamagishi et al. [34] and Vaz et al. [49,57,64] had investigated the dependence of the microstructure of UV-cured PDLC films on the volume concentrations of LC and polymer precursors and on T_{cure}. Variations in LC concentration led to either microdroplet (i.e., "Swiss-cheese") or reverse (i.e., "polymer-ball") morphologies (Section 5.3 and Fig. 5.4). "Polymer-ball" morphology does not confine the LC in microregions and, therefore, is less desirable than the microdroplet morphology [57]. Furthermore, films with this morphology often exhibit large hysteresis in their electrooptic response [34].

5.5.3 Operating Temperature Range

The operating temperature range of a PDLC film can be determined by calorimetric observation of its phase behavior, by optical microscopy, or by monitoring temperature-dependent transmittance changes when the film is switched between its off- and on-states. From such measurements, we have determined that the operating temperature range of a PDLC film is determined primarily by the choice of LC and the polymer matrix material and does not depend strongly on the polymerization rate [29,52,64,65]. A LC mixture with a wide nematic range is necessary to obtain a wide operating temperature range in a PDLC film. However, if certain components of a broad-range LC mixture preferentially remain dissolved in the polymer matrix or some uncured polymer precursors remain dissolved in the LC after cure (Section 5.4), the nematic range of the LC in the film will differ from that of the pure LC material. Studies with different matrices and liquid crystals have shown that changes in the polymer matrix can depress T_{NI} by as much as 30 °C [9,65]. Generally, however, the depression in T_{NI} is smaller than 5 °C.

Outdoor applications of PDLC films are far more demanding in terms of materials choice than are indoor applications. For example, the desired operating temperature range can exceed -30 °C to $+90$ °C. Although highly birefringent LC mixtures can be prepared to operate at the highest temperatures, improved formulations may be needed to obtain fast low-temperature response times.

5.5.4 Environmental Stability

The environmental stability of a PDLC film is especially important in outdoor applications. Without adequate protection, the stability can be sharply reduced by exposure to humidity, temperature, or UV radiation [66]. The primary degradation mechanisms are electrochemical reactions which are activated by moisture or UV radiation and accelerated by temperature [66].

Lackner et al. [66] studied the influence of short wavelengths on the stability of PDLC films by monitoring their electrooptic performance after exposure to radiation with different UV cutoffs. Films exposed for about 50 hours to 50 mW/cm^2 radiation with wavelengths as short as 351 nm exhibited poor electrooptic response. By contrast, samples protected by filtering out wavelengths below 370 nm exhibited good electrooptic response but had slightly ($\sim 10\%$) increased operating voltage after exposures of 1000 hours.

Protective filtering is especially important in UV-cured systems where the photoinitiator can accelerate electrochemical degradation of the LC. Electron-beam-curable resins may be less sensitive to UV degradatio because they do not need photoinitiators [38,39].

The lifetime of a PDLC film may also be extended by choosing LC materials with low susceptibility to electrochemical degradation. For example, while many pyrimidines, ester, azo, azoxy, and Schiff base linkage groups have limited stability, cyclohexyls and cyanobiphenyls and terphenyls have been found to be quite stable. However, since the cyclohexanes have low birefringence, cyanobiphenyls and terphenyls are currently the preferred materials.

Finally, PDLC lifetime may be extended by choosing stable polymer matrix materials (e.g., aliphatic rather than aromatic materials) and by including antioxidants, UV stabilizers, and other additives [67,68].

5.6 Film Fabrication and Morphology

5.6.1 Ultraviolet-Curing Conditions

Some aspects of film formation have already been discussed in Section 5.4. In this section we will describe the actual fabrication of UV-cured samples and discuss techniques to characterize the film microstructure.

PDLC films are prepared between two flat glass or plastic substrates coated with a thin layer of conducting indium tin-oxide (ITO) or other transparent conductor. They are cured using UV radiation and their thickness is controlled by 17.5 μm glass microfiber spacers. The three basic steps in film preparation consist of

(i) mixing the desired proportions of LC and polymer precursors;
(ii) spreading a small amount of the mixture between the substrates; and
(iii) exposing this "sandwich" to UV radiation to cure the polymer [5,7,8, 31,49,57].

As discussed in Section 5.4, T_{cure} and LC concentration are both important in controlling the cure process which, in turn, determines the size and distribution of the LC microdroplets [29,52,57,65]. Figure 5.16 indicates the range

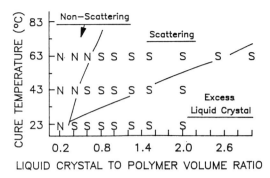

LIQUID CRYSTAL TO POLYMER VOLUME RATIO

FIGURE 5.16. Schematic diagram of the results obtained for PDLC films prepared with the LC ROTN404 and the NOA65 thiol-ene polymer matrix [49]. S denotes a scattering film and N a nonscattering film.

of temperatures and LC concentrations which yield [49] useful PDLC films based on the polymer matrix NOA65 and the LC ROTN 404 (see Table 5.1). In this figure S denotes a scattering film and N a nonscattering film. The diagram was obtained by curing many PDLC films at each of three temperatures (23°, 43°, and 63 °C) with various LC concentrations. Three distinct regions of the diagram are apparent:

(i) All samples with low LC concentrations failed to scatter light efficiently because the LC did not phase separate from the matrix. Films cured at higher temperatures required larger LC volume fractions for light scattering because of the increased solubility of the LC in the polymer at higher temperatures.
(ii) Samples prepared at low temperatures with large LC concentrations contained large drops of LC visible to the naked eye. Such films scattered light very inefficiently.
(iii) Films containing intermediate concentrations of LC scattered light. Films in this third region were characterized by a variety of droplet sizes and size distributions, as determined from SEM.

Figure 5.4(a) illustrates a relatively uniform distribution of microdroplets with an average diameter ≤ 1 μm. As noted in Section 5.4, other microstructures have been observed with large LC concentrations. For example, [57,64], some films made with EK11650 in NOA65 (Table 5.1) exhibited the "polymer-ball" morphology (Fig. 5.4(b)). The results illustrated in Figs. 5.4 and 5.16 show the importance of selecting the appropriate cure conditions to obtain the desired PDLC morphology (and, thus, the desired electrooptic performance) using PIPS. Micrographs like that of Fig. 5.4(a) also provide information which allows us to perform quantitative analysis of droplet-size distributions (next section).

5.6.2 Droplet Size Distribution

A quantitative characterization of the microdroplet size distributions of PDLC films is important for understanding their light scattering properties (Sections 5.7 and 5.8) [65,69,70]. This analysis is conveniently carried out using SEM photomicrographs of a film cross section (e.g., Fig. 5.4(a)). Figure 5.17 shows two-dimensional and three-dimensional histograms (2DH and 3DH, respectively) corresponding to the SEM photomicrograph of Fig. 5.4(a). The 2DH describes the number of droplet cross sections with diameters between d and $d + \Delta$, while the 3DH describes the number of actual, three-dimensional LC droplets with diameters between d and $d + \Delta$, where Δ is the width of each bin in the histograms. While the 2DH is obtained directly from the SEM photomicrograph, the 3DH is obtained from the 2DH histogram by matrix inversion techniques [38,39,71–77]. From the 3DH we find that the PDLC film of Fig. 5.4(a) has a mean droplet diameter of $\approx 0.75\ \mu$m and the volume fraction occupied by the droplets in that film is $\approx 32.3\%$.

5.6.3 Color

PDLC films can be tinted either by addition of dyes or pigments to the PDLC formulation or by use of an external color layer. There are advantages and disadvantages to each method.

Dyes or pigments must be added prior to the polymer cure. Care must be exercised to ensure that absorption by the colorants does not interfere with the absorption of UV radiation and the transfer of energy by the photoinitiator [78]. Good color contrast between the on- and off-states may be obtained with optically isotropic dyes as well as with dichroic dyes. Multiple light scattering within a PDLC film increases the effective optical path when the film is switched from the on-state to the off-state [79,80]. Since the color density varies exponentially with the optical path, it is easy to obtain efficient

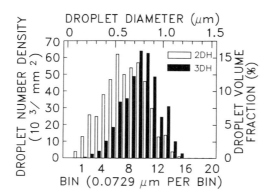

FIGURE 5.17. Microdroplet size histograms derived from the SEM of Fig. 5.4(a). Unfilled bars: 2D-histogram. Shaded bars: 3D-histogram.

coloring of the off-state even when the on-state is weakly colored. Thus, it may be practical to use isotropic dyes which are more readily available and often more stable than dichroic dyes.

External color can be added subsequent to cure, either by laminating a color filter to the PDLC device or by incorporating color into one or both of the substrates which support the transparent electrodes. Color may also be implemented to a limited degree by manipulating the formulation of the electrodes.

5.7   Electrooptic Properties

Electrooptic properties and potential applications of PDLC films have been reviewed recently [9,10,26,65]. Therefore, we will focus on selected electrooptic characteristics and how they are influenced by PDLC film composition and droplet morphology. We will discuss the relations between refractive indices and film transmittance, transmittance versus voltage characteristics, response times, and spectral transmittance characteristics.

5.7.1   Refractive Indices and Film Transmittance

The effects of refractive indices on film transmittance depend on viewing direction (angle of incidence), polarization, droplet shape, temperature, and matrix material.

Polarization Effects and On-State Transmittance: Spherical Droplets

In a PDLC film containing spherical droplets, the droplet directors are aligned normal to the film plane in the on-state and the film is azimuthally symmetric about its normal. In this case, the transmittance at normal incidence will be maximum in an index-matched film ($n_p = n_o$) regardless of the polarization of the incident light.

At oblique incidence, the on-state transmittance depends on polarization. For incident light polarized normal to the plane of incidence formed by the propagation direction of the incident light and the normal to the film, the droplet refractive index $n_d \approx n_o$ for any angle of incidence θ [42]. Increasing θ decreases transmittance by increasing both Fresnel reflection and the path length traversed by light in the film (see Section 8.3).

The on-state transmittance of light polarized in the plane of incidence is determined primarily by the increase in n_d with increasing θ [42], although Fresnel reflection and optical path length also play a role. If $n_p \leq n_o$, transmittance is maximum at normal incidence. As θ increases, scattering increases because of the increasing mismatch between n_d and n_p; consequently, the film appears hazy when viewed at large angles from its normal. The larger the birefringence of the LC, the more pronounced the haze. It is often necessary to compromise between maximizing off-state scattering (large birefringence) and minimizing haze (small birefringence) when choosing the LC for a PDLC film. Regardless of the birefringence, matching n_p to n_o will maximize the range of θ for which transmittance is high [42]. If $n_p > n_o$, transmittance will be maximized and haze minimized at some $\theta \neq 0$.

Polarization Effects and Film Transmittance: Ellipsoidal Droplets

If a shear stress is applied to a PDLC film during its formation, ellipsoidal droplets will be produced [27,55,81]. The major axes of the ellipsoids will be aligned (approximately) parallel to each other at some angle with the film plane. In the off-state, the droplet directors coincide with the major axes of the ellipsoids. In an applied electric field, molecular reorientation within the droplets causes the directors to rotate away from the major axes of the ellipsoids toward the film normal. At each field value, the droplet director orientation is determined by a balance between the electrostatic torque and the elastic torque due to the elongated shapes of the droplets [10].

Analysis of the transmittance characteristics of a PDLC film containing shaped droplets is complicated by the fact that the film normal is no longer an axis of symmetry. We discuss only the case where the plane of incidence contains the major axes of the ellipsoidal droplets of an index-matched film. For light polarized normal to this plane, $n_d \approx n_o$ for any θ and voltage and the film appears transparent. The magnitude of the transmittance depends on Fresnel reflection and optical path length as in the case of spherical droplets. For light polarized in the plane of incidence, the film will be transparent when $\theta = \theta_0$, the angle of incidence for which light propagates along the droplet directors. As θ moves away from θ_0, the film becomes increasingly cloudy. In the off-state, the angular range over which the film is transparent and the degree of transparency depend on how uniformly the droplets are aligned by shearing. The spread in droplet director orientation in the presence of the field determines the viewing angle range in the field-aligned state [55].

When illuminated with normally incident unpolarized light, a PDLC film with ellipsoidal droplets transmits the component polarized normal to the plane containing the droplet directors and scatters the component polarized in this plane. The degree of scattering and, hence, the polarization of the transmitted light, depends on the voltage across the film, i.e., the film is an electrically tunable polarizer.

Index Matching in Polymer-Dispersed Liquid Crystal Films with a Wide Operating Temperature Range

As discussed in the section on polarization effects, matching n_p and n_o maximizes on-state transmittance and viewing angle range. However, since n_o has a minimum value at some temperature below T_{NI} while n_p decreases monotonically with increasing temperature (Section 5.5.1), perfect index matching is achievable at only one or two temperatures. At other tempera-

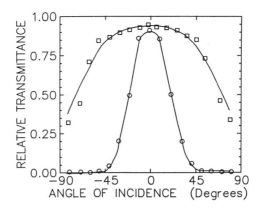

FIGURE 5.18. Relative transmittance versus angle of incidence for an index-matched PDLC film with an optically isotropic matrix material (circles) and for a "haze-free" PDLC film with a liquid crystalline polymer matrix [56].

tures, the effects of index mismatch on transmittance and viewing angle described above will come into play. For optimum PDLC operation over an extended temperature range, the LC should be chosen so that the operating temperature range of the PDLC film lies well below T_{NI}, preferably below the minimum in the n_o versus temperature curve. This choice will avoid the index mismatch caused by the rapid increase in n_o at temperatures near T_{NI}. This choice is not always possible if the PDLC film must operate over a very wide temperature range; in this case, operation over nearly the full nematic range of the LC is required. Nevertheless, reasonably good electrooptic performance over a wide temperature range has been demonstrated [9,65].

"Haze-Free" Polymer-Dispersed Liquid Crystal Film

Doane and West [56] have produced a "haze-free" PDLC film in which the matrix material is a side-chain liquid crystalline polymer whose ordinary and extraordinary refractive indices are close to those of the low molecular weight LC in the droplets. An applied electric field aligns the LC molecules in both the droplets and the matrix, resulting in a good refractive index match for all angles of incidence and a very wide viewing angle range, as illustrated in Fig. 5.18. When the field is removed, the droplets revert to a radial configuration (Fig. 5.2(c)) which scatters light for all angles of incidence.

5.7.2   Transmittance versus Voltage

Temperature Effects

Figure 5.19 illustrates the transmittance versus voltage ($\mathscr{T}$ versus V) curves for a 17.5μm thick UV-cured PDLC film at $-8°$, $30°$, and 60 °C. Transmit-

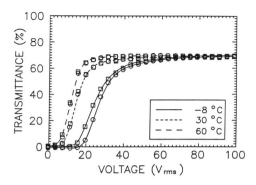

FIGURE 5.19. Transmittance versus voltage for a UV-cured PDLC film at different temperatures [32]. Circles denote data taken with increasing voltage; squares denote data taken with decreasing voltage.

tances were measured at 543.5 nm using a double-beam HeNe laser system with air in the reference beam [32]; thus, they include effects of reflection losses at the sample surfaces and at interfaces within the sample as well as absorption by the ITO electrode layers.

On-state transmittance near 70% and off-state transmittance below 1% are obtained at all temperatures above -8 °C for $V \geq 60V_{rms}$. The saturation voltage V_{sat} (the voltage required to obtain maximum transmittance) and the threshold voltage V_{th} (the voltage required to increase film transmittance above its off-state value) both increase as film temperature is lowered.

The $\mathcal{T}$ versus V curves of PDLC films generally show some hysteresis, i.e., the curves for decreasing voltage do not retrace those obtained for increasing voltage. Hysteresis is evident in the low-temperature data of Fig. 5.19. In general, if $\mathcal{T}$ versus V measurements are repeated, the same curves, including the hysteresis, will be obtained. Drzaic [25] has proposed a model for the hysteresis which is based on a two-stage response of the molecules within the droplet to the application or removal of an electric field. This model had been previously proposed [5] to explain the two-stage relaxation of PDLC films upon removal of a d.c. pulse. We note, however, that we have observed hysteresis in the $\mathcal{T}$ versus V characteristics of PDLC films which do not show two-stage relaxation but exhibit a simple exponential decay after removal of a driving electric field [9]. Therefore, there may be other factors accounting for the hysteresis which have not yet been identified.

$\mathcal{T}$ versus V measurements on a 60 μm thick PDLC film made from the same materials as the sample of Fig. 5.19 show that V_{th} is higher in the thick sample at all temperatures and maximum transmittance is not achieved at any temperature for $V \leq 150V_{rms}$. In general, both V_{th} and V_{sat} increase with film thickness because droplet reorientation is a function of applied electric field rather than applied voltage.

At any temperature, the shape of the $\mathcal{T}$ versus V curve depends on n_p and n_d. If $n_p = n_o$, $\mathcal{T}$ reaches a maximum for some value of V and remains at that value as V increases further. If $n_p < n_o$, $\mathscr{T}$ never reaches a maximum value. If $n_p > n_o$, $\mathscr{T}$ reaches a maximum value for some V and decreases as V increases further [42].

Effects of Film Morphology and Material Parameters

To reduce both electric power consumption and the cost of electronic drive circuitry, it is desirable to minimize the switching voltages of PDLC and NCAP films. Consequently, several groups have begun to investigate the parameters which control the switching voltage [3,10,26,35,55,81–84].

Switching these films involves the rearrangement of molecules within the droplets in response to an applied electric field. The rearrangement may involve a transition between different molecular configurations; such a transition occurs when droplets switch from a radial to an axial configuration (Fig. 5.2). In this case, the required electric field varies inversely with droplet diameter provided the droplet diameter is sufficiently large [10,26,55,81–84]. We will not discuss this case further since radial droplets have been found in few films of practical interest.

Switching a film with bipolar droplets involves the reorientation of the droplet directors rather than a configuration transition. Two stages in the reorientation have been observed by optical microscopy [10,84] and inferred from response time measurements [5,25], viz., reorientation of the droplet directors along the applied field and distortion of the molecular configuration within each droplet.

Droplet director reorientation occurs when the electric field within the droplet reaches a critical value which varies inversely with droplet size (radius for spherical droplets, semimajor axis for ellipsoidal droplets) [10,26,82,84]. The decrease in switching voltage with droplet size has been observed by several groups [3,35,45]. However, Nomura et al. have recently shown [45] that, for sufflciently large (about 3–4 μm) droplets, the switching voltage increases. They explain this behavior by modeling a dispersed LC droplet as a central core, in which the molecules are easily aligned by an applied field, and a surrounding shell in which the molecular alignment is controlled by surface forces. Changes in the relative volumes of the shell and core and the distribution of applied voltage between the droplet and the surrounding polymer as the droplet size changes combine to minimize the switching voltage for some droplet diameter.

The critical field is also proportional to a shape factor which vanishes for spherical droplets, indicating that no elastic distortion is necessary to reorient a spherical droplet director by an applied field while maintaining tangential alignment at the droplet wall [10,55]. As the droplet becomes more elongated, the critical field increases. Such an increase has been measured in NCAP films [85].

Finally, the critical field is proportional to a function of the complex dielectric constants of the LC and polymer matrix materials [10,26,55,81–84] which accounts for differences, between the applied electric field V/d (V is the applied voltage; d is the film thickness) and the field within the droplets. At low frequencies, this factor reduces to a function of the resistivities of the polymer and LC materials. The importance of resistivities has been demonstrated by several groups [55,86]. The resistivity of the LC material inside the droplets of a PDLC film can be substantially lower than that of the pure LC material since some polymer precursor material remains in the droplets after film formation [55]. Changing the resistivity of the polymer matrix by varying the amount of cross-linking material can change the switching voltage by nearly an order of magnitude [55]. In films formed by TIPS, addition of suitable additives (e.g., organometallic compounds) to the initial solution lowers the switching voltage [46]. Whether these additive effects are due to resistivity changes or to some other mechanism is not apparent at this time.

When the electric field inside the droplets exceeds the critical value for director reorientation, the director configuration within each droplet is distorted. This effect, which is small for bipolar droplets [10], has been predicted by computer simulations [10,26] and observed by optical microscopy [84].

5.7.3   Response Times

Temperature Dependence

We have measured the temperature-dependent rise time t_r and decay time t_d of PDLC films formed from different LC and polymer matrix materials. Typical results for UV-cured films [32] are:

(i) At room temperature t_r ranges from a few milliseconds to a few tens of milliseconds, depending on the rms excitation voltage V_{rms}; t_d is typically a few tens of milliseconds and does not depend on V_{rms}. As temperature increases, t_r and t_d both decrease. As temperature decreases, t_r and t_d both increase; however, even at temperatures below $-10\,°C$, t_r is generally below 250 ms and t_d below 600 ms at all reasonable voltages.

(ii) t_r depends strongly on voltage at all temperatures. As in conventional field-effect nematic LC devices, t_r^{-1} is a linear function of V_{rms}^2 provided V_{rms} is low enough that electric-field-induced distortion of the nematic director configuration may be assumed small [87]. At the highest voltages, the measured values of t_r^{-1} are lower than those obtained by extrapolating the linear curve from lower voltages, i.e., the actual rise times are longer than those expected from extrapolation. This is consistent with the idea that high electric fields produce a distortion of the molecular configuration within a droplet in addition to the rotation of the droplet directors; this additional distortion increases the response time.

(iii) t_d does not depend strongly on V_{rms} except at the lowest temperatures, where it appears to increase with increasing voltage. This unexpected voltage dependence may reflect the fact that, at low temperature, different degrees of molecular reorientation are obtained for different excitation voltages so that decay times are not being measured from the same initial state at all voltages.

At temperatures well below T_{NI}, the reported [55] temperature-dependent response times of thermally cured, epoxy-based PDLC films are similar to those of UV-cured films. However, the decay times of the thermally cured films increased as the temperature approached T_{NI}. We have observed similar effects in UV-cured films at temperatures near T_{NI} where the nematic and isotropic phases may coexist.

Size, Shape, and Material Effects

Since rotation of the nematic director configuration of a spherical droplet involves no change in the droplet elastic energy, the question of what makes a droplet return to its original state upon removal of an electric field is of considerable interest. Several authors [10,24–26,55,81–85] have noted that dispersed LC droplets are never really spherical. Even a small departure from sphericity means that elastic torques must be overcome to rotate the director configuration within a droplet by means of an electric field. When the field is removed, these torques restore the molecular configuration to its initial zero-field state.

The effect of droplet shape on t_d has been demonstrated in both epoxy-based PDLC films [55] and NCAP films [25]. This is illustrated in Fig. 5.20 which shows the response of two films to a d.c. pulse. Upon removal of the pulse, the transmittance of the film containing (very nearly) spherical droplets decreases sharply for a few milliseconds and then decays very slowly for several hundred milliseconds (Fig. 5.20(a)). The initial rapid decay has been attributed to the relaxation of the LC molecules to their equilibrium configuration within the droplets and the subsequent long decay to reorientation of the droplet directors to their original directions in the film [5,25]. Figure

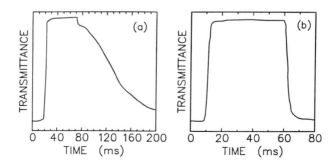

FIGURE 5.20. Transmittance curves showing room-temperature decay times of two epoxy-based PDLC films having the same composition [55]. (a) Film containing spherical droplets. (b) Film with shaped, aligned droplets formed by shearing the film during its fabrication.

5.20(b) shows the response of a film made from the same materials but sheared to produce aligned ellipsoidal droplets. t_d is now a few milliseconds. Similar effects have been observed over a wide temperature range [55].

Several groups [55,81] have derived equations for the response times of PDLC films with shaped droplets. t_r^{-1} is proportional to the sum of two terms: the first is independent of shape and depends on the electric field within the droplets; the second contains the effect of droplet shape and predicts that t_r will decrease as the droplets become elongated. t_d will increase as the droplet size increases but will decrease as the droplet becomes more elongated. These effects have been observed experimentally [55].

Response times of PDLC films can be lowered by adding a small amount of chiral mesogenic material to the LC material when the film is formed [46]. Addition of suitable additives, such as organometallic compounds, can further reduce the rise time [46].

5.7.4 Spectral Transmittance Characteristics

Figure 5.21 shows the spectral transmittance of a typical PDLC film. It was measured at normal incidence for quasi-collimated input illumination using a double-beam spectrophotometer with air in the reference beam [9,65]. This figure shows that PDLC films in their on-state are quite transparent at all visible wavelengths.

The off-state transmittance shown in Fig. 5.21 begins to increase at wavelengths above 700nm. The transmittance in the red portion of the visible spectrum and at near-infrared wavelengths can be reduced by increasing the droplet size, by using thicker films, or by adding dyes. Increasing film thickness decreases on-state transmittance and/or increases the switching voltage [3,9,32,65] while adding dyes decreases the on-state transmittance [80,88,89]. Increasing the droplet size decreases the switching voltage, as discussed

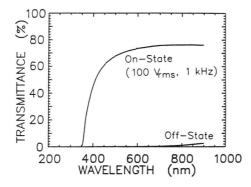

FIGURE 5.21. Transmittance versus wavelength for a UV-cured PDLC film [65]. These spectra were measured for quasi-collimated input illumination using a double-beam spectrophotometer with air in the reference beam.

above, and increases the total scattering cross section of the film, i.e., the fraction of the incident energy scattered out of the incoming beam provided the droplets do not become too large [9,45,65]. On-state transmittance can be increased and switching voltage reduced by using thinner films but this will also increase film transmittance in the off-state [9,32,65].

Transmittance curves like those in Fig. 5.21 are useful for predicting the behavior of PDLC films in applications where visual appearance is important. In applications where control of energy transmission through the film is of prime concern, hemispherical spectral transmittance and reflectance measurements with an integrating sphere are more relevant and have been used to assess the solar attenuation characteristics of PDLC films [9,65]. The effectiveness of PDLC films in energy control applications depends on the angular distribution of the scattered radiation, which is controlled by the concentration and size distribution of the droplets in the film [70].

5.8   Light Scattering

5.8.1   Theoretical Treatments

A complete theoretical description of light scattering in PDLC films would require a theory for multiple scattering of light by optically anisotropic particles in an optically isotropic matrix. Such a theory does not exist. Analyses of light scattering in PDLC films which have been published to date [45,69,90–93] have neglected multiple scattering by assuming that the light scattered by a LC droplet leaves the film before being scattered again by a second droplet. In this "single scattering approximation" the differential and total scattering cross sections for a PDLC film are obtained by multiplying the cross sections for a single droplet by the number of droplets which interact with the incident light beam.

Although there is no exact theory for scattering by an optically anisotropic particle, several approximations originally developed for isotropic particles have been extended to the anisotropic nematic droplets found in PDLC films. These approximations are based upon values of three key parameters: relative refractive index, size, and the maximum phase shift experienced by light in traversing a droplet.

The relative refractive index m is the ratio of the refractive index n_d of the LC droplet to the refractive index n_p of the polymer matrix material. n_d depends on the orientation of the droplet and the molecular configuration inside the droplet. For materials typically found in PDLC films, $1.52 \leq n_d \leq 1.75$ and n_p is usually ≈ 1.52 to match n_o. Thus, $1 \leq m \leq 1.15$ in typical films and it is reasonable to consider approximations for which $|m - 1| \ll 1$.

Size is conveniently specified by the dimensionless size parameter ka where $k = 2\pi n_p/\lambda$ (λ is the vacuum wavelength of the light impinging on the film) and a is a typical droplet dimension. In the remainder of this section we will restrict our discussion to spherical droplets, for which a is the droplet radius.

The maximum phase shift ρ is the difference between the maximum optical path traversed by a light ray in crossing a droplet and the optical path which it would have traversed in the matrix material in the absence of the droplet: $\rho = 2ka|m - 1|$.

Rayleigh–Gans (RG) scattering [94–96] spans the range between Rayleigh scattering ($ka \ll 1$, $\rho \ll 1$) and the "intermediate case" [94] for which ρ remains $\ll 1$ while ka becomes large. This case can occur only if the droplet is "optically soft," i.e., $|m - 1| \ll 1$. As noted earlier, this condition is satisfied for most PDLC film materials. Light scattering by a single nematic droplet in the RG approximation has been described by Žumer and Doane [90].

The anomalous diffraction (AD) approximation [94] also requires "optically soft" scatterers. It spans the range between the "intermediate case" and the limit ($ka \gg 1$, $\rho \gg 1$). Light scattering by a nematic droplet in the AD approximation has been described by Žumer [91].

If both ka and ρ are $\gg 1$, scattering can be treated as a combination of geometrical optics plus diffraction [94]. The geometrical optics approximation for scattering by a nematic droplet has been described by Sherman [93].

Both RG and AD calculations for nematic droplets are compared with angle-dependent light scattering measurements and transmittance measurements in the following sections.

5.8.2 Angle-Dependent Light Scattering Measurements

Montgomery [97] has measured the angular dependence of light scattering by PDLC films in both their on- and off-states as a function of the polarizations of the incident and scattered light and has compared his results with theory. Figure 5.22 shows the scattered intensity (normalized to the incident laser intensity) as a function of scattering angle (measured in air outside the film) for input light polarized parallel to the scattering plane formed by the propagation direction of the incident light and the observation direction. Figures 5.22 a and b show, respectively, $I_{\parallel}$ and $I_{\perp}$, the components of the scattered intensity polarized parallel and normal to the scattering plane. These data for a UV-cured film containing the LC ROTN403 in a NOA65 matrix are similar to data for biphenyl LC mixtures in both UV-cured and thermally cured matrices. Both the RG [90] and AD [91] calculations for a single nematic droplet predict that $I_{\perp} = 0$ in the on-state. Figure 5.22(b) (dashed curve) shows that, in a real PDLC film, $I_{\perp}$ is not only nonzero but comparable in magnitude to $I_{\parallel}$. Analysis of similar data for different materials and polarizations [97] has shown that, for the droplet concentrations and film thicknesses characteristic of PDLC films of practical interest, multiple scattering cannot be ignored in a theoretical description of angle-dependent light scattering, even in the on-state.

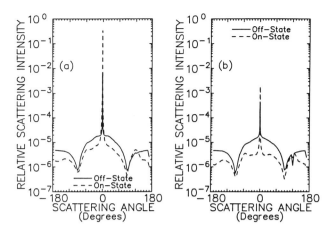

FIGURE 5.22. Scattered intensity at 632.8 nm versus scattering angle measured in air [97]. The detection system collected only light scattered within a cone of 0.5° half-angle about its optical axis. The incident light was polarized parallel to the scattering plane. The scattered light was polarized (a) parallel to the scattering plane; and (b) perpendicular to the scattering plane.

Whitehead et al. [98] have prepared PDLC samples, some having large and some having small droplets, and have measured their scattering characteristics. It is very difficult to prepare PDLC films with submicron droplets in which the droplet concentration is high enough to allow the angular dependence of scattering to be measured but low enough that multiple scattering effects will not affect the measurements. Consequently, the angular dependence predicted by RG theory has still not been tested under conditions where multiple scattering effects were unquestionably negligible. Whitehead et al. [98] did, however, prepare samples with very low concentrations of larger droplets which could be used to test the angular dependence predicted by AD theory. In these measurements, the samples were immersed in an index-matching fluid in a cylindrical container to minimize interface reflections which would introduce multiple scattering. Figure 5.23(a) shows the intensity of scattered light as a function of scattering angle inside such a film for incident light polarized parallel to the scattering plane. The measured scattered intensity is the sum of contributions from scattered light components polarized parallel and perpendicular to the scattering plane. The circles in Fig. 5.23(a) are measured intensities; the solid curve is calculated from the AD formulas of Žumer [91]. Because the intensity of the scattered light is very sensitive to droplet size in the AD regime, a Gaussian distribution of droplet radii about the mean 2.6 μm radius determined from optical microscopy was assumed in the AD calculations. Parameters for the distribution were obtained from least squares fits to transmittance measurements on the

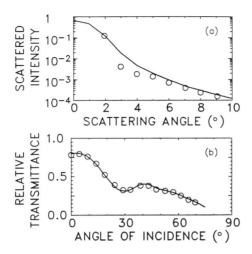

FIGURE 5.23. (a) Scattered intensity (arbitrary units) versus scattering angle internal to a PDLC film. (b) Relative transmittance of the same film versus angle of incidence [98]. The circles are experimental data. The solid curves were calculated using the anomalous diffraction approximation for a nematic droplet.

same sample (see next section). The agreement between theory and experiment is fairly good.

5.8.3 Transmittance Measurements

Angle-dependent scattering measurements directly measure the differential scattering cross section of a system and are strongly influenced by changes in the direction of scattered light produced by multiple scattering. The transmittance of a light-scattering film, measured with a detector having a small acceptance angle, is determined by the total light scattered out of the incident beam into all directions; i.e., transmittance measurements depend on the total, not the differential, scattering cross section. Light scattered out of the incident beam can be redirected by multiple scattering without affecting a transmittance measurement as long as the multiple scattering does not redirect a significant fraction of the scattered light back into the direction of the incident beam. In practice, multiple scattering has a negligible effect on transmittance provided less than 10% of the incident intensity is scattered out of the incoming beam [94]. Since transmittance measurements are less sensitive to multiple scattering effects than angle-dependent scattering measurements, they can be used to test both RG and AD light-scattering calculations of the total scattering cross sections of PDLC films for a wide range of droplet sizes and concentrations.

The basic relation linking the transmittance $\mathcal{T}$ of a PDLC film with σ, the total scattering cross section of a single LC droplet, is

$$\mathcal{T} = (1 - \mathcal{R})^2 \exp(-N_v \sigma d_{\text{eff}}). \tag{5.3}$$

In this equation $\mathcal{R}$ is the reflectance at an interface between the planar sample and the surrounding medium and N_v is the number of droplets per unit volume. The effective thickness $d_{\text{eff}} = d/\cos \theta'$ where θ', the angle between the normal to the film and the propagation direction of the light inside the film, is related to the external angle of incidence θ by Snell's law. Equation (5.3) has been used to study the transmittance characteristics of PDLC films in both the isotropic and the nematic phase.

Isotropic Phase: RG Regime

Montgomery and Vaz [69] have used the RG approximation to calculate the temperature-dependent transmittance of a PDLC film in its isotropic phase for unpolarized incident light and have compared their results with experiment. In their calculations they used measured values of $n_o(T)$, $n_e(T)$, and $n_p(T)$ and values of droplet size and concentration determined from SEM; these parameters were allowed to vary only over limited ranges determined by the precision with which they could be determined experimentally. The measured transmittances agreed with the RG calculations within experimental error over the entire temperature range of the measurements (40–85 °C).

Nematic Phase: AD Regime

The on-state transmittance of PDLC films with large nematic droplets has been measured as a function of the angle of incidence and compared with the predictions of AD theory [98]. The circles in Fig. 5.23(b) show the experimental $\mathcal{T}$ versus θ data for the sample whose angle-dependent scattering intensity was shown in Fig. 5.23(a). The input light was polarized in the plane of incidence. The solid curve in Fig. 5.23(b) was computed from (5.3) for a collection of droplets with a Gaussian distribution of droplet radii using Žumer's equation for σ in the AD approximation [91]. The parameters of the distribution were obtained by a least squares fit of the calculated transmittance curves to the measured data. The AD approximation gives an excellent description of the measured transmittance.

Nematic Phase: RG Regime

In the RG regime, when $ka \gg 1$, σ becomes

$$\sigma = 2\pi a^2 |m - 1|^2 (ka)^2. \tag{5.4}$$

For light polarized in the plane of incidence, m depends on the angle between the droplet director and the propagation direction of the incident light [42]. Hence, σ will vary with the angle of incidence. Figure 5.24 shows the transmittance of a PDLC film for light polarized in the plane of incidence [98].

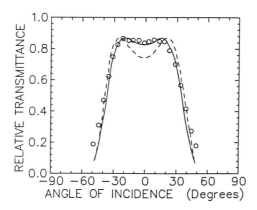

FIGURE 5.24. Relative transmittance of a PDLC film versus angle of incidence [98]. The circles are experimental data. The curves were calculated using the Rayleigh–Gans approximation for a nematic droplet.

The symbols are experimental data. The dashed curve is calculated [98] from (5.3) and (5.4) assuming that all droplet directors are aligned normal to the film by the applied electric field. Even allowing for uncertainty in n_o, n_e, and n_p, it was not possible to reproduce both the width of the transmittance curve and the magnitude of the depression at normal incidence. Whitehead et al. [98] hypothesized that the droplet directors may not have been perfectly aligned because the droplets were not perfectly spherical and electric-field alignment of a droplet depends strongly on its shape [26]. To account for this nonuniform alignment, they assumed a Gaussian distribution in the orientation of the droplet directors about the normal to the film. The solid curve in Fig. 5.24 was computed from (5.3) and (5.4) taking this distribution into account.

For light polarized normal to the plane of incidence, $n_d \approx n_o$ for all angles of incidence θ and σ does not depend on θ. Attenuation is due to the increase in d_{eff} with increasing θ. For a sample immersed in an index-matching fluid, $\theta' = \theta$ and $\mathscr{R} = 1$ in (5.3); under these conditions, a plot of $\ln \mathscr{T}$ versus $1/\cos \theta$ should yield a straight line whose slope is proportional to σ. For a known droplet radius, σ can be determined from the measured slope. For the sample of Fig. 5.24, the value of σ obtained by this procedure was about four times larger than that calculated from (5.4) using measured values of n_o, n_e, and n_p [98]. Several factors contributed to this discrepancy:

(i) uncertainty in the values of a and N_v determined from scanning electron microscopy;
(ii) a large uncertainty in n_p caused by residual LC dissolved in the matrix; and
(iii) the dependence of σ on θ introduced by the distribution in droplet director orientation.

Finally, we note that, if $a \leq \sim 0.1$ μm and the droplet volume fraction $\leq \sim 0.12$, the RG approximation can account for experimentally observed PDLC transmittance even for randomly oriented droplet directors (off-state) [92]. In this case, σ is very weak and scattered light is directed away from the forward direction; consequently, little scattered light is redirected back into the incident beam even for thick films with $\mathscr{T}$ as low as 20%.

5.8.4   Droplet Size Effects

Nomura et al. [45] have recently studied the effects of droplet size on PDLC transmittance. They approximated a PDLC film in the off-state as a collection of optically isotropic droplets whose refractive index was determined by averaging over director orientation. They then used Mie theory [94–96], which is exact for isotropic spherical particles in the absence of multiple scattering, to calculate film turbidity as a function of droplet diameter and wavelength. Calculated turbidities were compared with measured values of apparent absorbance (extinction) in their films. Both theory and experiment indicated that extinction initially increases with increasing droplet size, reaches a maximum value, and then decreases if droplet size increases further. However, the predicted diameter for maximum extinction (about 5 μm) and the measured value (about 3 μm) were different. Differences were attributed to the fact that the droplets were not perfect spheres, to the use of a single particle size instead of a size distribution, and to neglect of multiple scattering contributions to extinction.

5.8.5   Concluding Remarks Regarding Light Scattering

The on-state transmittance characteristics of PDLC films having low droplet concentrations are adequately described by the RG and AD approximations as modified for birefringent particles by Žumer [90,91]. Žumer's AD calculations also adequately describe the angular dependence of on-state light scattering by PDLC films with low concentrations of large droplets. Because of difficulties in measuring angle-dependent light scattering in films with low concentrations of small droplets, the predicted angular dependence of light scattering in the RG regime has not yet been completely verified. If the droplet size and volume fraction are sufficently small, Žumer's RG analysis also adequately describes the off-state transmittance of PDLC films with low droplet concentrations.

In PDLC films of practical interest, droplet volume fractions are too large to neglect multiple scattering effects and single-scattering analyses cannot quantitatively describe the available experimental data, particularly the measured angular dependence of light scattering [97]. Development of a multiple scattering formalism for systems containing optically anisotropic particles is a challenging and, as yet, unsolved problem.

5.9 Final Comments

In the 1980s PDLC films have emerged as potential competitors to other thin film electrooptic devices such as the twisted nematic cell. PDLCs have several advantages: they require no polarizers, have fast response times, and can easily be fabricated in a large, mechanically flexible format. Today about thirty research laboratories world-wide are studying PDLCs, principally in the United States, Japan, Europe, and Russia.

Recently considerable progress has been made in developing an understanding of mixing and phase separation, processes of great importance in PDLC formation. Hirai et al, [54,99] and Smith [100] have applied Flory–Huggins theory to describe the phase behavior of LC/polymer mixtures before, during, and after cure. They found that LC/matrix systems possess UCST (upper critical solution temperature) phase diagrams. Experimental phase diagrams of uncured LC/matrix mixtures [6,54,99,101] and cured PDLCs [6,102] have been determined and show semiquantitative agreement with the models. Smith [101] has investigated the dependence of LC phase separation on matrix cure and found that irreversible demixing occurs only when cure is almost complete. He has also described the temperature dependence of LC solubility in the cured polymer matrix [100], obtaining reasonable agreement with experiment [103,104]. The kinetics of phase separation have been studied both calorimetrically [29,30,52] and by light scattering [102,105], with the two methods giving compatible results.

Our present knowledge of PDLC formation mechanisms, although semiquantitative, allows us to recognize important parameters which optimize device performance. Among these are the size distribution of the LC microdroplets and the refractive indices of the polymer matrix and the LC. Droplet sizes in thermoset films can be controlled by the proper choice of PDLC cure rate, cure temperature, and LC concentration. Electrooptic properties of the films can be enhanced by suitable matching of the component refractive indices. Operating temperatures of the devices are determined primarily by the nematic range of the LC, which can sometimes be affected by the choice of matrix material. PDLCs operating over a temperature range from approximately $-20\,°C$ to $+100\,°C$ have been achieved. New LC mixtures with properties tailored to PDLC requirements are being developed. A new technology related to PDLCs has recently been developed: the polymer-stabilized cholesteric texture (PSCT) device [106,107]. PSCT displays provide both a gray scale and memory capability.

As a result of intensive research efforts, PDLC applications are being evaluated at a number of laboratories. Primary focus seems to be on large area devices, including flexible displays and switchable windows for architectural and automotive use. Commercialization of PDLCs may come soon, especially once large-scale fabrication techniques are perfected to fully optimize film performance and environmental life.

References

1. J.L. Fergason, U.S. Patent 4,435,047, March 6, 1984.

2. J.L. Fergason, SID Dig. Tech. Pap. **16**, 68 (1985).

3. P.S. Drzaic, J. Appl. Phys. **60**, 2142 (1986).

4. J.W. Doane, G. Chidichimo, and N.A. Vaz, U.S. Patent 4,688,900, August 25, 1987.

5. J.W. Doane, N.A. Vaz, B.-G. Wu, and S. Zumer, Appl. Phys. Lett. **48**, 269 (1986).

6. J.L. West, Mol. Cryst. Liq. Cryst. **157**, 427 (1988).

7. N.A. Vaz, G.W. Smith, and G.P. Montgomery, Jr., Mol. Cryst. Liq. Cryst. **146**, 17 (1987).

8. N.A. Vaz, G.W. Smith, and G.P. Montgomery, Jr., Mol. Cryst. Liq. Cryst. **146**, 1 (1987).

9. G.P. Montgomery, Jr., in *Large Area Chromogenics: Materials and Devices for Transmittance Control*, edited by C.M. Lampert and C.G. Granqvist (SPIE Optical Engineering Press, Washington, 1990).

10. J.W. Doane, in *Liquid Crystals—Applications and Usages*, edited by B. Bahadur (World Scientific, Teaneck, 1990).

11. G.P. Montgomery, Jr. and N.A. Vaz, U.S. Patent 4,775,266, October 4, 1988.

12. D.K. Yang and P.P. Crooker, Proc. SPIE **1257**, 60 (1990).

13. E.V. Generalova, A.S. Sonin, Ya.S. Freidzon, V.P. Shibaev, and I.N. Shibaev, Vysokomolek. Soedin. A **25**, 2274 (1983) [Polymer. Sci. USSR **25**, 2641 (1983)].

14. E. Dubois-Violette and O. Parodi, J. Phys. (Paris) Coll. **C4**, 57 (1969).

15. R.D. Williams, Rutherford Appleton Laboratory Report RAL-85-028, unpublished.

16. R.D. Williams, Rutherford Appleton Laboratory Report RAL-85-062, unpublished.

17. P.S. Drzaic, Mol. Cryst. Liq. Cryst. **154**, 289 (1988).

18. H. Yang, D.W. Allender, and M.A. Lee, Bull. Am. Phys. Soc. **33**, 275 (1988).

19. A.E. Köhler, Z. Phys. Chemie (Leipzig) **269**, 196 (1988).

20. Y.-D. Ma, B.-G. Wu, and G. Xu, Proc. SPIE **1257**, 46 (1990).

21. H. Kohlmüller and G. Siemsen, Siemens Forsch. Entwickl. **11**, No. 5, 229 (1982).

22. J.L. Fergason, U.S. Patent 4,616,903, October 14, 1986.

23. K.N. Pearlman, European Patent Appl. 85/301,888.5, March 19, 1985.

24. P.S. Drzaic, Proc. SPIE **1080**, 11 (1989).

25. P.S. Drzaic, Liq. Cryst. **3**, 1543 (1988).

26. J.W. Doane, A. Golemme, J.L. West, J.B. Whitehead, Jr., and B.-G. Wu, Mol. Cryst. Liq. Cryst. **165**, 511 (1988).

27. J.L. West, J.W. Doane, and S. Žumer, U.S. Patent 4,685,771, August 11, 1987.

28. N.A. Vaz, *Workshop on Polymer-Dispersed Liquid Crystal Films*, Warren, MI, October 30, 1986.

29. G.W. Smith and N.A. Vaz, Liq. Cryst. **3**, 543 (1988).

30. G.W. Smith, Mol. Cryst. Liq. Cryst. **180B**, 201 (1990).

31. N.A. Vaz and G.W. Smith, U.S. Patent 4,728,547, March 1, 1988.

32. G.P. Montgomery, Jr., N.A. Vaz, and G.W. Smith, Proc. SPIE **958**, 104 (1988).

33. A.M. Lackner, J.D. Margerum, E. Ramos, S.-T. Wu, and K. C. Lim, Proc. SPIE **958**, 73 (1988).

34. F.G. Yamagishi, L.J. Miller, and C.I. van Ast, Proc. SPIE **1080**, 24 (1989).

35. A.M. Lackner, J.D. Margerum, E. Ramos, and K.C. Lim, Proc. SPIE **1080**, 53 (1989).

36. M. Mucha and M. Kryszewski, J. Therm. Anal. **33**, 1177 (1988).

37. T. Gunjima, H. Kumai, S. Tsuchiya, and K. Masuda, U.S. Patent No. 4,834,509, May 30, 1989.

38. N.A. Vaz, G.W. Smith, and G.P. Montgomery, Jr., Proc. SPIE **1257**, 9 (1990).

39. N.A. Vaz, G.W. Smith, and G.P. Montgomery, Jr., Mol. Cryst. Liq. Cryst. **197**, 83 (1991).

40. A. Golemme, S. Žumer, D. W. Allender, and J.W. Doane, Phys. Rev. Lett. **61**, 2937 (1988).

41. J.L. West, A. Golemme, and J.W. Doane, U.S. Patent 4,673,255, June 16, 1987.

42. B.-G. Wu, J.L. West, and J.W. Doane, J. Appl. Phys. **62**, 3925 (1987).

43. A. Golemme, S. Žumer, J.W. Doane, and M.E. Neubert, Phys. Rev A **37**, 559 (1988).

44. B.-G. Wu and J. William Doane, U.S. Patent 4,671,618, June 9, 1987.

45. H. Nomura, S. Suzuki, and Y. Atarashi, Japan. J. Appl. Phys. **29**, 522 (1990).

46. P.W. Mullen and F.E. Nobile, U.S. Patent No. 4,888,126, December 19, 1989.

47. G.W. Smith, "PDLC Formation Kinetics and Phase Behavior," lecture at Kent State University, June 3, 1987.

48. G.W. Smith and N.A. Vaz, unpublished.

49. N.A. Vaz, Proc. SPIE **1080**, 2 (1989).

50. G. Pasternak, "Fundamental Aspects of Ultraviolet Light and Electron Beam Curing," *Radiation Curing Workshop*, the Association of Finishing Processes of the Society of Manufacturing Engineers, Chicago, IL, September 15, 1981.

51. C.E. Hoyle, R.D. Hensel, and M.B. Grubb, J. Polymer Sci., Polymer. Chem. Ed. **22**, 1865 (1984).

52. G.W. Smith, Mol. Cryst. Liq. Cryst. **196**, 89 (1991).

53. Perkin-Elmer Corp., Norwalk, CT.

54. Y. Hirai, S. Niiyama, H. Kumai, and T. Gunjima, Proc. SPIE **1257**, 2 (1990).

55. J.H. Erdmann, J.W. Doane, S. Žumer, and G. Chidichimo, Proc. SPIE **1080**, 32 (1989).

56. J.W. Doane and J.L. West, SID Dig. Tech. Pap. **21**, 224 (1990).

57. N.A. Vaz, paper No. 328, American Chemical Society, 21st Central Regional Meeting, Cleveland, OH, 31 May–2 June, 1989.

58. N.A. Vaz and G.P. Montgomery, Jr., J. Appl. Phys. **62**, 3161 (1987).

59. L.J. Miller, C. Van Ast, and F.G. Yamagishi, WO89/06,264, July 13, 1989.

60. L.J. Miller, C. Van Ast, and F.G. Yamagishi, U.S. Patent No. 4,891,152, January 2, 1990.

61. T. Gunjima, H. Kumai, S. Tsuchiya, and K. Masuda, European Pat. Appl. No. 87/118,579.9, December 15, 1987.

62. T. Gunjima, H. Kumai, M. Akatsuka, and S. Tsushiya, U.S. Patent No. 4,818,070, April 4, 1989.

63. T. Korishima, Y. Hirai, and E. Aoyama, Jap. Pat. Appl. Kokai No. 01-61,238, March 8, 1989.

64. N.A. Vaz and G.P. Montgomery, Jr., J. Appl. Phys. **65**, 5043 (1989).

65. G.P. Montgomery, Jr., Proc. SPIE **1080**, 242 (1989).

66. A.M. Lackner, E. Ramos, and J.D. Margerum, Proc. SPIE **1080**, 267 (1989).

67. C.G. Roffey, *Photopolymerization of Surface Coatings* (Wiley, New York, 1982).

68. H. Lee and K. Neville, *Handbook of Epoxy Resins* (McGraw-Hill, New York, 1982).

69. G.P. Montgomery, Jr. and N.A. Vaz, Phys. Rev. A **40**, 6580 (1989).

70. G.P. Montgomery, Jr., J.L. West, and W. Tamura-Lis, J. Appl. Phys. **69**, 1605 (1991).

71. J.R. Havens, D.B. Leong, and K.B. Reimer, Mol. Cryst. Liq. Cryst. **178**, 89 (1990).

72. E. Scheil, Z. Metallk. **27**, 199 (1935).

73. S.A. Saltykov, *Stereometric Metallography*, 2nd ed. (Metallurgizdat, Moscow 1958).

74. R.T. DeHoff, Trans. Met. Soc. AIME **224**, 474 (1962).

75. R.T. DeHoff and F.N. Rhines, Trans. Met. Soc. AIME **221**, 975 (1961).

76. R.L. Fullman, Trans. AIME J. Metals **197**, 447 (1953).

77. R.L. Fullman, Trans. AIME J. Metals **197**, 1267 (1953).

78. *Radiation Curing*, edited by the Education Committee of the Radiation Curing Division, the Association for Finishing Processes of the Society of Manufacturing Engineers, Dearborn MI, 1984.

79. N.A. Vaz, unpublished.

80. J.L. West, W. Tamura-Lis, and R. Ondris, Proc. SPIE **1080**, 48 (1989).

81. B.-G. Wu, J.H. Erdmann, and J.W. Doane, Liq. Cryst. **5**, 1453 (1989).

82. J.H. Erdmann, S. Žumer, B.G. Wagner, and J.W. Doane, Proc. SPIE **1257**, 68 (1990).

83. M.V. Kurik, A.V. Koval'chuk, O.D. Lavrentovich, and V.V. Sergan, Ukrain Fiz. Zh. **32**, 1211 (1987).

84. A.V. Koval'chuk, M.V. Kurik, O.D. Lavrentovich, and V.V. Sergan, Zh. Èksper. Teoret. Fiz. **94**, 350 (1988) [Soviet Phys. JETP **67**, 1065 (1988)].

85. P.S. Drzaic, Proc. SPIE **1257**, 29 (1990).

86. G. Chidichimo, G. Arabia, A. Golemme, and J.W. Doane, Liq. Cryst. **5**, 1443 (1989).

87. L.M. Blinov, *Electro-Optical and Magneto-Optical Properties of Liquid Crystals* (Wiley, New York, 1983).

88. G.P. Montgomery, Jr. and N.A. Vaz, Appl. Opt. **26**, 738 (1987).

89. P.S. Drzaic, R.C. Wiley, and J. McCoy, Proc. SPIE **1080**, 41 (1989).

90. S. Žumer and J.W. Doane, Phys. Rev. A **34**, 3373 (1986).

91. S. Žumer, Phys. Rev. A **37**, 4006 (1988).

92. S. Žumer, A. Golemme, and J.W. Doane, J. Opt. Soc. Amer. A **6**, 403 (1989).

93. R.D. Sherman, Phys. Rev. A **40**, 1591 (1989).

94. H.C. van de Hulst, *Light Scattering by Small Particles* (Wiley, New York, 1957).

95. M. Kerker, *The Scattering of Light and Other Electromagnetic Radiation* (Academic Press, New York, 1969).

96. C.F. Bohren and D.R. Huffman, *Absorption and Scattering of Light by Small Particles* (Wiley, New York, 1983).

97. G.P. Montgomery, Jr., J. Opt. Soc. Am. B **5**, 774 (1988).

98. J.B. Whitehead, Jr., S. Žumer, and J.W. Doane, Proc SPIE **1080**, 250 (1989).

99. Y. Hirai, S. Niiyama, H. Kubai, and T. Gunjima, Reps. Res. Lab. Asahi Glass Co. **40**, 285 (1990).

100. G.W. Smith, Mol. Cryst. Liq. Cryst. **225**, 113 (1993).

101. G.W. Smith, Phys. Rev. Lett. **70**, 198 (1993).

102. T. Kyu, M. Mustafa, J. Yang, J.Y. Kim, and P. Palffy-Muhoray, in *Polymer Solutions, Blends, and Interfaces*, edited by I. Noda and D.N. Rubingh (Elsevier, Amsterdam, 1992).

103. G.W. Smith, G.M. Ventouris, and J.L. West, Mol. Cryst. Liq. Cryst. **213**, 11 (1992).

104. P. Nolan, M. Tillin, and D. Coates, Mol. Cryst. Liq. Cryst. Lett. **8**, 129 (1992).

105. J.Y. Kim and P. Palffy-Muhoray, Mol. Cryst. Liq. Cryst. **203**, 93 (1991); J.Y. Kim, Ph.D. Thesis, Kent State University, May 1992.

106. D.K. Yang and J.W. Doane, SID Digest of Technical Papers **XXIII**, 759 (1992).

107. J.W. Doane, D.K. Yang, and Z. Yaniv, Japan Display '92, Hiroshima 73 (1992).

6

Nuclear Magnetic Resonance Spectroscopy of Thermotropic Liquid Crystalline Polymers

L. Stroganov

6.1 Introduction

The science of liquid crystalline (LC) polymers has today accumulated a lot of data. However, these seem to be only a tremendous amount of mozaic pieces that still have to be fitted together to understand the structural organization of LC polymers as a whole.

Keeping this situation in mind we conceive that our primary task is:

(i) to select the pieces of most reliable data obtained by NMR; and
(ii) to give a compendium of these data that would be sufficiently clear and complete for the reader to learn whether the information presented matches his own experience and knowledge.

The subject to be predominantly discussed involves the consideration of the ordering and dynamics in the main-chain and side-chain LC polymers. The studies of polymeric LC discotics [1], combined main-chain/side-chain LC polymers [2], and the majority of very important publications treating the dynamics of the director reorientation [3] are beyond the scope of this chapter because of the lack of space.

As regards the NMR per se, the discussion will be the briefest possible to ensure a true understanding of the results presented. This seems to be quite natural, since a number of sophisticated reviews on the methodology of the NMR spectroscopy of LC polymers was published by Böffel and Spiess, Müller, Meier, and Kothe; these and others are referred to in the text, however, we will not repeat here some of the results discussed in these papers.

6.2   Thermotropic Liquid Crystalline Polymers with Mesogenic Groups in the Main Chain

This section deals with the NMR investigations of thermotropic LC main-chain polymers comprising regularly alternating rigid mesogenic fragments and flexible spacers. Three types of polymers differing in the structure of a rigid fragment R, the length of an aliphatic spacer F, and molecular masses were predominantly involved in the relevant studies. These polymers are listed in Table 6.1.

6.2.1 ^{1}H and ^{2}H Nuclear Magnetic Resonance in Fast Motion Limit

The PMR spectra of R1Fn LC polymers were subjected to detailed analysis by Volino et al. [4,5]. The spectra of aligned nematics have the form of a symmetrical signal approximately 50 kHz wide displaying a triplet structure characteristic of low-molar-mass nematics [6]. Martins et al. [4] described a model for the simulation of these spectra. Figure 6.1 shows the specific features of the spectra associated with the contributions of mesogenic units and of the spacers.

(i)   The two peaks at ± 10 kHz correspond to dipole–dipole interactions between ortho protons of the rings.
(ii)   The two shoulders at ± 5 kHz are associated with the ring methyl groups. The intensities and positions of these shoulders are strongly dependent on the dihedral angle between the two rings and on the order parameter S.
(iii)   The two broad shoulders on both sides of the spectrum and the broad line at the center are the contributions of the spacer in a rather extended conformation.

The splitting of the external lines can be given by $2\delta_v = [\frac{3}{2}\pi\gamma^2\hbar/r^3]\,|S|/2$, where $|S|/2$ is the degree of order of an isolated pair of methylene protons aligned, on average, perpendicularly to the long axis of the mesogenic moiety (S is the order parameter of the mesogens). The best fit of the model with respect to the experimental line shape is obtained when the mean dihedral angle between the two rings is $36°$ and the order parameter is $S = 0.8$. The order parameter S and the main splitting $2\delta_N$ are related by a simple relationship

$$2\delta_N/\text{kHz} \approx 24.08S. \qquad (6.1)$$

This assignment is supported by the studies performed with compounds selectively deuterated in the spacer. Figure 6.1 shows the spectra of the polymer R1F10 (a) and of the same polymer but deuterated in the spacer (b). The two shoulders are seen to have disappeared in the spectrum of the deuterated polymer (b). Hence, in spectrum (a) these are unequivocally associated with

TABLE 6.1. Structures of main chain LC polymer.

Code	Polymer	Spacer (n) and polymeric chain length ($\bar{x}$)	Reference
R1Fn		$5 < n < 18$ $1.5 < \bar{x} < 42$	[14, 16]
R2Fn		$n = 9; 10$	[7]
R3Fn		$n = 9; 10$ $10 < \bar{x} < 63$	[8, 32]

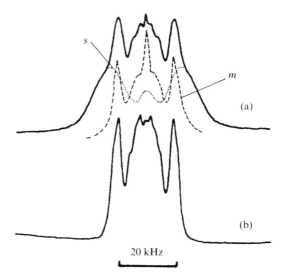

FIGURE 6.1. PMR spectra of the main-chain LC polymer [4]. (a) is the nematic phase of the polymer R1F10; (b) is the polymer R1F10 deuterated in the spacer; ($\cdots$) is the subspectrum of the spacer; and ($---$) is the subspectrum of the mesogenic group.

the spacers. Similarly, by measuring the PMR spectrum of a model compound without the ring methyl groups, the shoulders at ± 5 kHz were shown to be due to these ring methyl groups. Hence, analysis of the PMR spectra of main-chain LC polymers makes it possible to extract from the relevant data the order parameters of the mesogenic units, of the spacers, and the dihedral angle between the rings.

DMR has nowadays become a powerful technique for investigating the structural order in LC polymers. Indeed, the pattern of the DMR spectrum is determined by the interaction between the nuclear quadrupole moment of the deutrons and the local electric field gradient (efg) tensor. In aliphatic molecules the principle axis of the efg tensor is directed along the C—D bond vector, and DMR provides a tool for direct monitoring of the reorientation of this vector in specficially labeled alkyl chain segments. For each C—D bond within the aligned nematic phase the DMR splitting can be assessed in the fast motion limit with the aid of the following equation:

$$\Delta v_i = \tfrac{3}{2}(e^2qQ/h)S_{zz}\langle \tfrac{3}{2}\cos^2 \Phi - \tfrac{1}{2}\rangle \tag{6.2}$$

where e^2qQ/h ($=174$ kHz) is the quadrupole coupling constant, S_{zz} is the orientational order parameter of chain segments with respect to the direction of the nematic domain, Φ refers to the angle between the C—D bond and the axis of aligned chain segments, and the angle brackets denote averaging over all the allowed conformations.

Figure 6.2 shows the DMR spectra of the R2F9 polymer deuterated in the

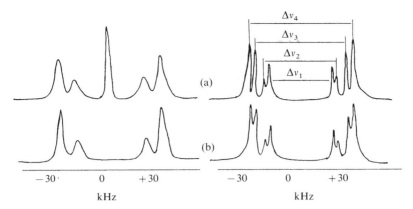

FIGURE 6.2. ^{2}H NMR spectra of the main-chain LC polymer [7]. (a) is the spectra of the polymer R2F9 and of the model dimer

just below the clearing temperature. (b) is the temperature 12° below the clearing point. The quadrupolar splittings Δv_i correspond to different CD_2 groups of the spacer: $\Delta v_1 - \alpha + \beta$, $\Delta v_2 - \delta$, $\Delta v_3 - \varepsilon$, and $\Delta v_4 - \gamma$.

spacer [7]. A sharp peak at the center of the spectrum is associated with the isotropic phase of the biphase region that still remained under the experimental conditions. The spectrum patterns of the dimer and the polymer appear to be very similar. However, since the dimer samples are less viscous the resolution of the signal is better. Besides, the quadrupolar splitting of the polymer was found to be somewhat larger than that of the corresponding dimer. The four splittings in the spectrum of the dimer were discriminated between the different CD_2 groups of the spacer ($\Delta v_1 = \alpha + \beta$, $\Delta v_2 = \delta$, $\Delta v_3 = \varepsilon$, $\Delta v_4 = \gamma$). In the spectrum of the polymer these splittings are grouped into two doublets.

Thus, DMR spectroscopy in a fast motion limit makes it possible to measure quadrupolar splittings Δv_i, which are explicitly related to the order parameters of chain segments via (6.2), the mean angles between the C—D bonds and the alignment axis, and the order parameters of mesogenic groups.

6.2.2 Multipulse Dynamic Nuclear Magnetic Resonance

The methods discussed in the preceding section can be used to examine fast molecular motions only. However, correlation times for the molecular motions in LC polymers can vary within a very broad range (Section 6.2.3) extending far beyond the fast motion limit. Müller et al. [8] suggested a complex of experimental techniques offering most detailed information concerning micro- and macroscopic ordering and dynamics of the processes with correlation times within the range 10^{-10}–100 s. To give an adequate description of the experimental results a dynamic NMR model for $I = 1$ spin system involving the density matrix formalism was proposed.

Within the framework of this model the intermolecular motion is assumed to proceed via a diffusive process. Two rotational correlation times $\tau_{R_\perp}$ and $\tau_{R_\parallel}$ were assumed to be sufficient to describe all intermolecular transition rates, where $\tau_{R_\perp}$ is the correlation time for reorientation of the symmetry axis of a molecule, whereas $\tau_{R_\parallel}$ refers to rotation around this axis. For the intramolecular motion a random jump process is assumed. According to this model τ_j refers to the average lifetime in one conformation, and n_k refers to the relative occupational probability. For the phenyl ring conformations r^+ and r^-, related by $180°$ ring flips, are equally populated ($n_{r^+} = n_{r^-}$). The *trans* (t) and *gauche* (g$^+$, g$^-$) conformations of alkyl chain segments are characterized by n_t-*trans* population. Then the *gauche* populations can be calculated from the relationship $2n_g + n_t = 1$.

Two exponential orientational distribution functions are introduced in the model discussed: one of them specifies the orientation of the local director axes through the same (Section 6.2.6, (6.3), (6.4)) and the second gives the distribution of the long molecular axes with respect to the local director. The constants in the power coefficients of the exponents are associated with the macro- and micro-orientational order parameters $S_{z''z''}$ and $S_{z'z'}$, respectively.

Choosing the distinct multipulse experimental procedure for a distinct application depends on the order of magnitude of the correlation times. In the range 10^{-8} s $< \tau_R < 10^{-4}$ s the quadrupole echo sequence is used [9]. Studies of faster motion, corresponding to 10^{-10} s $< \tau_R < 10^{-8}$ s, in the $5T$ magnetic field require the inversion recovery sequence technique [10]. The Jeener–Broekaert sequence [12] makes it possible to study extremely slow molecular motion with correlation times as long as $\tau_R = 100$ s [11].

6.2.3   Molecular Dynamics

Let us survey the results of the multipulse dynamic NMR studies of inter- and intramolecular motions in the main chains of LC polymers, by way of the example involving R3F10 polymers of different chain length and with alkyl spacer fragments deuterated in the α position [8]. The correlation times of inter- and intramolecular motions are shown in Fig. 6.3. The first includes rotation around the chain axis and fluctuation of this chain axis, while the second refers to the *trans–gauche* isomerization of the first segment of the flexible spacer. The values of activation energies for these motions are listed in Table 6.2.

In anisotropic melt the correlation times of chain rotation and chain fluctuation are around 10^{-8} s by the order of magnitude, whereas *trans–gauche* isomerization proceeds even more rapidly. The value of the activation energy of *trans–gauche* isomerization (15.9 kJ/mol) is comparable with the energy

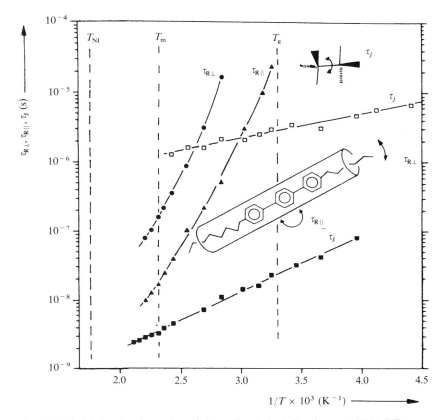

FIGURE 6.3. Molecular dynamics of the main-chain LC polymer R3F10 [8]

$$\text{--}[\text{O}\text{--}\bigcirc\text{--}\text{COO}\text{--}\overset{\text{Cl}}{\bigcirc}\text{--}\text{OOC}\text{--}\bigcirc\text{--}\text{O}\text{--}\text{CD}_2\text{--}(\text{CH}_2)_8\text{--}\text{CD}_2\text{--}]_x\,.$$

τ_j and $\tau_{R_\parallel}$, $\tau_{R_\perp}$ are the correlation times of the intramolecular (kink motion of the α segment of the spacer) and the intermolecular (chain rotation and chain fluctuation) motions. Dashed lines indicate different phase transition temperatures: T_{NI} is the isotropic–nematic transition, T_m is the melting point, and T_g is the glass transition. Liquid crystalline and crystalline components, observed below the melting point, are distinguished by full and open symbols.

barrier for the rotation about the C—C bond in alkanes. High activation energies for rotation and fluctuation of the chain axes (48.3 kJ/mol and 52.6 kJ/mol) presumably reflect the cooperative character of these processes.

Below the melting point a new type of intramolecular motion (open squares in Fig. 6.3) appears, which is several orders of magnitude slower than *trans–gauche* isomerization in the LC phase. This motion is associated with

TABLE 6.2. Activation energies for various motions of R3F10 polymer [8].

Temperature range (°C)	Activation energy			
	Intermolecular motion		Intramolecular motion	
	$E_{R\perp}$ (kJ/mol)	$E_{R\parallel}$ (kJ/mol)	E^{LC} (kJ/mol)	E^{cryst} (kJ/mol)
$T_m = 160 < T < T_{NI} = 260$	52.6	48.3	15.9	—
$T_g = 30 < T < T_m = 160$	52.6–85.0	48.3–71.0	15.9	6.7
$T < T_g = 30$			15.9	6.7

the crystalline phase of the biphase region (Section 6.2.6). Decomposing the relaxation curves into two components makes it possible to evaluate the degree of crystallinity (60%), which appears to be actually independent of temperature. The activation energy of this process (6.7 kJ/mol) is close to the corresponding value for solid paraffins.

The chain rotation and chain fluctuation correlation times versus the reciprocal temperature dependences appeared to be nonlinear. Hence, the apparent activation energy increases with decreasing temperature. Below the glass-transition temperature all intermolecular motions are "freezed" and intramolecular motions dominate in the retaining dynamic processes. In fact, intramolecular motions are still feasible even at very low temperatures ($T = 136$ K, $\tau_j = 10^{-4}$ s).

The rate of *trans–gauche* isomerization is somewhat higher for CD_2 groups that are closer to the center of the flexible fragment. Within the chain lengths of $\bar{x} = 10$–63 repeating units, intermolecular motions are independent of the polymer chain length. Molecular reorientation appears to occur essentially by single-step processes; in this case, any distribution of correlation times must be restricted to less than one decade.

6.2.4   Orientational Ordering

It should be noted that the order parameters measured in the NMR procedures refer to the nematic phase aligned in the strong magnetic field of the NMR instrument. In this regard, NMR methods are preferable to other methods, such as diamagnetic anisotropy, IR dichroism, etc. The latter ones average over the entire sample. Hence, they may fail to distinguish the coexisting isotropic phase and the nematic domains which have not been properly aligned, thus leading to lower estimates of nematic order parameters.

Figure 6.4 [13] serves to illustrate the precision of the method. It depicts the temperature dependences of the order parameter of the R2F10 polymer measured by distinctly different procedures, viz., deduced from the splittings of the *ortho*-protons of phenylene rings registered by PMR and from the quadrupolar splittings of deuterons in the same polymer deuterated in the phenylene rings. As is seen, the discrepancy between the PMR and DMR

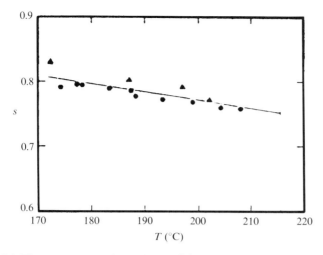

FIGURE 6.4. The temperature dependence of the order parameter, measured by the ^{1}H and ^{2}H NMR methods for the LC polymer R2F10 [13]. ● is ^{1}H NMR and ▲ is ^{2}H NMR.

order parameter is about 0.02. Martins et al. [4] evaluate the error in S measured by PMR as 0.05.

Ordering of Rigid Fragments

The order parameter of polymer nematics with mesogens in the main chain reaches values as high as 0.85–0.9 [8]. This is much larger than the order parameters of comb-shaped nematics as well as common low molecular weight nematics (0.4–0.6). This situation is probably due to the restriction on orientational fluctuations imposed by the mesogenic units connected by the spacers.

The rise of the order parameter accompanying the isotropie–nematic transition is about 0.75 [13,14]. As predicted by Ronca and Yoon [15], for the isotrope–nematic transition of semiflexible polymers in the limit of high molecular weight, this brings the order parameter to its upper limit. Unfortunately, the error of such measurements is rather large because of the large error in the detection of the transition temperature (Section 6.2.6).

Figure 6.5 illustrates the S versus reduced temperature dependences for R1F10 of different polymerization degree. We can see that the order parameter of mesogenic groups increases rapidly for short chains, whereas it levels off at about 8–10 repeating units [14]. A similar situation has also been observed for other polymers [8,16].

For the main-chain LC polymers, an odd–even alternation of mesogen ordering is observed similar to low-molar-mass nematics:

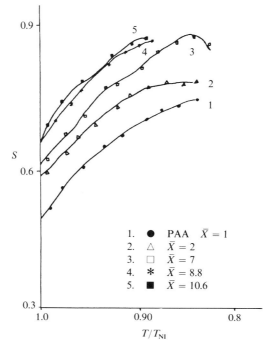

FIGURE 6.5. Temperature dependence of the order parameter as a function of the polymer chain length ($\bar{X}$) for R1F10 polymers [14]

$$\text{PAA}\quad \text{MeO}-\langle\!\!\!\bigcirc\!\!\!\rangle-N\!\!\overset{\overset{\displaystyle O}{\uparrow}}{=}\!\!N-\langle\!\!\!\bigcirc\!\!\!\rangle-\text{OMe}\qquad (\bar{X}=1)$$

The undulations are explained by changes of phase compositions at different temperatures in the sample with broad molar mass distribution (see Section 7.2.5).

(i) For n-even the mesogen order is much larger than for n-odd, whereas between n-odd and n-even it is regularly alternating. This inference is unequivocally illustrated by the data in Fig. 6.6 [16] comparing the PMR spectra of R1F10 and R1F9 for which $\delta_N \propto S_{zz}$ (Section 6.2.1). Table 6.3 compiles the data reflecting the dependence between the mesogen order parameter and the spacer lengths for polymer R1Fn in the range of 5–12 CH_2 groups.

(ii) Contrary to the low-molar-mass nematics, in which an odd–even effect is restricted to alternating the magnitude of the order parameter and dampens out rapidly at $n \geq 4$, in LC polymers a normal nematic level of order was presumed for the n-odd series [19], while for n-even (through $n = 16$) the proposed structural organization is a "micellar cybotactic nematic"

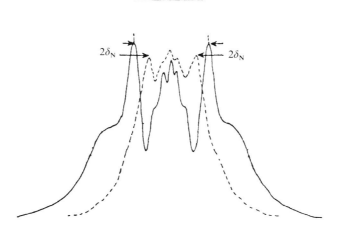

FIGURE 6.6. The odd–even effect in the main-chain LC polymers [16]. (———)
PMR spectra of the even polymer R1F10 and (– – –) the odd one R1F9, $T/T_{NI} = 0.92$.

(N_c), in which the polymer chains are extended and confined to layers
skewed by 41–45° with respect to the director of nematics.

Ordering of Flexible Fragments

Analysis of the splitting of the subspectrum of the protons of the flexible
fragments (Fig. 6.1) provokes us to assume that the absolute value of the
order parameter of CH_2 groups is approximately half the value of the order
parameter of rigid fragments. This case corresponds to the situation when
orientation of the interproton vector is, on average, perpendicular to the
direction of the mesogen orientation. In other words, on average, the chains
are aligned along the long axis of the mesogenic moiety. Figure 6.7 shows the
DMR spectra of the nematic phase of polymers R1F10 and R1F7 with completely deuterated spacers. As is seen, for polymer R1F10, only two doublets

TABLE 6.3. Orientational order parameter S as a function of the spacer length of
polymers

$$-[O-\bigcirc-N\overset{\overset{\textstyle O}{\uparrow}}{=}N-\bigcirc-O-CO-(CH_2)_n-CO-]-$$

n	5	7	8	9	10	11	12
S	0.35	0.49	0.72	0.35	0.71	0.44	0.75

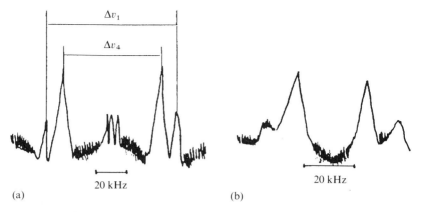

FIGURE 6.7. ^{2}H NMR spectra of the spacer-deuterated LC polymers [17]. (a) R1F10; (b) R1F7; Δv_1 is the quadrupolar splitting corresponding to the deuterons in the α position of the ester linkage, and Δv_4 are the remaining deuterons.

Δv_1 and Δv_4 with their relative intensities 1 : 4 ratio are observed. The largest splitting Δv_1 is associated with the α-methylene groups adjacent to the mesogens, whereas Δv_4 results from the methylene groups comparising the "internal" fragment of the spacer. Quite naturally, all of these groups follow nearly the same "orientational order" [18]. By using (6.2) and the values for the orientational order parameter S_{zz} extracted from the PMR spectra, we can calculate the conformational order parameters $S_i = |\langle P_2(\cos \beta_i)\rangle|$ (where β_i is the angle between the C—D bond and the long axis of the molecule) corresponding to orientation of the C—D bond in the α position $|S_1|$ and in all other positions $|S_4|$. Figure 6.8 illustrates the results. The following comments are necessary:

(i) $|S_1|$ is almost independent of temperature. This implies that, on average, the first CH$_2$ group is always oriented similarly with respect to the mesogen and may thus be observed as the inherent part of it. At $|S_1| \approx 0.40$ the C—D bond is, on average, almost perpendicular to the long axis, since for $|S_1| = 0.5$, $\beta = 90°$.

(ii) $|S_4|$ decreases with the rise in temperature. Since behavior corresponds to the increase in the amplitude of rotational fluctuations about the C—C bonds of the spacer and its gradual leveling out in the $N + I$ biphase region. In other words, the spacer undergoes disordering more rapidly than the rigid fragments.

In a number of papers [8,16,17] it was demonstrated that (see also Section 6.2.6):

(i) at the same reduced temperature the spacer order is smaller for n-odd;

(ii) in the vicinity of the T_{NI} transition the relative ordering of the spacers is

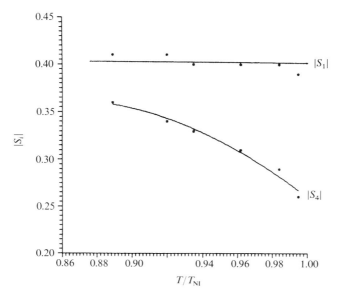

FIGURE 6.8. The conformational order parameter of various CD_2 groups of the spacer of the R1F9 polymer as the function of temperature [18]. $|s_1|$ is the conformational order parameter of the CD_2 groups in the α position of the ester linkage, and $|s_4|$ is the conformational order parameter of the remaining CD_2 groups.

always the same regardless of the chain length and parity of the spacer. This fact can be explained assuming that there exists a certain minimal degree of order corresponding to the appearance of the nematic phase.

Contrary to the Δv_i splittings providing information on S_{zz} through (6.2), the relative ^2H splittings $\Delta v_i / \Delta v_1$, as treated in the rotational isomeric states (RIS) approximation, are determined exclusively by the population of various rotational states of the bonds of a flexible spacer f_i^t. The values for f_i^t can be calculated in the following way. The $\langle P_2(\cos \beta_i) \rangle$ is averaged over all conformers. These values are fitted to the experimental data taking the statistical weights of various conformers as adjustable parameters. The trans fraction f_i^t for each bond is thus obtained. The results of RIS simulation of the DMR spectra of R2F9 and R2F10 LC polymer are summarized in Table 6.4 [7]. We can see that for R2F10 the odd–even alternation of the f_i^t value along the chain is very distinctive, whereas for R2F9 such a trend becomes less conspicuous.

An entirely different approach to the problem of LC order, called the dynamical racemization, was recently proposed by Volino et al. [19–21]. Within the frame of this approach the molecules are assumed to exist in essentially two equivalent conformations. These are the mirror image of one

TABLE 6.4. The *trans* fractions of each bond of the flexible spacer as estimated from ^{2}H NMR data [7].

Bond	Polymer	
	R2F9	R2F10
C_1-C_2	0.57	0.36
C_2-C_3	0.77	0.97
C_3-C_4	0.51	0.48
C_4-C_5	0.67	0.91
C_5-C_6	—	0.48

another with respect to a plane in which the long molecular axis lies. Thus, for instance, a set of angles $\{\varphi_i\}$ around successive single bonds used to describe a conformation would be transformed to the set $\{-\varphi_i\}$ for the mirror image conformation (Fig. 6.9). As in the RIS model, values of $\langle P_2(\cos \beta_i)\rangle$ can be easily calculated and fitted to the experimental data on ^{2}H quadrupole couplings, taking the set of angles $\{\varphi_i\}$ as adjustable parameters. The data in Fig. 6.9 pertaining to the R1F10 dimer demonstrate that the two halves of the even spacer are all *trans* and are linked by a central C—C bond with a significant defect. For the odd spacer the central part is all *trans* followed by C—C bonds on both ends of the spacer with defects of the *gauche* type ($|\varphi_3| = |\varphi_6| > 60°$); these bonds ensure the linkages with the ester groups presumably via defects of the *trans* type. However, the RIS-type model predicts quite different physical consequences, viz., *gauche* states ($f^t \neq 1$) for all bonds with an oscillation in f_i^t along the chain (see Table 6.4). The defects in the RIS model are delocalized by the conformational changes, whereas the dynamical racemization model predicts that these defects appear at the well-localized sites in the spacer.

Volino et al. [19] gave a detailed analysis of the entropy changes at the clearing transition within the frameworks of RIS and dynamical racemization models. However, neither was shown to be advantageous. The final choice between these two models requires additional experimental data.

Biaxiality of Molecular Ordering

In almost all up-to-date studies on LC polymers the generally probable deviation of the LC polymer symmetry from cylindrical is neglected, the first-order parameter S_{zz} (uniaxial term) being the only one taken into account. However, description of the ordering in rigid, noncylindrical elongated molecules requires that the second-order parameter, $\delta = S_{xx} - S_{yy}$, reflecting the difference in ordering of two minor principal (short) axes OX and OY would be introduced. The quantity $\eta_S = \delta/S_{zz}$ defines the anisotropy parameter of

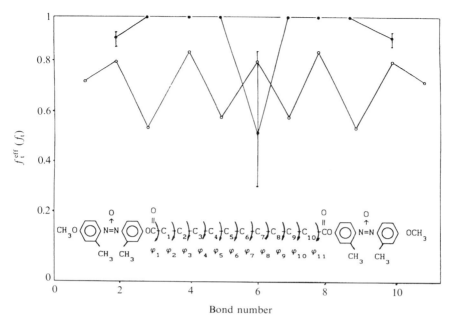

FIGURE 6.9. Comparison of the RIS model prediction with that of the dynamic racemization model for dimer [19]

the molecular order; it is directly related to the difference between the amplitudes of librations about the OX and OY axes.

For main-chain LC polymers Samulski [22] suggested recently that "linking successive cores together via a flexible spacer could very well amplify biaxial librations of the mesogenic cores." In [23] the biaxiality of molecular order in the polymers R1F7 and R1F10 was comprehensively investigated by applying PMR, DMR, and magnetic susceptibility measurements. The results of these studies can be summed up as follows:

(i) $S_{xx} - S_{yy}$ for polymers are essentially larger than for model compounds.
(ii) S_{xx} reveals no dependence on the chain length, whereas S_{yy} decreases with increasing chain length.
(iii) An odd–even effect is observed with the parity of spacers for $S_{xx} - S_{yy}$.

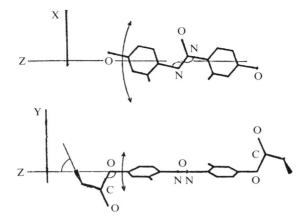

FIGURE 6.10. Sketch of the two orthogonal views of a possible structure of R1Fn polymers [23]. The plane of the ester group makes a large angle with the mesogenic group (roughly located between the two phenyl rings) and a small angle with the symmetry plane of the first methylene group. The arrowed arc-circles picture the different libration amplitudes in the two perpendicular planes from which the biaxiality originates.

Hence, in this case, biaxiality appears to be specifically related to the polymeric nature of the substance. Figure 6.10 pictures the model explaining the difference between the fluctuation amplitudes in the XOZ and YOZ planes by the particular structure of the repeating unit of a polymer chain. The fluctuation in the YOZ plane is presumed to involve larger displacements of the chain fragment next to the mesogenic unit; these are supposed to be more and more constrained as the chain length increases. This explains the decrease in S_{yy} and the increase in $S_{xx} - S_{yy}$ in polymers, contrary to their invariability in low molecular weight model compounds.

6.2.5   Biphase Region

Isotropic–Nematic Biphase Region

The NMR spectra of all main-chain LC polymers display a feature appearing to be quite specific, viz., the y comprise a narrow component whose width corresponds to that of the isotropic liquid [4,5,7,16,24–28]. Splitting of the narrow components, into signals corresponding to aliphatic and aromatic protons [25], as well as the independent assessment of the phase composition of the biphase region from the DMR spectra of selectively deuterated polymers [5] makes it possible to specify the molecular fragments involved in the isotropic motion responsible for the narrow component. It was found that in all investigated cases the ratio between the spectral components precisely correlated with the chemical structure (see also Section 6.3.3) This fact may interpreted as the molecule as a whole belongs to the different phases of the biphase system.

At the same time, the narrow component is detected by NMR at temperatures which are essentially lower than that at which the biphase region is no longer detectable by optical microscopy. The fraction of this "isotropic phase" was estimated from the signal area to amount to 38%. This features has led Stupp [25] to suggest that the biphase region is limited by the temperature determined by optical microscopy. At a lower temperature, preceding the phase separated state, isotropic fluctuations are liable to occur. To explain these isotropic fluctuations, relying on the results of conformational calculations, it was suggested [26] that some fragments in linear polymeric nematics may possess sufficient conformational versatility to commute dynamically between the nematic and pseudoisotropic fields.

There is a number of specific features characteristic of the biphase region that should be emphasized.

(i) The span of the biphase region of both odd- and even-numbered samples attains a maximum at $n \approx 6$–8 (Table 6.5) [16] and then starts to diminish progressively [29].

(ii) The span of the biphase region is larger the wider the molecular mass distribution is [24], and the greater the flexibility of the nonuniformity of the copolymers [25]. A specially synthesized "chemically uniform" sample showed a 25–40-fold narrowing of the biphase region.

(iii) The nematic phase of the biphase region is enriched with the longest chains while the isotropic phase is enriched with the shortest ones [24]. This is due to the fact that the temperature of the I–N transition increases with the molecular mass of a polymer. Hence, the molecular mass composition of the LC domains formed at different temperatures within the I–N biphase can be quite different.

TABLE 6.5. Biphase range as a function of the spacer length of polymers [16]

$$-[O-\!\!\!\bigcirc\!\!\!-N\!\!=\!\!N-\!\!\!\bigcirc\!\!\!-O-CO-(CH_2)_n-CO-]-\,.$$
$$\qquad\qquad\quad Me\qquad\quad Me$$

n	5	7	8	9	10	11	12
Biphase range (°C)	10.5	6*	23	12	18	11	18

* Polymer fraction with narrow ratio $\overline{M}_w/\overline{M}_n \cong 1.06$.

The thermal history and the processes taking place in the I–N biphase are reflected in the morphology and properties of the resulting isotropic, nematic, and solid phases. That is why the segregation by the chain length processes were subjected to detailed investigation by polarization microscopy, DSC, and NMR [29–31]. Isothermal evolution of the biphase region proceeds via two stages:

(i) rapid (several minutes) phase separation involving orientational and conformational ordering, and distribution of the chains of different length between the I and N components resulting in the formation of metastable "domains" segregated by chain length and the level of order; and

(ii) a very slow coalescence–homogenization stage involving coalescence and growth of the domains, and homogenization of their chain-length distributions.

The time scale for this stage was assessed by using DMR of PAA–D_{14} to probe heterogeneity in the pure N phase and comparing it to that of the low molecular weight model. For R1F7, $\tau_{1/2} \approx 100$ h, the order of magnitude for the dimensional scale of heterogeneity in the system is about 20 μm. The extent of molecular segregation and initial heterogeneity is invariably stronger for n-even than for n-odd.

Liquid Crystalline—Crystalline Biphase Region

Biphase systems are usually formed on the cooling of a polymer below melting point. The DMR spectra of biphase systems display the components corresponding to mobile and rigid deuterons, which are assigned to liquid crystalline and crystalline phases, respectively. The specific features of such biphase systems are as follows:

(i) The time scale for *trans–gauche* reorientation in the crystalline component is three orders of magnitude less than in the LC one (Fig. 6.3).

(ii) On passing through the melting point the conformational order for α and δ CH_2 groups of the flexible spacer jumps to a constant value of $n_t = 0.8$ regardless of the parity of the spacer (Fig. 6.11), whereas for the LC component the conformational order continuously increases from 0.69 to 0.76.

(iii) In the R3F10 polymer the order parameter is quantitatively retained (mobile and rigid components show the same limiting order parameter $S_{zz} = 0.9$) when the polymer is cooled below the melting point and the glass transition temperature. Thus, in this case, long-range orientational order is preserved even upon crystallization, which is quite uncommon for conventional low-molar-mass mesogens. In contrast, the R3F9 polymer, i.e., the one with the odd-numbered spacer, shows two limiting order parameters, viz., $S_{zz} = 0.82$ for LC and $S_{zz} = 0.55$ for crystalline components (Fig. 6.12).

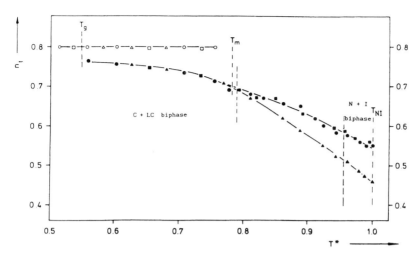

FIGURE 6.11. Temperature dependence of the conformational order of the spacer in the biphase region of selectively deuterated main-chain LC polymers R3F10 and R3F9 [8]

Full and open symbols correspond to liquid crystalline and crystalline components, respectively.

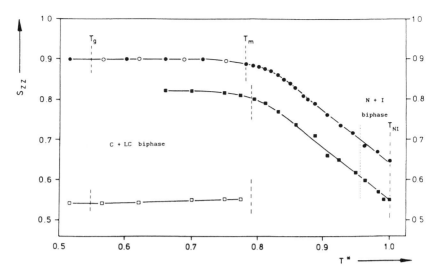

FIGURE 6.12. Orientational order in the biphase regions of main-chain LC polymers R3F10 and R3F9 [8]

Full and open symbols correspond to liquid crystalline and crystalline components.

6.2.6   Macroscopic Ordering

Practically all imaginable applications of the LC polymer involve the creation of sufficiently perfectly oriented structures. Moreover, most of the methods for investigating the LC polymer require that the monodomain orientational structure has to be preliminarily established. Therefore, it seems to be a matter of exclusive importance that a method be found for the reliable assessment of the perfection of the orientational structure. NMR appears to be one of the methods to fit the purpose. For a description of the macroscopic order the distribution function of the local directors of the sample is used

$$f(\varphi, \theta) = f(\theta) = N \exp(k \cos^2 \theta), \tag{6.3}$$

where k specifies the perfection of the orientational structure, N is the normalization constant, and φ and θ are the azimuthal and polar angles of the local director of the sample, respectively. Also used is the macroscopic order parameter $S_{z''z''}$ defined as follows:

$$S_{z''z''} = \int_0^\pi (\tfrac{3}{2} \cos^2 \theta - \tfrac{1}{2}) f(\varphi, \theta) \sin \theta \, d\theta. \tag{6.4}$$

For a random distribution of the director axes, $S_{z''z''} = 0$. In the opposite case of completely uniform alignment, $S_{z''z''} = 1$.

^{2}H Nuclear Magnetic Resonance Method

Table 6.6 lists the values of the macroorder parameter of the R3F10 LC polymer obtained by the DMR method (Section 6.2.2). We can see that there is no alignment in the magnetic field of 0.35 T. For $B = 7$ T the uniform alignment is obtained. Since the dielectric anisotropy of the polyester is negative, the distribution of the local director is two dimensional ($S_{z''z''} = -0.5$). The evolution of macroscopic alignment in the thermotropic polyester is slow, so that even in the magnetic field of 3 T it takes 300 s for the saturation of the macroorder.

DMR was also used to investigate the structure of melt spun fibers from R3F10 polyester in the N phase [32]; the polymer was selectively deuterated in the α position in the spacer. The molecular and mechanical characteristics of the fibers are summarized in Table 6.7.

^{1}H Nuclear Magnetic Resonance Method

The values for the macroorder parameter of Tables 6.6 and 6.7 are somewhat overestimated. This is due to the fact that in nematic liquid crystals thermally excited orientational fluctuations (elastic modes) necessarily arise. These are determined by the relation between Leslie coefficients and Frank constants. In low molecular mass nematics the majority of these fluctuations are fast on

TABLE 6.6. Parameters characterizing the macroscopic alignment of the R3F10 polymer [8].*

Orientation method	Alignment conditions	Reduced temperature (T/T_{NI})	Macroorder parameter
Electric field	$E = 48$ kV/cm $f = 50$ kHz	0.82	-0.5
Magnetic field	$B = 7.0$ T	0.82	1.0
Magnetic field	$B = 0.35$ T	0.82	0.0
Extrusion	Draw ratio = 49	0.67	0.9

* Error in determining $S_{z''z''}$ is < 8%.

TABLE 6.7. Molecular and mechanical properties of the fibers. Samples obtained from polymer R3F10 (degree of crystallinity, $\sim 55\%$; draw ratio, ~ 50) [32].

Molar mass ($\bar{M}_\omega$)	Spinning* temperature ($T_s - T_m$)	Order parameters†			Mechanical properties‡		
		$S_{z'z'}$	S_{zz}	$S_{z''z''}$	E (GPa)	σ (GPa)	ε (%)
10,000	15	0.7 (0.75)	0.9	0.9	11	0.14	2.5
10,000	10	0.7 (0.75)	0.9	0.9	11.4	0.15	2.5
30,000	33	0.7 (0.75)	0.9	0.9	10.7	0.31	2.5
30,000	18	0.7 (0.75)	0.9	0.9	11.4	0.32	2.5
30,000	8	0.7 (0.75)	0.9	>0.9	22.0	0.34	2.5

* T_s is the spinning temperature and T_m is the melting point.
† $S_{z'z'}$ is the order parameter of the α-segment of the spacer; value in brackets refers to crystalline component (biphase), S_{zz} is the orientational order parameter, and $S_{z''z''}$ is the macroorder parameter.
‡ E is Young's modulus, σ is the tensile strength, and ε is the elongation at break.

the NMR time scale (10^{-4}–10^{-6} s). In the viscous nematic polymer an appreciable fraction of these modes are presumably slow. This implies that there exists a certain distribution of local directors by orientation (6.3) and the overall macroordering can be described using $S_{z''z''}$. In PMR spectra the distribution of local directors is manifested in the deviation of the angular dependence of the spectrum, $F_\beta(v)$, from that for the nematic monodomain ($S_{z''z''} = 1$)$f_\beta(v)$, in the fast motion limit [34]

$$f_\beta(v) = P_2^{-1}(\cos \beta)f_0(v/P_2(\cos \beta)), \qquad (6.5)$$

where β is the angle between the director and the magnetic field, f_0 is the spectrum for $\beta = 0$ in the absence of the statical distribution of directors, $P_2(\cos \beta) = \frac{3}{2} \cdot \cos^2 \beta - \frac{1}{2}$, and v is the frequency. For the computation of the spectrum, taking account of the statistical distribution of the directors $F_\beta(v)$, (6.5) must be integrated by the angles φ and ϑ of the distribution (6.3) "shifted" by the angle β

$$F_\beta(v) = \int_0^{2\pi} \int_0^\pi P_2^{-1}(\cos \gamma)f_0(v/P_2(\cos \gamma))f(\varphi, \theta) \sin \theta \, d\varphi \, d\theta, \qquad (6.6)$$

where $\cos \gamma = \sin \theta \cos \varphi \sin \beta + \cos \theta \cos \beta$, and γ is the angle between the local director axis and the magnetic field direction. At $S_{z''z''}$, only slightly different from 1, $f_0(v)$ is close to $F_0(v)$. Figure 6.13 [35] shows the experimental curve for the angular dependence $F_\beta(v)$ of the PMR spectra for polyester R1F7 deuterated in the spacer; the curves fitted to the described model with $S_{z''z''} = 0.962$ and $f_0(v) = F_0(v)$ are also shown. As is seen, the conformity between the two patterns is quite satisfactory.

Therefore, at slight deviations of the static director distribution from the ideal one characteristic of the monodomain structure ($S_{z''z''} \approx 1$), this ap-

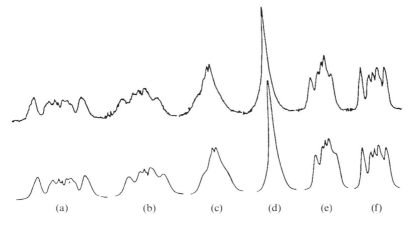

FIGURE 6.13. Line shape of the experimental (upper) and fitted (lower) angle dependences of the PMR spectra of a nematic monodomain for the R1F7 polymer [35]: $T = 393$ K; (a) 0°; (b) 25.5°; (c) 37°; (d) 47°; (e) 76.5°; (f) 90°.

proach to PMR spectra simulation may be efficiently used to assess the macroscopic order parameter $S_{z''z''}$.

6.3 Liquid Crystalline Polymers with Mesogenic Groups in the Side Chains

The NMR researchers were predominantly involved in the studies of polymeric LC with mesogens in the main chain. LC comb-shaped polymers of comb-shaped structure with mesogens in the side chains were in much less favor. Such a situation was largely due to the complexity of the task of examining LC comb-shaped polymers [36]. These are much less-ordered systems displaying essentially greater possibilities for the formation of various LC phases [37]. The group of polymers most thoroughly studied by today comprises siloxane, acrylate, and methacrylate polymers, with phenylbenzoate or cyanobiphenyl mesogenic groups attached to the backbone via an ester linkage and a flexible aliphatic spacer.

6.3.1 Molecular Dynamics and Ordering in the Glassy State

Below the glass transition temperature comb-shaped LC polymers form polymeric glasses fully retaining the mesophase order. All intermolecular motions are frozen at the glass transition temperature, and the dynamics in the glass are controlled by intramolecular phenyl flips and a kink motion of the spacer's bonds.

TABLE 6.8. Molecular order and motion of a labeled phenylene unit in glassy comb-shaped LC polymers [38].

$$
\begin{array}{ccccc}
-[-CH_2-CH-]_n- & -[-CH_2-CH_n-]- & & & -[-CH_2-\overset{\displaystyle CH_3}{\underset{\displaystyle |}{C}}-]_n- \\
| & | & & & | \\
C=O & C=O & CH_3 \qquad\qquad CH_3 & & C=O \\
| & | & -[-Si-O-]_n-\quad -[-Si-O-]_n- & & | \\
O & O & | \qquad\qquad\quad | & & O \\
(CH_2)_2 & (CH_2)_6 & (CH_2)_3 \qquad\quad (CH_2)_6 & & (CH_2)_6
\end{array}
$$

(phenylene–C=O–O–C6D4–OCH3 mesogenic units in all five columns)

Mesophase type	N	S_A	N	S_C	N
Order paramter	0.65	0.88	0.65	0.85	0.65
Width parameter $\bar{\beta}$ of orientational distribution function, grade	18	10.5	19	11.5	18.5
Width 2σ of correlation time distribution function (in decades)	2.2	2.5	2.6	2.4	2.3
Activation energy E_a, of the 180° phenyl flip (kJ/mol).	42	47	43	48	46

In [38,39] the dynamics of phenyl flips and the orientational ordering of mesogenic groups were subjected to detailed analysis. For siloxane, acrylate, methacrylate comb-shaped LC polymers deuterated in phenyl nuclei ^{2}H, NMR solid echo spectra were registered at various rotational angles of the sample director with respect to the magnetic field and at various pulse spacings of the solid echo sequence. The respective data are compiled in Table 6.8. For the calculation of the line shapes of the spectra, the function of the orientational distribution $P(\beta)$ of the long axes of mesogenic groups and the log–Gaussian function $K(\ln \Omega)$ for the flips correlation frequency distribution were used

$$K(\ln \Omega) = (\sigma\sqrt{2\pi})^{-1} \exp[-(2\sigma^2)^{-1} \ln(\Omega/\Omega_0)], \qquad (6.7)$$

where Ω is the correlation frequency, Ω_0 is the center of the distribution of the correlation frequency, and σ is the distribution variance. The subspectrum for the given orientational angle and a single value of the jump frequency Ω

was calculated within the framework of the two-site jump exchange model [40]. Assessment of the parameters of the model from experimental spectra (Table 6.8) led to the following conclusions:

(i) The function of the orientational distribution is, with a high degree of precision, a Gaussian one, $N \exp(-\sin^2 \beta/2 \sin^2 \bar{\beta})$, its width being characterized by $\bar{\beta}$ and the order parameter S (6.4).

(ii) The order parameter is fully retained on cooling to the glassy state.

(iii) The correlation frequency distribution is satisfactorily described by the log–Gaussian distribution function with the distribution width somewhere around two decades.

(iv) The temperature dependences of the center of the flip correlation frequency distribution Ω_0 is described by the Arrhenius plot (values for E_a are listed in Table 6.8). The activation energies and the width of the flip correlation time distribution are comparable to that of conventional amorphous polymers [38], although for highly ordered systems a drastic narrowing of the distribution width might have been expected.

The ordering in frozen LC phases of polyacrylates and polymethacrylates were comprehensively studied by Böeffel and Spiess [38,41,42]. The angle dependences of ^2H NMR spectra were interpreted in terms of the orientational distributions [43] of the C—D bonds in polyacrylates and the C—CD$_3$ bonds in polymethacrylates. Analysis of the orientational distributions revealed the following features:

(i) The long axis of the mesogenic fragment and the spacer are oriented along the director. The order parameter in the frozen smectic phase of polymethacrylate

$$
\begin{array}{c}
\text{Me} \\
| \\
-[-\text{C}-\text{CH}_2-]- \\
| \\
\text{O}=\text{CO}-(\text{CH}_2)_6-\text{O}-\langle\bigcirc\rangle-\text{COO}-\langle\bigcirc\rangle-\text{OMe}
\end{array}
$$

is 0.88 for the mesogenic group and 0.52 for the spacer.*

(ii) Orientation of polymeric backbones in the case of polyacrylates is different from that of polymethacrylates. In polymethacrylates, regardless of their phase state (N or S), the chains are oriented on the average normal to the director. The most probable angle β between the director and the C—CD$_3$ bond of 54° (half of the tetrahedral angle) corresponds to such a situation. Consequently, the bisector of the (RO$_2$C)—C—(CD$_3$) angle is oriented along the director (Fig. 6.14(a)). In acrylates the backbone itself is oriented parallel to the director. In this case, the most probable angle between the C—D bond and the director is 90°.

---

* Similar results were also obtained by the 2D ^{13}C MAS method [44].

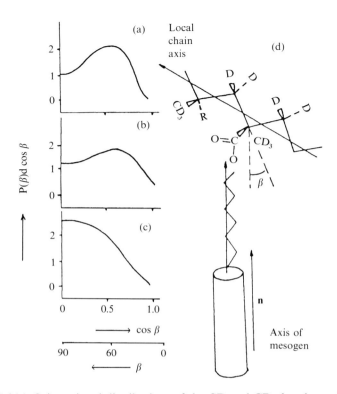

FIGURE 6.14. Orientational distributions of the CD and CD_3 bond axes in comb-shaped LC polymers [38,41]:

(a) smectic mesophase of

$$-[-\underset{\underset{O=CO-(CH_2)_6-O-\bigcirc-COO-\bigcirc-OC_4H_9}{\overset{|}{C}}}{\overset{Me}{\overset{|}{C}}}-CH_2-]-$$

(b) nematic mesophase of

$$-[-\underset{\underset{O=CO-(CH_2)_6-O-\bigcirc-COO-\bigcirc-OMe}{\overset{|}{C}}}{\overset{Me}{\overset{|}{C}}}-CH_2-]-$$

(c) smectic mesophase of

$$-[-CH_2-\underset{\underset{-O=CO-(CH_2)_6-O-\bigcirc-COO-\bigcirc-OMe}{\overset{|}{C}H}]}{}$$

$P(\beta) \cos \beta$ is the value proportional to the number of CD or CD_3 bonds located at the angle β with respect to the director.

(iii) Quantitative characteristics describing the ordering of the main chains in the smectic phases of polyacrylates and polymethacrylates are essentially different as well. The by-the-angle distribution function for mesogenic fragments of smectic polymethacrylate

$$\begin{array}{c} Me \\ | \\ -[-C-CH_2-]- \\ | \\ O=CO-(CH_2)_6-O-\bigcirc-COO-\bigcirc-OC_4H_9 \end{array}$$

is characterized by the mean width $\bar{\beta} = 20°$; this corresponds to the order parameter of the bisector axis of 0.6 and the anisotropy of the radius of gyration, as studied by SANS, $4:1$ [45]. Smectic polyacrylate

$$\begin{array}{c} -[-CH_2-CH-]- \\ | \\ O=CO-(CH_2)_6-O-\bigcirc-COO-\bigcirc-OMe \end{array}$$

is ordered to a substantially lesser extent, $\bar{\beta} = 50°$. In this case, the orientational distribution is broad and the anisotropy of the radius of gyration is as small as $1:1.25$ [46], the long dimension being in alignment with the director.

(iv) There is no difference in the ordering of the CH_2 groups of the spacer differently distanced from the mesogen. Such a behavior pattern for comb-shaped LC polymers is essentially different from those observed for polymers with mesogens in the main-chain and conventional low molecular mass LC. The difference in the ordering of the spacer and subsequent carbonyl carbon is very small (according to [44] the order parameters are 0.52 and 0.45, respectively), whereas already beyond the next bond linking the carbonyl group to the main chain ($-CO-CH-$)

the ordering decreases dramatically ($S = 0.25$). Thus the conformational state of the backbone-carbonyl bond appears to be one of the key features controlling the LC order. The differences in the orientational behavior of polyacrylates and polymethacrylates is presumably associated with the steric hindrance of the rotation around this bond in polymethacrylates [38].

6.3.2 Ordering and Dynamics in Liquid Crystalline Phases above the Glass Transition Temperature

In LC the uniaxial phase quadrupolar splittings of the ^2H spectra and dipole splittings of the proton spectra are proportional to the order parameter (Sections 6.2.1 and 6.2.2), provided that molecular reorientations are fast on the NMR scale. Figure 6.15 shows the proton spectra of nematic polyacrylates

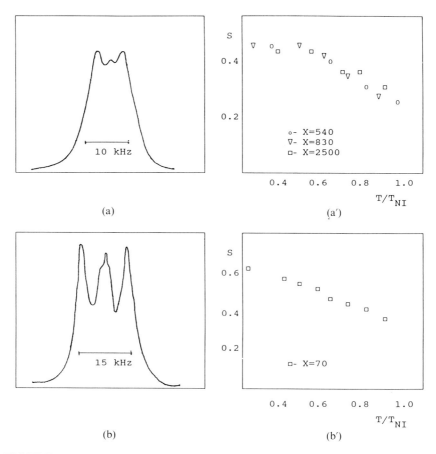

FIGURE 6.15. PMR spectra (a, b) and temperature dependences of the orientational order parameter (a', b') for the comb-shaped nematic polyacrylates [47,48]

(a) $-$[$-$CH$_2$$-CH-$]$-$
          |
          O=CO$-$(CH$_2$)$_4$$-O-$⬡$-$⬡$-$CN

(b) $-$[$-$CH$_2$$-CH-$]$-$
          |
          O=CO$-$(CH$_2$)$_5$$-O-$⬡$-COO-$⬡$-$OMe

with phenylbenzoate and cyanobiphenyl mesogenic groups and the temperature dependences of the order parameter [47,48]. As compared to the spectra of polymers with mesogenic groups in the main chain the spectra of comb-shaped LC polymers are narrower, especially in the range of the protons of the spacer (see Fig. 6.1).

Such a pattern is unambiguously associated with the essentially looser ordering of the CH_2 groups of the spacer in comb-shaped LC polymers. As evaluated by Böeffel et al. [50], the order parameter of the spacer in the nematic phase of the polyacrylate

$$-[-CH_2-CH-]-$$

$$O=CO-CD_2(CH_2)_5-O-\text{⟨ring⟩}-COO-\text{⟨ring⟩}-OMe$$

amounts to about half of the order parameter of the mesogen.

The splitting of phenyl protons, which is proportional to the order parameter of the *para*-axis of phenyl rings, is smaller than for low molecular mass analogues (Fig. 6.16) [49]. Böeffel et al. [50] observed a similar effect for quadrupolar splittings in the spectra of deuterated polymers of different chain length and low molecular mass analogue. In the array

$$C_4H_9O-\overset{D\ \ D}{\underset{D\ \ D}{\text{⟨ring⟩}}}-COO-\text{⟨ring⟩}-OC_{10}H_{21}$$

$$-[-CH_2-CH-]_{10}-$$
$$O=CO-(CH_2)_6-O-\text{⟨ring⟩}-COO-\overset{D\ \ D}{\underset{D\ \ D}{\text{⟨ring⟩}}}-OMe$$

$$-[-CH_2-CH-]_{100}-$$
$$O=CO-(CH_2)_6-O-\text{⟨ring⟩}-COO-\overset{D\ \ D}{\underset{D\ \ D}{\text{⟨ring⟩}}}-OMe$$

quadrupolar interaction diminishes and, hence decreases in the same order as the order parameter of mesogen. The abrupt change of the order parameter accompanying the isotrope–nematic transition is 0.2–0.3; this is substantially smaller than the value predicted by the Maier–Saupe theory.

In comb-shaped LC polymers, at a relatively small decrease of the reduced temperature ($T/T_{NI} \approx 0.85$ or even greater), the increase in quadrupolar splitting in ^2H NMR spectra occurs more rapidly than the possible increase in the order parameter [38], it being an important specific feature of comb-shaped LC polymers. It is interpreted [38,50] as the slowing down of molecular motions associated with the high viscosity of comb-shaped LC polymers and insufficient decoupling of the motions involving a polymer backbone, on the one hand, and mesogenic groups, on the other (this is especially important in polymethacrylates). As a result, the "fast motion limit" is no longer valid and quadrupolar splittings, as well as the moments of the PMR spectra line shape, are no longer related directly to the order parameter. Although both

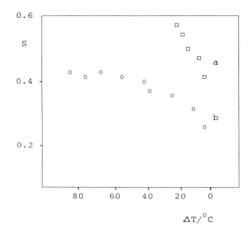

FIGURE 6.16. Temperature dependence of the order parameter of the comb-shaped nematic polymer and of the low-molar-mass model compound [49]

(a) $CH_3(CH_2)_5$—⟨⟩—⟨⟩—CN

(b)   —[CH$_2$—CH]—
       |
       O=CO—(CH$_2$)$_5$—O—⟨⟩—⟨⟩—CN

$\Delta T = T_{NI} - T \ (°C)$.

of the described processes somehow contribute to the decrease of mobility, a quite different mechanism unrelated to the decrease of the local mobility and, hence, especially efficient at high temperatures may be suggested to account for the increase in quadrupolar splittings. This mechanism makes use of the equilibrium distribution of the director orientation fluctuations; it will be discussed below (Section 6.3.4).

6.3.3   Biphase Region

The biphase region in comb-shaped LC polymers was examined much less extensively than in linear LC polymers (Section 6.2.5). This fact is due to the fact that in this case the biphase is less common and its temperature range is substantially narrower. Usually, it expands to several degrees around the isotrope–nematic transition temperature. Besides this, the jump in the ordering accompanying the transition of comb-shaped LC polymers to the LC phase is quite small and, consequently, the difference in the line widths of isotropic and LC phases is not large as well. This fact makes the NMR investigation of the biphase region in comb-shaped LC polymers a complicated task. At the same time Stroganov et al. [51] reported on a narrow component reliably registered by PMR in the S_A and S_B phases of comb-shaped smectic polyacrylate containing the mesogenic Schiff base

$$-[-CH_2-CH-]-$$
$$O=CO-(CH_2)_{11}-O-\langle\rangle-CH=N-\langle\rangle-C_4H_9$$

groups in the side chains. However, there are two considerations that raised doubt that the analysis of the biphasity can actually be carried out by assessing the narrow component of the spectrum.

(i) At a given cooling rate a perfect LC phase fails to form. This fact appears to be natural because of the high viscosity of polymeric liquid crystals; it is also confirmed by the dependence between the perfection of orientational structures in low molecular mass smectics [48] and cooling rates.

(ii) It is not the molecules of the isotropic phase that give rise to the narrow component, but only certain more mobile fragments of these, for instance, the aliphatic ones.

Both considerations were comprehensively treated and finally refuted [51]:

(i) The cooling rate was chosen on the basis of a quantitative assessment of the perfection of orientational structures from the angular dependences of the second moments of PMR spectra (see below).

(ii) The type and number of protons contributing to the narrow component, as disclosed in the analysis of the high-resolution spectra, were fully consistent with the stoichiometry of the compound (Fig. 6.17). Thus, it is exactly the coexistence of the S_A and S_B phases with the isotropic one that gives rise to the narrow component; the fraction of the isotropic phase as a function of temperature is shown in Fig. 6.17.

It should be noted that the studies of the biphase transitional region of comb-shaped LC polymers can add substantially to the understanding of polymeric mesophase organization. Such important mesophase characteristics as the perfection of orientational structure, the molecular mass segregation of LC domains, and others are presumably "pre-shaped" in the biphase region, as is the case of polymers with mesogens in the main chain (Section 6.2.6). We believe that this trend was unduly deprived of attention, and that these studies will result in a sufficiently thorough picture of the processes taking place in the biphase region of comb-shaped LC polymers.

6.3.4   Macroscopic Ordering

As in Section 6.2.6, the director orientational distribution function is used for a quantitative description of the macroscopic ordering. The deviation of the macroscopic orientational structure from the ideal one is manifested via the *angular dependences of NMR spectra* (Fig. 6.13). At the same time, the meth-

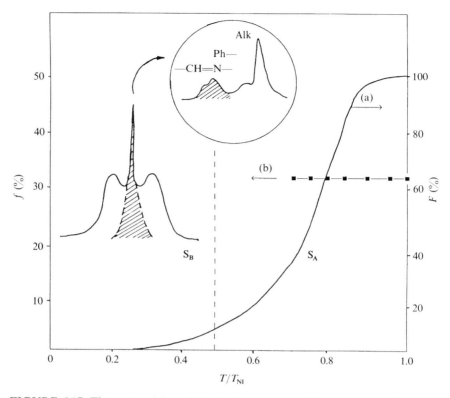

FIGURE 6.17. The composition of the biphase region of the comb-shaped smectic polymer [54]

$$-\!\!\!\begin{array}{c}\text{[} CH_2\!-\!CH \text{]} \\ | \\ O\!=\!CO\!-\!(CH_2)_{11}\!-\!O\!-\!\bigcirc\!-\!CH\!=\!N\!-\!\bigcirc\!-\!C_4H_9 \end{array}$$

(a) is the temperature dependence of the isotropic phase fraction, F, deduced by integration of the narrow component of the wide-line PMR spectrum, and (b) is the temperature dependence of the phenyl and azomethine fraction, f, deduced by integration of the phenyl and azomethine part of the high-resolution PMR spectrum. The dashed line indicates the S_A–S_B phase transition.

od for assessing the perfection of the orientational structure, which (perfection) is identified with the parameter k of the orientational distribution function, is valid for large k only (orientational structure is close to the ideal one). For systems that are loosely ordered on a macroscopic scale, angular dependences of the moments of PMR spectra provide information concerning the values of k. Molchanov et al. [53] calculated k from the dependences of the second moment of PMR spectra on the angle β between the director of a LC and of the magnetic field direction

$$M_2(\beta, k) = M_2(0, \infty) \int_0^{2\pi} \int_0^{\pi} P(\theta, k)(\tfrac{3}{2}\cos^2\gamma - \tfrac{1}{2})^2 \sin\theta \, d\varphi \, d\theta, \quad (6.8)$$

where

$$\cos\gamma = \sin\theta \cos\varphi \sin\beta + \cos\theta \cos\beta,$$

$$P(\theta, k) = (1 - \alpha)/2 + \alpha N^{-1} \exp(k\cos^2\theta),$$

$$N = \int_0^{\pi} \exp(k\cos^2\theta) \sin\theta \, d\theta.$$

Notation is the same as in (6.6) except for the distribution function $P(\theta, k)$, which here includes an additional constant term $(1 - \alpha)$ referring to the unoriented fraction of the liquid crystal sample. This approach was used to investigate the formation of the orientational structure in comb-shaped smectic polyacrylate

$$-[-CH_2-CH-]-$$
$$O=CO-(CH_2)_{11}-O-\langle\!\langle\bigcirc\rangle\!\rangle-CH=N-\langle\!\langle\bigcirc\rangle\!\rangle-C_4H_9$$

induced by the magnetic field. It was demonstrated that various distribution functions in (6.8)—e.g., $P(\theta, k)$ proportional to $\exp(k\cos^2\theta)$, $\cosh(k\cos\theta)$ or $k^3/[1 + (k^3 - 1)\sin^2\theta]^{3/2}$—used to characterize orientation in the fibers [55], produced an equally satisfactory description of the angular dependences of the second moments [51,54], whereas the attempts to describe orientational distribution assuming a mixture of ideally oriented and nonoriented structures lacked consistency with experimental data. Furthermore, orientational perfection was shown to exhibit substantial dependence on the cooling rate and magnetic field intensity (Roth and Krücke [56] obtained similar results for comb-shaped siloxanes). At small fields (0.89 T) and high cooling rates (3–5 deg/min) orientational perfection is small, $k = 4$–5, and irreproducible (see Fig. 6.18). At a cooling rate of 0.25 deg/min and a magnetic field intensity of 1.4 T, orientational perfection reaches a saturation level. It is noteworthy that, for a linear polyester with mesogenic groups in the main chain, saturation of orientational perfection is attained at almost the same field intensity, viz. 1.6 T [57]. The ease with which the polymeric smectic reaches the orientational limit, as compared to the low molecular mass smectic, is quite unfamiliar as well [52]: orientation reaches the saturation level at the field intensity of 2.2 T, orientational perfection being actually the same for the polymer ($k = 12$) and the low molecular mass smectic ($k = 15$).

A more precise description of the macroscopic order was achieved when the first moments (centers of gravity) of the halves of PMR spectra $\int_0^{\infty} vf(v) \, dv / \int_0^{\infty} f(v) \, dv$ were analyzed. Kireev et al. [58,59] examined the angular dependences of the first moments for various models of the orientational

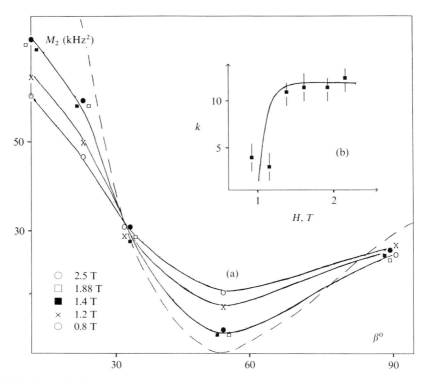

FIGURE 6.18. Macroordering of the comb-shaped smectic polymer in a magnetic field [54]

$$\text{-[-CH}_2\text{--CH-]-}$$
$$\text{O=CO--(CH}_2)_{11}\text{--O-}\langle\bigcirc\rangle\text{--CH=N-}\langle\bigcirc\rangle\text{--C}_4\text{H}_9$$

(a) is the angle dependence of the second moments of the PMR spectra in fields of various intensity; (b) is the magnetic field intensity dependence of the orientation quality; and (---) is the angle dependence of the second moments for the ideal orientational structure ($k = \infty$).

distribution of the director. For a mixture of ideally oriented ($k = \infty$) and nonoriented ($k = 0$) structures the angular dependence is given by the following relationship:

$$M_1(\beta) = M_1(0, \infty)|\tfrac{3}{2}\cos^2\beta - \tfrac{1}{2}| + M_H. \tag{6.9}$$

When the director distribution is described by the function $P(\theta, k)$ the relationship is

$$M_1(\beta, k) = M_1(0, \infty)\int_0^{2\pi}\int_0^{\pi} P(\theta, k)|\tfrac{3}{2}\cos^2\gamma - \tfrac{1}{2}|\sin\theta\, d\varphi\, d\theta. \tag{6.10}$$

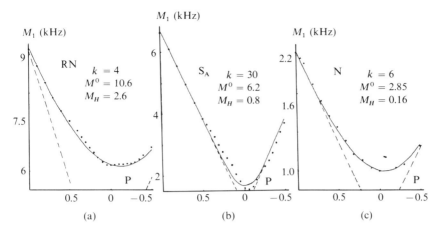

FIGURE 6.19. Quality of the macroscopic orientational structure of the comb-shaped LC polymer in different mesophases [59].

• are the experimental values of the first moments of the halves of PMR spectra ($\int_0^\infty \nu f(\nu)\, d\nu$) for various angles β of the macroscopic director to the magnetic field ($p = \frac{3}{2}\cdot\cos^2\beta - \frac{1}{2}$); (——) is the fitted dependence for the orientational distribution; $P(\theta, k) \propto \exp(k\cos^2\theta)$; M_H is the unoriented structure moment; $M^0 = M(0; \infty)$ is the ideal orientational structure moment; and (---) is the computed dependence for ideal orientational structure ($k = \infty$).

Notation is the same as in (6.8) except for the additional constant term M_H referring to the unoriented fraction of the sample. Orientational structures in different phases of the comb-shaped polyacrylate with cyanobiphenyl meso-genic groups and a six CH_2 units long aliphatic spacer were examined by the described method. Shown in Fig. 6.19 are the dependences of the first moment calculated using (6.9), (6.10) and the experimental data for nematic, smectic, and reentrent nematic (RN) phases. High macroscopic order in the S_A phase (b) seems to be quite natural. At the same time, low macroscopic order in both nematic phases appears to be quite unexpected. This result is in contradiction to the well-known property of nematics to form monodomain orientational structures at magnetic field intensities exceeding a certain thres-hold. It also cannot be accounted for either by the slowing down of molecular motion due to high viscosity or by the incomplete decoupling of the motions involving polymer backbone and mesogenic groups (Section 6.3.2), because the transition from the S_A phase to the N phase taking place on increasing the temperature can affect molecular motions only so as to intensify them. This must lead to deviations from the monodomain model, arising from the low intensity of molecular motions, becoming smaller and, consequently, to the increase of k. However, in reality, k decreases abruptly (b, c). At the same time, analysis of the angular dependences of the first moment (a, c) unambiguously indicates on the existence of the static, on the NMR time scale, and on the orientational director distribution in the RN and N phases with $k = 4$ and 6. These results have been interpreted [54,55] in terms of the equilibrium fluctuations of the director orientation (see also Section 7.2.6). Since the relaxation time of the fluctuation is proportional to the viscosity of the medium, in a viscous LC polymer melt relaxation of a sufficiently large fraction of fluctuations is slow on the PMR time scale. This situation may be equivalently described by the static orientational distribution of the director leading, in its turn, to the deviation of the angular dependences of the spectra from the ideal pattern.

References

1. B. Hüser and H.W. Spiess, Macromol. Chem., Rapid Commun. **9**, 337 (1988).

2. K. Kohlhammer, B. Reck, H. Ringsdorf, and G. Kothe, 18 Freiburger Arbeitstagung Flüssige Kristallen (1988).

3. L.B. Stroganov, A.N. Prokhorov, R.A. Galyullin, E.V. Kireev, V.P. Shibaev, and N.A. Platé, Polym. Sci. USSR. **34**, 89 (1992).

4. A.F. Martins, J.B. Ferreira, F. Volino, A. Blumstein, and R.B. Blumstein, Macromolecules **16**, 279 (1983).

5. F. Volino and R.B. Blumstein, Mol. Cryst. Liq. Cryst. **113**, 147 (1984).

6. St. Limmer, H. Schmiedel, B. Hilner, A. Löshe, and S. Grande, J. Phys. (Paris) **4**, 69 (1980).

7. A. Abe and H. Furuya, Macromolecules **22**, 2982 (1989).

8. K. Müller, P. Meier, and G. Kothe, in *Progress in NMR Spectroscopy*, Vol. 17, edited by J. Emslly, J. Feeney, and L. Sutcliffe (Pergamon Press, Oxford, 1985).

9. J.G. Powles and J.H. Strange, Proc. Phys. Soc. **82**, 6 (1963).

10. K.R. Jeffrey, Bull. Magn. Reson. **3**, 69 (1981).

11. H.S. Spiess, J. Chem. Phys. **72**, 6755 (1980).

12. J. Jeener and P. Broekaert, Phys. Rev. **157**, 232 (1967).

13. D.Y. Yoon, S. Bruckuer, W. Volksen, and J.C. Scott, Faraday Discuss. Chem. Soc. **79**, 41 (1985).

14. R.B. Blumstein, E.M. Stickles, M.M. Gauthier, A. Blumstein, and F. Volino, Macromolecules **17**, 177 (1984).

15. G. Ronca and D.Y. Yoon, J. Chem. Phys. **76**, 3295 (1982).

16. A. Blumstein, M.M. Gauthier, O. Thomas, and R.B. Blumstein, Faraday Discuss. Chem. Soc. **79**, 33 (1985).

17. R.B. Blumstein and A. Blumstein, Mol. Cryst. Liq. Cryst. **165**, 361 (1988).

18. E.T. Samulski, M.M. Gauthier, R.B. Blumstein, and A. Blumstein, Macromolecules **17**, 479 (1984).

19. F. Volino, J.A. Ratto, D. Galland, P. Esnault, and A.J. Dianaux, Abstract of the 8th Liquid Crystal Conference of Socialist Countries (8LCCSC), Krakow, Poland (1989).

20. D. Galland and F. Volino, J. Phys. (Paris) **50**, 1743 (1989).

21. F. Volino, D. Galland, J.B. Ferreira, and A.J. Dianoux, J. Mol. Liq., Special Issue in Honour of Professor C. Brot (in press).

22. E.T. Samulski, Faraday Discuss. Chem. Soc. **79**, 7 (1985).

23. P. Esnault, D. Galland, F. Volino, and R.B. Blumstein, Macromolecules **22**, 3734 (1989).

24. F. Volino, J.M. Alloneau, A.M. Giroud-Godquin, R.B. Blumstein, E.M. Stickles, and A. Blumstein, Mol. Cryst. Liq. Cryst. Lett. **102**, 21 (1984).

25. S.I. Stupp, Polymer Preprints **23**, 509 (1989).

26. S.I. Stupp, J.S. Moore, P.G. Martin, J. Wu, and F. Chen, "Molecular Organisation in Liquid Crystal Polymers," submitted for publication.

27. J.S. Moore and S.I. Stupp, Macromolecules **20**, 273 (1987).

28. J.S. Moore and S.I. Stupp, Macromolecules **20**, 282 (1987).

29. J.F. d'Allest, P. Sixou, A. Blumstein, and R.B. Blumstein, New Orleans, Mol. Cryst. Liq. Cryst. **157**, 229 (1988).

30. D.Y. Kim, J.F. d'Allest, A. Blumstein, and R.B. Blumstein, ibid., p. 253.

31. P. Esnault, F. Volino, J.F. d'Allest, and R.B. Blumstein, ibid., p. 273.

32. K. Müller, A. Schleicher, E. Ohmes, A. Ferrarini, and G. Kothe, Macromolecules **20**, 2761 (1987).

33. M. Warner, Mol. Phys. **52**, 677 (1984).

34. H. Schmiedel, B. Hillner, S. Grande, A. Lösche, and St. Limmer, J. Magn. Res. **40**, 369 (1980).

35. P. Esnault, J.P. Casquillo, and F. Volino, Liq. Cryst. **3**, 1425 (1988).

36. V.P. Shibaev and N.A. Platé, Adv. Polymer Sci. **60/61**, 173 (1984).

37. M. Warner, in *Side Chain Liquid Crystal Polymers*, edited by C.B. McArdle (Blackie, Glasgow, 1988).

38. Ch. Böeffel and H.W. Spiess, in *Side Chain Liquid Crystal Polymers*, edited by C.B. McArdle (Blackie, Glasgow, 1988).

39. U. Pschorn, H.W. Spiess, B. Hisgen, and H. Ringsdorf, Makromol. Chem. **187**, 2711 (1986).

40. H.W. Spiess and H. Sillescu, J. Magn. Res. **42**, 381 (1981).

41. Ch. Böeffel and H.W. Spiess, Macromolecules **21**, 1626 (1988).

42. Ch. Böeffel and H.W. Spiess, Makromol. Chem., Rapid Commun. **7**, 777 (1986).

43. H.W. Spiess, in *Developments in Oriented Polymers—1*, edited by J. Ward (Applied Science, London, 1982).

44. B. Blümich, Ch. Böeffel, G. Harbison, Y. Yang, and H.W. Spiess, Ber. Bunsenges. Phys. Chem. **91**, 1100 (1987).

45. P. Keller, B. Carvalno, J. Cotton, M. Lambert, F. Moussa, and G. Pepy, J. Phys. (Paris) Lett. **46**, L-1065 (1975).

46. R. Kirste and H. Ohm, Makromol. Chem., Rapid Commun. **6**, 179 (1985).

47. E.V. Kireev, E.B. Barmatov, L.B. Stroganov, S.G. Kostromin, and R.V. Talrose, Abstracts of the All-Union Symposium, "Polymer—90", Leningrad (1990), p. 105.

48. L.B. Stroganov, E.V. Kireev, M.A. Rogunova, S.G. Kostromin, and V.P. Shibaev, Abstracts of the All-Union Symposium on Rheology, Odessa (1990), p. 197.

49. M.V. Piskunov, S.G. Kostromin, L.B. Stroganov, V.P. Shibaev, and N.A. Platé, Makromol. Chem., Rapid Commun. **3**, 443 (1982).

50. Ch. Böeffel, B. Hisgen, U. Pschorn, H. Ringsdorf, and H.W. Spiess, Israel J. Chem. **23**, 388 (1983).

51. L.B. Stroganov, A.E. Prizment, R.V. Talroze, V.P. Shibaev, and N.A. Platé, Preprints of *IUPAC International Symposium on Characterisation and Analysis of Polymers*, Melbourne (1985), p. 530.

52. Yu.V. Molchanov, P.M. Borodin, S. Grande, and A. Löshe, in *Proceedings of the Third Liquid Crystalline Conference of Socialist Countries*, Vol. 1, Budapest (1979).

53. Yu.V. Molchanov, A.F. Privalov, Yu.B. Amerik, V.G. Grebneva, and I.I. Konstantinov, Vysokomolek. Soedin. **A27**, 2206 (1985).

54. L.B. Stroganov, A.E. Prizment, M.V. Piskunov, A.N. Olonovski, and R.V. Talroze, in *Proceedings of XXII ALL-Union Conference "Vysokomolekularnye Soedineniya,"* Alma-Ata (1985).

55. O. Kratky, Kolloid Z. **64**, 213 (1934).

56. H. Roth and B. Krücke, Makromol. Chem. **187**, 2655 (1986).

57. G. Sigaud, Do.Y. Yoon, and A.C. Griffin, Macromolecules **16**, 875 (1983).

58. E.V. Kireev, Graduation paper, Moscow State University, 1986.

59. E.V. Kireev, L.B. Stroganov, T.I. Gubina, R.V. Talrose, and S.G. Kostromin, in *Liquid Crystals and Their Practical Application* (Abstracts of the All-Union Conference), Chernigov (1988).

7

Mesophase of Graphitizable Carbons

H. Marsh and M.A. Diez

7.1 Introduction

Carbon materials are classified as being graphitizable or nongraphitizable. Nongraphitizable carbons are produced from such nonfusing materials as wood, nut-shell, carbons from highly cross-linkaged polymers (e.g., cellulose), thermosetting resins (e.g., poly(furfuryl)alcohol), and nonusing coals. The macromolecular skeletal structure of these materials remains during heat treatment (pyrolysis) because only small molecules are lost during pyrolysis. Resultant carbons are characterized by isotropic properties, high microporosity, and surface areas in the range $500-1500$ m^2 g^{-1}. On the other hand, graphitizable carbons are formed from parent materials, e.g., pitches from petroleum and coal, which pass through a fluid stage during pyrolysis (carbonization). This fluidity facilitates the appropriate molecular mobility of the aromatic molecules of the system, resulting in intermolecular dehydrogenative polymerization reactions to create aromatic, lamellar (disc-like) molecules of size $200-1000$ amu. These molecules "associate" to create a new *liquid crystal* phase, the so-called mesophase. A fluid phase is the dominant requirement for the production of graphitizable carbons. Definitions for carbon materials have been published in 1982 and 1983 [1].

In studies of both types of carbons, optical microscopy, using polarized light with a half-wave retarder plate inserted between the surface of the polished sample and the analyzer, has been recognized as a very useful "tool" and has played a major role in studies of carbonization mechanisms.

Nongraphitizable carbons have a short-range order of stacked constituent lamellar molecules (< 5 nm) arranged randomly in space. They appear isotropic (without optical activity) in polarized light microscopy. Graphitizable carbons have a longer range of order extending from < 1 μm to about 200 μm. These materials appear as anisotropic with optical activity, and also exhibit yellow, blue, and purple areas (optical texture) with an interchange of color on rotation of the specimen, in association with the use of the half-wave

TABLE 7.1. Definitions of the nomenclature to describe the size and shape of the optical texture in polished surfaces of anisotropic cokes [3].

Isotropic	(I)	No optical activity
Very fine-grained mosaics	(VMf)	< 0.5 μm in diameter
Fine-grained mosaics	(Mf)	$< 1.5 > 0.5$ μm in diameter
Medium-grained mosaics	(Mm)	$< 5.0 > 1.5$ μm in diameter
Coarse-grained mosaics	(Mc)	$< 10.0 > 5.0$ μm in diameter
Supra mosaics	(SM)	Mosaics of anisotropic carbon orientated in the same direction to give a mosaic area of isochromatic color.
Medium-flow anisotropy elongated	(MFA)	< 30 μm in length; < 5 μm in width
Coarse-flow anisotropy elongated	(CF)	$< 60 > 30$ μm in length; $< 10 > 5$ μm in width
Acicular flow domain anisotropy	(AFD)	> 60 μm in length; < 5 μm in width
Flow domain anisotropy	(FD)	> 60 μm in length; > 10 μm in width
Small domains, $\sim$ isochromatic	(SD)	$< 60 > 10$ μm in diameter
Domains, $\sim$ isochromatic	(D)	> 60 μm in diameter

D_b is from the basic anisotropy of low-volatile coking vitrains and anthracite.
D_m is by growth of the mesophase from the fluid phase.
Ribbons (R) strands of mosaics inserted into an isotropic texture.

retarder plate. Each color represents a given orientation of the aromatic lamellar molecules. The size and shape of these areas have nomenclatures which are analyzed and compared by Coin [2]. A nomenclature to describe these optical textures is given in Table 7.1 [3]. The origin of this anisotropy, developed in pyrolyzing pitch, is attributed by Brooks and Taylor [4,5] to the formation of nematic liquid crystals during pyrolysis of these coals, pitches, and model polyaromatic compounds, following the initial work of Taylor [6]. He made initial observations in the Wongawillie coal seam in New South Wales, Australia. He studied induced effects in the coal seam of an igneous dyke which passed vertically through the seam subjecting the coal to a slow carbonization process over distances approaching one kilometer. In coal samples approaching the dyke, Taylor [6] observed the development of small anisotropic spheres, initially micrometer size, in the vitrinite. The spheres, in polished section, exhibited highly reflective optical properties which were consistent with the anisotropic stacking of the constituent lamellae parallel to an equatorial plane. These spheres grew in size at the expense of the vitrinite and coalesced to form anisotropic-coke on approaching the dyke. From these observations, Taylor [6] attributed the origin of the spheres to aromatic, discotic, nematic liquid crystals (mesophase). Further, he recognized that these anisotropic spheres offered the first evidence to understand the dependence of anisotropic carbon formation from coals and pitch materials upon *liquid crystals* as precursor [5].

7.2   Discotic Nematic Liquid Crystals

Liquid crystals systems have been recognized since 1888 although, before 1968, they were often just a curiosity in the laboratory. The term was originally used to describe systems obtained by melting substances such as cholesteryl benzoate which had unusual fluid properties. Such systems possessed more stuctural order than found in normal isotropic liquids but were not "true" crystals. This term "liquid crystal" is used to describe these systems. Friedel [7] criticized this term in a debate on nomenclature and suggested "mesophase" as a better term describing this intermediate state. But, mesophase is rather too imprecise a term itself. However, its use as a description of an intermediate phase in carbon formation is now firmly established.

Of the several categories, into which liquid crystals are classified, it is the aromatic, discotic (lamellar) nematic systems which have relevance in carbonization processes. To understand the formation and properties of the mesophase, it is necessary to consider briefly the properties of "conventional" liquid crystals and to establish the main differences with the liquid crystals of the carbonaceous mesophase.

Whereas with conventional liquid crystals, heating and cooling bring about a phase change, at reproducible temperatures, of all the material in the sample, e.g., nematic to isotropic or nematic to smectic, for mesophase from pitch the mesophase is actually manufactured within the pyrolyzing pitch and is still chemically very active. Thus these systems are never in a state of chemical equilibrium; conversion of isotropic pitch to mesophase occurs over a range of temperature.

The exact mechanism of the phase change from pitch to the "crystallinity" of graphitizable carbon is difficult to study in detail because of the high temperatures at which it occurs (673–773 K); the chemical instability of the polyfaromatic organic compounds, which are formed during the process, as well as the extreme complexity of the molecular composition within the pitch. The use of pure compounds as standards or models is very useful in understanding the mechanism of the phase change. The first example of a model discotic compound forming liquid crystals was observed for a hexasubstituted ester of benzil [8].

From optical, thermodynamic, and x ray studies, a new type of liquid crystal was suggested that was different from the classical or conventional nematic liquid crystals. The structure proposed is shown in Fig. 7.1 in which *discs* are stacking aperiodically in columns and these columns have a two-dimensional array [8]. This model has a two-dimensional order but with a liquid-like statistical disorder in the third dimension.

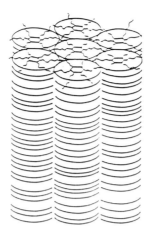

FIGURE 7.1. Schematic representation of the structure of a columnar liquid crystal. The discs are irregularly spaced in columns [8].

Later, the model was described as "canonic" [9,10], "semicolumnar" [11], or "discotic" [12]. Other liquid crystal systems, based on disc-shaped mesogens are reported [13–15]. The canonic or columnar structure has been confirmed using x ray analysis [16]. Chandrasekhar has reviewed the physics of these new systems [17].

Growth units of discotic, nematic liquid crystals which are formed within the fluid phase on the carbonization process are detected using transmission electron microscopy (TEM) at a size of < 0.1 μm [18]. The initial stacking of the molecules has been examined using TEM and the selected area diffraction (SAD) [19,20]. From high magnification, lattice-fringe images of carbonaceous systems [20], the aromatic molecules can be seen to stack in columns which are in close contact with each other and in compact arrangements. These columns are not interconnected and are curved and distorted. This observation appears to indicate that carbonaceous nematic liquid crystals could be columnar. However, this order is fragile and is only temporal. On further heat-treatment of the pyrolyzing system, they adopt the structure of discotic, nematic liquid crystals by the molecular flow involved during the subsequent coalescence and deformation processes. They become spherical because of the minimum surface energy requirements. The transition can also be seen using a hot-stage polarizing light microscopy as the formation of small optically anisotropic spheres (1 μm approx.) [21].

It is now recognized that the mesophase is a plastic, anisotropic phase which grows from the fluid pyrolysate and solidifies to give an anisotropic coke, i.e., a graphitizable carbon. The mesophase possesses a nematic liquid crystal structure and this imparts its plastic properties [22], its high degree of anisotropy [5], the effects of magnetic fields which align constituent molecules of the mesophase [23], the orientation of molecules parallel to contained surfaces [24], and of the eutectic effects [25].

These properties are well established for conventional liquid crystals. Other evidence indicates that the mesophase (intermediate phase formed during the pyrolysis of pitch and coal prior to coke formation) has physical properties that are similar, but has very different chemical properties. For example, the conventional liquid crystals have reversible "thermotropic" properties. They exhibit a reversible phase change from an isotropic liquid phase to the nematic liquid crystal phase on cycling around the phase-transition temperature; the mesophase shows only some initial reversibility. However, in the long term, this process becomes irreversible. The liquid crystals formed in heat-treated pitch are chemically reactive and ultimately produce solid carbon (coke).

Lewis [26], using the technique of hot-stage microscopy, demonstrated the initial thermotropic behavior of liquid crystals in a pitch produced by the high-pressure pyrolysis of naphthalene. On maintaining an isothermal heat-treatment temperature of 623 K, small anisotropic spheres (0.5 μm in diameter) of mesophase were formed in the isotropic matrix pitch. On rapid heating from 623 K to 723 K, the spheres disappeared into the isotropic matrix pitch, and only when the system was cooled to 623 K did the spheres reappear. In this experiment, Lewis demonstrated an important property of nematic liquid crystals, i.e., reversible formation, because the discotic molecules are kept together only by physical forces. However, this phenomenon could be observed only for a few cycles of heating and cooling. When the carbonization process progressed, the system lost its reversibility, because of chemical reactions between molecules in one layer and probably between molecules in different layers, i.e., cross linking.

In summary, the three principal differences between the carbonaceous discotic, nematic liquid crystals and conventional rod-like nematics is summarized as follows:

(i) Conventional liquid crystals are chemically stable whereas carbonaceous discotic, nematic liquid crystals (mesophase) are chemically reactive. The constituent molecules undergo a process of polymerization between molecules in one layer and possibly by molecules in different layers (cross linkage).

(ii) Conventional liquid crystals consist of aligned rod-like molecules whereas the mesophase is composed of disc-like polyaromatic molecules.

(iii) For conventional liquid crystals, the phase change from isotropic liquid to liquid crystal occurs as a consequence of the reduction of the kinetic energy (molecular mobility) by lowering the temperature; in carbonaceous systems of mesophase, with an increasing pyrolysis temperature and an increasing average molecular size, a point is reached where the kinetic energy of the molecule is insufficient to prevent the "association" of these large molecules (by van der Waals forces) into the format of a nematic liquid crystal. Transition temperatures are not constant for the mesophase.

7.3   Formation of the Mesophase

Materials which form the mesophase have a wide variation of initial chemical composition and a complexity of molecular structure. Each of these materials, e.g., petroleum and coal-tar pitches, contains several hundreds of individual compounds. Initial extents of aromaticity vary considerably from about 7% of aromatic hydrogen in Gilsonite pitch [27] to over 90% in coaltar pitch [28].

To simplify the study of mesophase development, in relation to the chemistry of carbonization, experiments have been carried out using single organic compounds and their mixtures, as model compounds [29,30], as well as carbonizations of separated fractions of petroleum feedstocks [31,32]. Although the carbonizations of single organic compounds are as complicated in terms of molecular structure as are coal-tar and petroleum pitches, these studies do give a better understanding of the carbonization mechanisms.

It is established that the first stage in the formation of both isotropic and anisotropic carbons from carbonaceous precursors is pyrolysis. When systems are pyrolyzed in an inert atmosphere, the rupture of C—C, C—O, and C—H bonds and the loss of noncarbonizing volatile matter occur. The recombination of reactive fragments in the fluid phase give high molecular weight planar polyaromatics. This stage of pyrolysis is accompanied by the presence of free radicals. The relative proportions of the two types of free radicals formed (reactive and stable) influence the development of the order and alignment of planar aromatic molecules [33]. The structure and reactivity of the reactive free radicals control the pyrolysis process. Their influence on the pyrolysis process is described below.

Mesophase is formed only when quite restrictive chemical and physical conditions are operating within a pyrolyzing system. Initially, the translational energies of constituent molecules exceed cohesion energies with the isotropic fluid phase approaching Newtonian properties. When the size of the constituent polyaromatics (containing aliphatic side-chains and unpaired electrons) is ~ 2.5 nm in diameter, the molecular weight is 500–2000 amu [41], the C/H ratio is close to 2 [22], and they can be considered as "discs." Figure 7.2 shows the model of constituent molecules in the mesophase pitch as proposed by Zimmer and White [35]. When the association of constituent molecules becomes energetically favorable, molecules of this type, and larger, become bound to each other, surface to surface, by the van der Waals dispersion forces. Thus, the origin of the discotic, nematic, liquid phase is one of homogenous nucleation. The development process is not a "seeded" one, as with conventional inorganic/organic crystals.

In the early stages of pyrolysis, the "association" times are very short

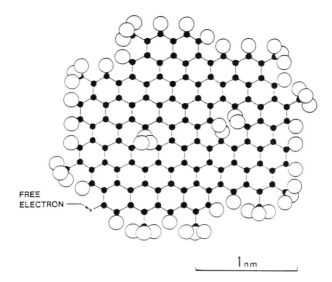

FIGURE 7.2. Model of a planar aromatic stable free radical of the type that forms lamellar nematic liquid crystals [35].

($\approx 10^{-8}$ s) due to the mobility of the molecules (kinetic energy), and no liquid crystals are formed. With continuing pyrolysis, the time of association of the mesogen molecules of similar size and shape increases and, consequently, growth units of molecules are formed which are eventually seen in the optical microscope as anisotropic, reflecting spheres of 1 μm in diameter. The thermotropic properties of "classical" liquid crystals are not usually observed. This is because polymerizations occur within mesophases such that cohesion energies always exceed translational energies. In addition, chemical cross linkages between constituent mesogens of the growth units also thermally stabilize them against dissociation.

Figure 7.3 shows a model diagram representing mesogen molecules and structure of the mesophase (Fig. 7.3(a)), an optical micrograph of mesophase spheres in an isotropic matrix (Fig. 7.3(b)), and a scanning electron microscopy (SEM) micrograph (Fig. 7.3(c)) of mesophase spheres from acenaphthylene, HTT 550 °C.

Greinke and Singer [36] reported molecular weight changes during the pitch pyrolysis and mesophase growth. Using high-temperature centrifugation to separate the coexisting phases during the transformation to mesophase, they studied the separated phases by size-exclusion chromatography and other analytical techniques. They established that the molecular weight distribution (MWD) of the mesophase did not change during formation of the mesophase and that both phases contain similarly sized molecules in different proportions. This phenomenon can be explained by the polymerization of smaller molecular weight materials (400–1000 amu) within the iso-

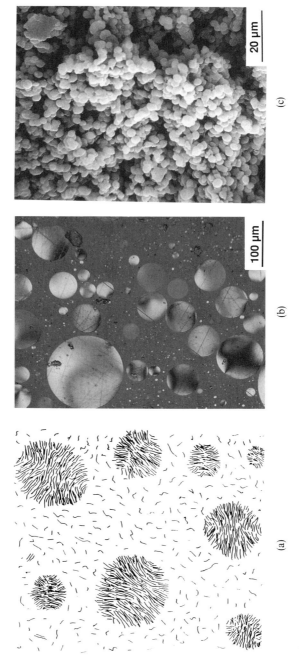

FIGURE 7.3. (a) Schematic diagram of a discotic nematic liquid crystal sphere in a liquid pyrolysate; (b) an optical micrograph of mesophase spheres; and (c) a SEM micrograph of clustered mesophase spheres.

tropic phase, the selective transfer of these molecules (400–1000 amu) into the mesophase, and also by the stability within the isotropic phase of larger sized molecules (> 1000 amu). However, the mesophase, as estabished, is chemically reactive and further polymerization/condensation/cross-linkage reactions occur, with increasing time and temperature, until the mesophase ceases to be a fluid and becomes plastic and finally a viscoelastic solid (coke).

7.4   Structure in the Mesophase

During the very initial growth stages of mesophase development, the discotic mesogen stacks initially in columns [37]. As a requirement of minimum surface energy, a spherical shape to the growing mesophase then becomes established. The constituent lamellar molecules of the sphere are parallel to each other and parallel to an equatorial plane as indicated by electron diffraction [5]. Optical microscopy indicates that the layers become orientated toward the poles of the spheres, so that they are perpendicular to the interface with the isotropic matrix (Fig. 7.4). This morphology has been identified in spheres as small as a few microns, but these spheres seldom develop beyond 50 μm diameter without developing more complex internal structures [35] as a result of coalescence (Section 7.5). The nematic liquid crystal structure is clearly confirmed by phase contrast, high-resolution electron microscopy (HREM) studies of the mesophase and graphitic carbon [20].

Although the Brooks and Taylor model [5] is the most common morphology for mesophase spheres, other mesophase spheres with different internal molecular orientations of constituent molecules have been described. These different morphologies are dependent on carbonization conditions, the presence of inerts and/or additives, and other factors which influence the process of mesophase growth. Unusual optical effects in mesophase spheres contained in pitches from pyrolysis of model compounds (anthracene and naphthalene) with anhydrous aluminum chloride catalyst have been observed [38,39]. Here, the lamellae are orientated parallel to the equatorial plane

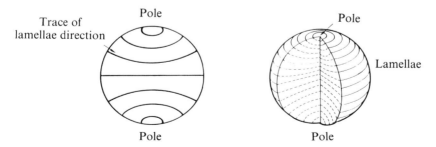

FIGURE 7.4. Morphology of mesophase spheres of the Brooks and Taylor model [5].

within the center of the sphere and are parallel to the surface of the sphere. This difference in structure is believed to be due to interactions at the interface of the isotropic pitch/mesophase.

Other types of morphology have been proposed by Honda and coworkers [40–42] as observed in heat-treated products of quinoline-soluble portions of coal-tar pitch containing carbon black particles. Observations suggested the existence of spheres which have the lamellae stacked perpendicular to a polar diameter and aligned to the surface of the sphere and parallel to the interface with the isotropic matrix. Honda and coworkers. [40–42] suggested that this new type of mesophase sphere may be produced via the influence of, for example, particles of carbon black. So, particles which adhere to surfaces have an influence on the orientation of layers of the aromatic molecules.

A fourth type of mesophase morphology was reported by Hüttinger [43]. He found them in coal-tar pitches containing small amounts of carbon black, like the Honda model. Initially, these particles act as a nucleus (an unusual phenomenon) and are the origin for the mesophase formation (Fig. 7.5). The lamellae are orientated concentrically around the carbon black particle, having the symmetry of an onion [40,41,44,45].

The structures within the mesophase described above are for the sphere. But, under specific carbonization conditions, the mesophase adopts other different shapes. A "spaghetti" mesophase is formed from the carbonization of anthracene at the high pressure of 250 MPa and at 763 K [46]. The effect of this pressure is to increase the viscosity of the mesophase and to prevent coalescence, Fig. 7.6(a). These elongated units change shape to spheres as the viscosity is lowered by increasing the carbonization temperature, passing through an intermediate ovoid shape. Figure 7.6(b) shows the mesophase from a carbonized mixture of anthracene and phenanthrene [3:7]

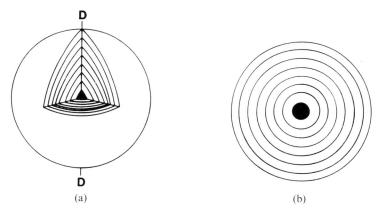

(a)   (b)

FIGURE 7.5. Morphology of mesophase spheres of the Hüttinger model [43]: (a) a sectioned sphere showing lamellae and carbon black nucleus; and (b) a cross section on DD.

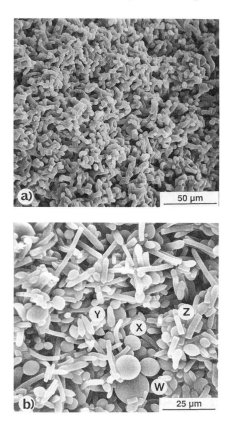

FIGURE 7.6. SEM micrograph of the mesophase: (a) "spaghetti" shape, anthracene, HTT 763 K, 5 K min^{-1}, 0.5 h soak time, 250 MPa; and (b) mixed shape, anthracene:phenanthrene (3:7), HTT 823 K, 5 K min^{-1}, 300 MPa [46]. [Z] "spaghetti" shape; [Y] "ovoid" shape; [X] "lemon" shape; and [W] spherical shape.

300 MPa and 823 K, with elongated ("spaghetti") (position Z), ovoid (position Y), "lemon" (position X) and spherical shapes (position W). The mesophase with different shapes, including "cylinders" which have converted into noncoalesced spherical shapes within the sample (of lower surface energy), arises because of small differences in temperature that critically affect viscosity.

In summary,

(i) the morphology of the mesophase (different alignment of the molecular layers and shape) depends on the experimental conditions of the carbonization process and also on the presence of foreign material particles in the system; and

(ii) the mesophase spheres with the Brooks and Taylor morphology [5] are the most common from a wide range of conditions of temperature, pressure, and time (duration) of carbonization.

7.5   Growth and Coalescence

The stages following the formation of anisotropic mesophase spheres within a pyrolyzing pitch are the growth and coalescence of these spheres. With increasing time and/or temperature of carbonization, the size and molecular weight of the planar polyaromatic molecules increase. These then become incorporated into mesophase spheres which increase in size under normal conditions of pressure and temperature. If the carbonization process is relatively rapid, then the mesogen molecules form new spheres. Otherwise, the growth of spheres continues until contact is made with others, when they "instantaneously" form a larger sphere. This phenomenon is known as "coalescence" (Fig. 7.7).

Coalescence is a function of the viscosity of the polymeric mesophase which must remain sufficiently deformable (low viscosity/high plasticity) to respond to the requirements of minimum surface energy. Smith et al. [47] estimate the interfacial energy of the mesophase/isotropic phase, from the Frenkel relationship, to be about 0.02 dyn cm^{-1}.

When spheres of low viscosity mesophase coalesce, the structure within the resultant unit is rapidly reorganized. The new mesophase units adopt the spherical shape with larger size but do not retain the detail of the Brook and Taylor morphology [5]. As coalescence progresses, the mesophase spheres become more viscous. Viscosities possessed by the mesophase at temperatures where coalescence occurs are a function of two properties of the mesophase:

 (i) the extent of polymerization or cross-linkage; and
 (ii) the extent of the nonplanarity of the constituent molecules.

The higher the extent of polymerization, the higher the viscosity of the mesophase. The extent of polymerization is dependent on the chemical reactivity of the constituent molecules. If this reactivity is high, then at comparatively low temperatures the extent of polymerization is extensive and the mesophase has little plasticity (high viscosity).

When two spheres coalesce, each with the initial growth structures of equatorial stacking, there occurs mutual deformation of this equatorial stacking which is not reestablished in the enlarged sphere (Fig. 7.7). Examples are in Fig. 7.7, which are optical micrographs of polished surfaces from the carbonization of petroleum feedstock at 713 K. Figure 7.7 shows small spheres with the Brooks and Taylor morphology and also small spheres coalescing with this morphology (position Y). In position X (Fig. 7.7), larger spheres have lost their initial structure becoming, structurally, more complicated coalesced spheres [48].

When the coalescence of spheres takes place toward the end of the process

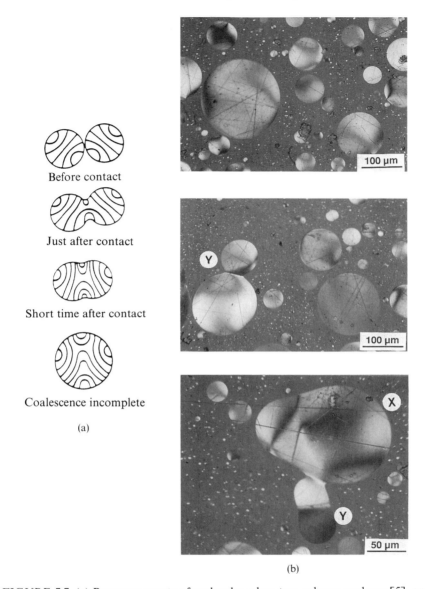

Before contact

Just after contact

Short time after contact

Coalescence incomplete

(a)

(b)

FIGURE 7.7. (a) Rearrangements of molecules when two spheres coalesce [5]; and (b) optical micrographs of mesophase spheres from the carbonization of petroleum feedstock at 713 K [48]. [Y] spheres with the Brooks and Taylor morphology during the coalescence process and [X] spheres with complexity structural.

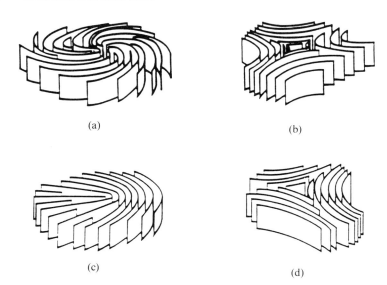

FIGURE 7.8. Wedge disclinations found in coalesced mesophase pitch and aniso-tropic carbons: (a) $+2\pi$; (b) -2π; (c) $+\pi$; and (d) $-\pi$ [49].

of conversion from the pitch to bulk mesophase, lamellar stacking imperfec-tions (disclinations) are introduced into the bulk fully coalesced mesophase "phase." Such disclinations are mobile within the plastic mesophase. As a result of plastic deformation within the mesophase (e.g., during flow) some of these disclinations are eliminated. However, others remain in the structure of the coke throughout further calcination and graphitization treatments. Their type and frequency govern the resultant physical and mechanical properties of carbons, i.e., the influence on the strength of the resultant cokes/graphites in terms of crack propagation mechanisms [22]. Disclinations in the car-bonaceous mesophase are reported by White and Zimmer [49]. Figure 7.8 illustrates wedge disclinations found in coalesced mesophase and anisotropic carbons [22]. As an example, Fig. 7.9 is a SEM micrograph of etched surfaces of petroleum coke in which "node," denoted $(+\pi, -\pi)$ (positions Z and X) and "crosses" as $(+2\pi, -2\pi)$ (position Y), can be observed.

7.6   Formation of Anisotropic Carbons

The structure within coke (graphitizable carbon) is governed by the proper-ties of the mesophase at the time of its solidification and these are dependent on the parent feedstocks. As commented above, the generation of mesophase within a pyrolyzing system occurs if several requirements are met:

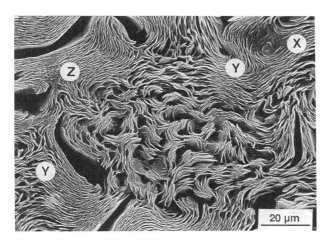

FIGURE 7.9. SEM micrograph of the etched surface of a petroleum coke showing wedge disclinations: $[Z]\ -\pi$; $[Y]\ -2\pi$; and $[X]\ +\pi$.

(i) the intermolecular reactivity of the constituent molecules is constrained to limit the molecular growth of polyaromatic molecules, not exceeding ~ 900 amu, on average [36];

(ii) the system must remain fluid to temperatures of $400°-450\ °C$, such that the constituent polynuclear aromatic molecules have enough mobility (high fluidity within the system) to enable the establishment of a physical liquid crystal system; and

(iii) the intermolecular reactivity of constituent molecules of the mesophase must facilitate growth, coalescence, and movement within the fluid phase prior to solidification.

If the reactivity of the molecular systems is too high, randomly orientated polymerization and cross-linkage reactions occur rapidly at temperatures which are too low for the establishment of a liquid crystal system. It then solidifies to an isotropic carbon [50,51], e.g., carbons prepared from phenolic resin.

Although the pyrolysis chemistry of parent carbonaceous materials in forming anisotropic carbons is complicated, factors describing the control of the mesophase are known. Variations in the viscosity of the fluid system with increasing pyrolysis temperature explain differences between the optical texture of resultant cokes. Figure 7.10 shows this relationship [52]. In systems which have a relatively high viscosity (PQ) and a short temperature range at minimum viscosity (QR), the mesophase spheres are too viscous to grow or coalesce and so fuse on contact with rapid solidification (RS) to produce a coke of fine-grained mosaics ($< 5\ \mu m$). On the other hand, highly aromatic

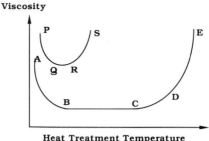

FIGURE 7.10. Variation between the viscosity of a pyrolyzing system and the heat treatment temperature [52].

feedstocks possess relatively low viscosity (AB) and a constant minimum viscosity is maintained over a wide range of temperature (BC) during the first stages of pyrolysis. At this stage, the mesophase spheres adopt the spherical shape with the morphology of Brooks and Taylor [5]. With increasing temperature the coalesced "bulk" mesophase becomes the dominant phase and cross-linking between lamellae and an increase in viscosity occur (CD), leading to solid coke (DE) which shows larger sizes of optical texture as domains (> 50 μm).

Figure 7.11 shows the relationship between minimum viscosity during the pyrolysis of pitch and the size of optical texture in the resultant cokes.

In summary, transitions from the smallest textures (mosaics) to the largest textures (domains) are associated with the intermolecular reactivity which is dependent upon:

  (i)   more aromatic feedstocks;
 (ii)   the absence of alkyl side chains;
(iii)   the absence of reactive functional groups; and
 (iv)   the absence of heteroatoms within the molecules.

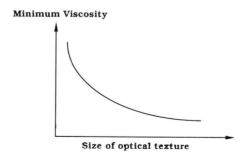

FIGURE 7.11. Variation in size of the optical texture of coke with minimum viscosity in pyrolisis of pitch.

7.7 Factors Influencing the Formation of Anisotropic Carbons

7.7.1 Chemical Factors

Role of Free Radicals

Electron spin resonance (ESR) indicates the presence of unpaired electrons in the pyrolysis chemistry which involves the formation of free radicals as intermediates. The role of "reactive" transient free radicals in pyrolysis is of interest because of their influence on the kinetics and mechanism of the process as reviewed by Lewis and Singer [33]. These authors proposed that "reactive" transient free radicals with low molecular weights act as intermediates in the formation of relatively stable, higher molecular weight free radicals and planar aromatic molecules. Stable free radicals are incorporated into the mesophase whereas transient free radicals influence the development of the mesophase.

Consequently, the transition from a smaller to a larger size of optical texture is associated with a decrease in activity of the "reactive" transient free radicals [53]. Consequently, consideration of the role of the hydrogen transfer reactions and the availability of transferable hydrogen in terms of the stabilization of reactive radicals in mesophase development have proved successful. Figure 7.12 shows a scheme of the influence of the hydrogen transfer reactions during pyrolysis of petroleum residues resulting in either isotropic or anisotropic carbon [53].

Evidence exists indicating that a relative increase in stable free radical concentration during pyrolysis results in less order in resultant carbons [54,55]. A linear increase in free radical concentration with time, as polymerization continues, indicates that free radical concentrations are directly related to the molecular weight of the polymerized product. In attempts to

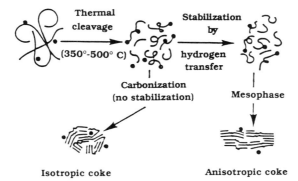

FIGURE 7.12. Schematic diagram of the carbonization process leading to isotropic and anisotropic carbons. The influence of hydrogen transfer reactions [53].

relate ESR data with mesophase formation, little change was found in free radical concentration during the stage of mesophase formation. The most rapid increase in free radical concentration occurs when the bulk mesophase is formed [56]. This is confirmed by others [27] in studies using such techniques as mass spectrometry (MS) and nuclear magnetic resonance (NMR). At this stage, two competitive processes coexist:

(i) mesophase formation following recombination of molecular fragments into polynuclear aromatic mesogens (decreased free radicals concentration); and

(ii) simultaneous dehydrogenation reactions (increased free radicals concentration).

Influence of Heteroatoms

The effect of heteroatoms on the pyrolysis processes, using ESR, has been studied. The presence of heteroatoms, such as oxygen, increases the spin concentration with increasing oxygen content and the size of the resultant optical texture is reduced [57].

Heteroatoms (S, O, and N) and functional groups of molecules in the pyrolysis system influence the chemical reactivity and, consequently, the mesophase formation [29]. Extensive studies of the effects of heteroatoms on the development of optical texture exist and are described in terms of the chemical reactivity modifications to the carbonizing system. Following small additions of sulphur (< 5% mole), a dehydrogenative process occurs between molecules and hydrogen is removed as H_2S. The resultant product polymerizes to form a mesophase with a higher coke yield, without any significant decrease in the graphitizability of the resultant anisotropic coke. However, if the addition of sulphur is higher (> 5% mole) the ability to form cross-linkages increases and thio-bridges with sulphur retention within the new mesogens is extensive and random. The fluidity of the system is reduced, preventing the growth of the large mesogen molecules and, consequently, reducing the optical texture of the resultant coke which becomes more disordered. When about 10% mole of sulphur is added, a nongraphitizable isotropic carbon results [29].

The effects of additions of model compounds containing heteroatoms (N, O, and S) to pitches and acenaphthylene on carbonization processes have also been studied [58]. When N- or S-containing compounds are incorporated to the system, mesophase growth was inhibited. Similarly, additions of compounds containing oxygen such as phenols result in smaller optical textures of the product carbons. Marsh et al. [58] proposed a mechanism involving the formation of reactive free radicals or the formation of ether and peroxide linkages. On the other hand, some compounds containing oxygen, such as quinones, were found to enhance mesophase growth. The mechanism could involve hydrogen transfer and radical stabilizing reactions.

Recently, studies of the oxidation of coal asphaltenes concluded that amounts < 6 wt.% could increase the coke yield without causing any decrease in terms of the optical texture size [59]. The asphaltenes were oxidized by two methods:

(i) heat treatment in air at 200 °C for 24 h; and
(ii) oxidation over chromyl chloride.

However, if higher amounts than this limit (6 wt.%) are added, the coke yield is increased but the graphitizability of the precursor is destroyed [59].

Influence of Metals

Metals catalyze pyrolysis reactions and so influence the development of the optical texture of the resultant coke.

Mochida et al. [60] showed that additions of alkali metals (lithium, sodium, and potassium) in high concentrations inhibit the development of the mesophase and produce an isotropic carbon. The metals catalyze dehydrogenation reactions to such an extent that the resulting condensed polyaromatic molecules increase the melting point of the pyrolzing system, and so reduce the temperature range of the liquid phase required to form graphitizable carbon. That is, the necessary mesogens are established at temperatures too low to give sufficient mobility to these mesogens to orientate themselves into nematic liquid crystal structures.

Aluminum chloride [38,61] and ferrocene [62] have the same effect on the development of the optical texture of cokes. However, for some systems of low intrinsic reativity, the use of catalysts is beneficial to create mesogens at temperatures appropriate to liquid crystal generation.

Inerts: Quinoline Insolubles

The properties of parent materials such as the extent of aromaticity, molecular size distribution, and primary quinoline insolubles (QI) influence carbonization behavior. QI material can conveniently be classified into several types. Primary QI are usually found in coal-tar pitches in different amounts and they are produced in the carbonization process. Coal-tar pitches contain up to 12 wt.% of primary QI; petroleum pitches do not contain QI. QI material contains small particles, usually < 1 μm in diameter, and resembles carbon black. The morphology of QI material has been described by Marsh et al. [63].

These primary QI materials are usually referred to as the "inerts" content of pitches. However, the so-called "inerts" must not be considered to be "inert" during the carbonization process of the pitches. They have a very important and significant role, because the coalescence and growth of the mesophase is very sensitive to the presence of these particles during the process. During the growth of the mesophase from the pyrolyzing system, these primary QI particles adhere to the surface of the mesophase sphere. This phenomenon modifies the shape and prevents growth and coalescence, and smaller optical textures are generated in the resultant coke.

7.7.2   Physical Factors

Not only do chemical factors influence the development of the mesophase, other physical factors exist such as: the rate of heating, soak time, temperature, and pressure. Such factors are intimately related to the final coke structure. Variations in these carbonizing parameters influence the evolution of the volatile matter produced during the process and this controls the viscosity of the system.

An increase in the heating rate influences coke quality and results in greater perfection of the graphitic structure [64]. Higher heating rates allow higher temperatures to be reached in a shorter time, and this influences the rates of evolution of volatile matter. Some volatiles are retained in the system and promote additional fluidity (decreased viscosity), and hence allow a better mobility of the molecules leading to promoted growth and coalescence of the mesophase.

Similar explanations are applied to the effects of such factors as time and temperature which are not interdependent [65]. The maximum size of the optical texture and coalescence results if the mesophase is formed under conditions which provide a minimum viscosity in a short time. Studies of the

FIGURE 7.13. SEM micrograph of the botryoidal spherical mesophase, acenaphthylene, HTT 823 K, 5 K min^{-1}, 0.5 h soak time, 260 MPa.

influence of carbonization parameters (temperature, time, and pressure) on petroleum residues have been described by several authors [66–70].

A pressurized carbonization creates a closed system, so preventing loss of volatile matter. Hence, variations of carbonization pressure also modify the final coke structure. Forrest and Marsh [71] found that it increased the size of the mesophase spheres due to the retention of the organic volatile material which would normally be lost as volatiles. These volatiles increase fluidity in the system and then facilitate the growth and coalescence as explained above.

However, too great an increase in pressure (e.g., 300 MPa) reduces anisotropy due to hindrance to the mesophase growth and coalescence. Mesophase spheres tend to form agglomerations. The resultant appearance of the carbon has been described as "botryoidal" [72] (Fig. 7.13). Pressures less than 0.1 MPa produce a reduced size of optical texture because of increased viscosity in the pyrolyzing system as a result of loss of small molecules as volatiles [73].

7.8   Graphitization

Heat treatment of graphitizable carbons to high temperatures (in excess of 3000 K) results in the formation of three-dimensional polycrystalline graphite, with an essentially hexagonal structure.

Anisotropic carbons (cokes) heat treated typically, 1000–1500 K, contain

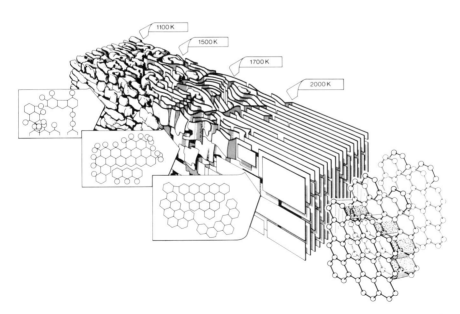

FIGURE 7.14. Diagram indicating changes in the structure of lamellae of anisotropic carbons on heat treatment [74].

many structural defects and heteroatoms, mainly hydrogen and oxygen. During the graphitization process, the nematic stacking of structures established in these types of carbon is progressively perfected. By imaging, using phase contrast electron microscopy [74,75] for a range of carbons at different heat-treatment temperatures up to 2000 K, changes in the stacking sequences of aromatic lamellae with graphitization are indicated. The model proposed to describe changes in internal and intralamellar reordering of the polyaromatic lamellae is shown in Fig. 7.14 [74].

With a heat-treatment temperature of 1100–1500 K, the carbon formed via the mesophase consists of imperfect lamellae with many defects such as side chains, heteroatoms, and stacking defects. At temperatures of 1500–1700 K, the lamellae are more ordered and perfect than initially, but they are still not perfect (see Fig. 7.13). By a heat treatment of 1700 K, the lamellae are now parallel. The lamellae are 10–22 nm in length and the interlayer spacing approaches that of the graphite. This type of material has been described as "graphitic carbon." On heating graphitic carbon to temperatures in excess of 2000 K the lamellae increase in length, the perfection of stacking also increase, and the three-dimensional ordering of graphite is developed.

7.9   Industrial Relevance of the Mesophase: Discotic, Aromatic, and Nematic Liquid Crystals

The awareness that structure in the many forms of commercial graphitizable carbons is a consequence of different development modes of the mesophase in pyrolyzing liquid crystal systems [4,5] made a tremendous impact on the carbon/graphite-producing industries [77]. Not only was the scientific control of established procedures improved, but new processes were created. The history of these effects is of textbook proportions but can be outlined as follows.

7.9.1   Delayed Coking

From an industrial point of view, the piece of equipment used to convert pitch feedstocks to a graphitizable carbon (green-coke) is the delayed-coker. This consists basically of a drum which is charged with several hundred tonnes of preheated petroleum feedstock ($> 300\,°C$), subsequently heated further to $450°–475\,°C$ and kept at this temperature for about 24 h. After this time, the feedstock becomes converted, via a liquid crystal mesophase, to green-coke. The quality of green-coke, in terms of optical texture is of critical importance. The larger the size of the optical texture, the higher the commercial value of the green-coke. Hence, there exists a strong incentive to optimize the quality and treatment given to the feedstock. The application of "mesophase science" to establish the optimum "window" of the delayed coking process is a success story.

There exist two general qualities of green-coke:

(i) regular- or sponge-coke with optical textures, generally of coarse-grained mosaics (10 μm in diameter) and medium-flow anisotropy (< 30 μm in length); and

(ii) needle-cokes with acicular flow domain anisotropy (> 60 μm in length, < 5 μm in width).

The regular cokes find use as filler-cokes in anode production for use in "Hall cells" to produce aluminum from the electrolytic reduction of aluminum oxide. Calcined needle-cokes act as filler-cokes in the manufacture of high-performance graphite electrodes used to transmit electrical power to crucibles for steel production.

7.9.2   Metallurgical Coke

Metallurgical coke is used in blast furnaces. Its role there is:

(i)   to support the iron oxide ore burden (high strength);

(ii)   to act as the carbon source for the reduction of iron oxide ores (suitably low reactivity); and

(iii) to provide thermal energy.

Metallurgical coke must have an optical texture of fine-grained mosaics (1.5 μm in diameter). The theory of the carbonization of coals, including the use of coal blends and of additives (hydrogen donors [53]) has benefited the coal-coking industry considerably [78–80].

7.9.3   Carbon Fibers

The last two decades have seen the development of a new form of carbon, i.e., the carbon fiber. Polyacrylonitrile (PAN) was developed intensively, initially in [81]. This gives nongraphitizable, isotropic carbon fibers which have very successful applications in composite systems, both with polymers and with carbon as matrix materials. The concept of spinning mesophase into carbon fibers [81], to take advantage of the prealignment of molecules in the liquid crystal structure to obtain high strength, high-modulus fiber, has been actively pursued, mainly in Japan.

7.9.4   Graphite Production

The graphite industry produces many forms of graphite for many types of application, varying from electrodes (brushes) for electrical motors, to electrodes for batteries, and to moderators and structural components in the nuclear power industry. The carbonization of coal-tar pitch, with its quinoline-insoluble content and the influence of filler-coke properties has benefited from advances in knowledge of the mesophase, liquid crystal chemistry.

7.9.5   Carbon Blacks

These soot-like particles are formed by the vapor phase pyrolysis/partial combustion of hydrocarbons. The mechanism of formation involves the generation of liquid droplets (clusters of atoms). The establishment of the concentric, onion-like arrangement of constituent molecules within the droplet, with parallelism of the structure to the surface of the droplet, is most probably via a liquid crystal mechanism.

References

1. International Committee for Characterization and Terminology of Carbon, Carbon **20**, 445 (1982); Carbon **21**, 517 (1983).

2. C.A. Coin, Fuel **66**, 702 (1987).

3. H. Marsh, *Ironmaking Proceedings of the Iron and Steel Society of AIME*, Warrendale, PA, **41**, 2 (1982).

4. J.D. Brooks and G.H. Taylor, Carbon **3**, 185 (1965).

5. J.D. Brooks and G.H. Taylor, in *Chemistry and Physics of Carbon*, edited by P.L. Walker, Jr. (Marcel Dekker, New York, 1968), Vol. 4, p. 243.

6. G.H. Taylor, Fuel **40**, 465 (1961).

7. G. Friedel, Ann. Physique **18**, 273 (1922).

8. S. Chandrasekhar, B.K. Sadashiva, and K.A. Suresh, Pramana **9**, 471 (1977).

9. F.C. Frank and S. Chandrasekhar, J. Phys. (Paris) **41**, 1285 (1980).

10. W. Helfrich, in *Liquid Crystals*, edited by S. Chandrasekhar (Heyden, London, 1979), p. 7.

11. W. Helfrich, J. Phys. (Paris) Coll. **40**, C3-105 (1979).

12. J. Billard, J.C. Dubois, N.H. Tinh, and A. Zann, Nouv. J. Chimie **2**, 535 (1978).

13. J.C. Dubois, Ann. Phys. **3**, 131 (1978).

14. C. Destrade, M.C. Mondon, and J. Malthete, J. Phys. (Paris) Coll. **40**, C3-17 (1979).

15. N.H. Tinh, J.C. Dubois, J. Malthete, and C. Destrade, C.R. Acad. Sci. (Paris) **C286**, 463 (1978).

16. A.M. Levulet, J. Phys. (Paris) Lett. **40**, L81 (1979).

17. S. Chandrasekhar, Phil. Trans. Roy. Soc. London A **309**, 93 (1983).

18. M. Inhatowicz, P. Chiche, J. Deduit, S. Pregermain, and R. Tournant, Carbon **4**, 41 (1966).

19. M. Monthioux, M. Oberlin, A. Oberlin, and X. Bourrat, Carbon **20**, 167 (1982).

20. D. Augie, M. Oberlin, A. Oberlin, and P. Hyvernat, Carbon **18**, 337 (1980).

21. D.S. Hoover, A. Davis, A.J. Perrotta, and W. Spackman, Ext. Abst. 14th Biennial Conference on Carbon, American Carbon Society, Pennsylvania State University (1979), p. 393.

22. J.L. White, Progr. Solid State Chem. **9**, 59 (1974).

23. Y. Sanada, T. Furuta, and H. Kimura, Carbon **10**, 664 (1972).

24. H. Marsh, Fuel **52**, 205 (1973).

25. H. Marsh, C. Cornford, and G. Hermon, Fuel **53**, 168 (1974).

26. R.T. Lewis, Ext. Abst. 12th Biennial Conference on Carbon, American Carbon Society, Pittsburg (1975), p. 215.

27. H. Marsh, J.W. Akitt, J.M. Hurley, J. Melvin, and A.P. Warburton, J. Appl. Chem. **21**, 251 (1971).

28. D. McNeil, in *Bituminous Materials, Asphalts Tars, and Pitches*, edited by A.J. Hoiberg (Interscience, New York, 1966), Vol. 3, p. 139.

29. E. Fitzer, K. Mueller and W. Schaefer, in *Chemistry and Physics of Carbon*, edited by P.L. Walker Jr. (Marcel Dekker, New York, 1971), Vol 7, p. 237.

30. H. Marsh and P.L. Walker, Jr., in *Chemistry and Physics of Carbon*, edited by P.L. Walker Jr. and P.A. Thrower (Marcel Dekker, New York, 1979), Vol. 15, p. 230.

31. V.A. Weinberg, J.L. White, and T.F. Yen, Fuel **62**, 1503 (1983).

32. S. Eser and R.G. Jenkis, Carbon **27**, 877 (1989).

33. I.C. Lewis and L.S. Singer, in *Chemistry and Physics of Carbon*, edited by P.L. Walker Jr. and P.A. Thrower (Marcel Dekker, New York, 1981), Vol. 17, p. 1.

34. J.L. White, Progr. Solid State Chem. **9**, 59 (1975).

35. J.E. Zimmer and J.L. White, *Disclinations Structures in Carbonaceous Mesophase and Graphite*, Aerospace Report (1976).

36. R.A. Greinke and L.S. Singer, Carbon **26**, 665 (1988).

37. A. Oberlin, Carbon **22**, 521 (1984).

38. C.A. Kovac and I.C. Lewis, Ext. Abst. 13th Biennial Conference on Carbon, American Carbon Society, Irvine, CA (1977), p. 199.

39. C.A. Kovac and I.C. Lewis, Carbon **16**, 433 (1978).

40. H. Honda, Y. Yanada, S. Oi and K. Fukuda, Ext. Abst. 11th Biennial Conference on Carbon, American Carbon Society, Gatlinburg, TE (1973), p. 219.

41. T. Imamura, Y. Yanada, S. Oi, and H. Honda, Carbon **16**, 481 (1978).

42. S. Matsumoto, S. Oi, T. Imamura, N. Nakamiza, Y. Yanada, and H. Honda, Ext. Abst. 13th Biennial Conference on Carbon, American Carbon Society, Irvine, CA (1977), p. 201.

43. K.J. Hüttinger, in *Proceedings 1st International Conference on Carbon* (Deutschen Keramischen Gesellschaft, Baden-Baden, 1972), p. 5.

44. C.A. Atkinson, Ph.D., University of Newcastle-upon-Tyne, UK, 1983.

45. J.R. Lander, Ph.D., University of Newcastle-upon-Tyne, UK, 1984.

46. H. Marsh, F. Dachille, J. Melvin, and P.L. Walker, Jr., Carbon **9**, 159 (1971).

47. G.W. Smith, J.L. White, and M. Buechler, Carbon **23**, 117 (1985).

48. S. Martínez and R. Santamaria, University of Newcastle-upon Tyne, UK, Private communication, 1990.

49. J.L. White and J.E. Zimmer, Ext. Abst. 13th Biennial Conference on Carbon, American Carbon Society, Irvine, CA (1977), p. 203.

50. H. Marsh and C.S. Latham, in *Petroleum Derived Carbons*, Symposium 303 (ACS, Washington, DC, 1986), p. 1.

51. I. Mochida and Y. Korai, in *Petroleum Derived Carbons*, Symposium 303 (ACS, Washington, DC, 1986), p. 29.

52. Y. Yokono and H. Marsh, in *Coal Liquefaction Products*, edited by H.D. Schultz (Wiley, New York, 1983), Vol. 1, p. 125.

53. T. Yokono, T. Obara, Y. Sanada, S. Shimomura, and T. Imamura, Carbon **24**, 29 (1986).

54. C. Jackson and W.F.K. Wynne-Jones, Carbon **2**, 227 (1964).

55. L.S. Singer and I.C. Lewis, Carbon **16**, 417 (1978).

56. K.J. Huttinger, Chem.-Ing. Tech. **43**, 1145 (1971).

57. L.S. Singer and I.C. Singer, US Air Force Materials Laboratory, Reports ML-TDR-65-125, Vol. 3, Union Carbide Corporation, Carbon Products Division, AD482, 152 (1965).

58. H. Marsh, J.M. Foster, G. Hermon, M. Iley, and J.N. Melvin, Fuel **52**, 243 (1973).

59. M.A. Sadegui, K.M. Sadegui, and T.F. Yen, Carbon **27**, 233 (1989).

60. I. Mochida, E. Nakamura, K. Maeda, and K. Takeshita, Carbon **13**, 489 (1975).

61. I. Mochida, T. Ando, K. Maeda, and K. Takeshita, Carbon **14**, 123 (1976).

62 H. Marsh, J.M. Foster, G. Hermon, and M. Iley, Fuel **52**, 234 (1973).

63. H. Marsh, C.S. Latham, and E.M. Gray, Carbon **23**, 555 (1985).

64. H. Marsh, I. Gerus-Piasecka, and A. Grint, Fuel **59**, 343 (1980).

65. I. Mochida and H. Marsh, Fuel **58**, 809 (1979).

66. I. Mochida, T. Oyama, Y. Korai, and Y. Qu. Fei, Fuel **67**, 1171 (1988).

67. I. Mochida, Y. Korai, T. Oyama, Y. Nesumi, and Y. Todo, Carbon **27**, 359 (1989).

68. H. Marsh, C. Atkinson, and C. Latham, Ext. Absts. 16th Biennial Conference on Carbon, American Carbon Society, San Diego, (1983), p. 74.

69. E. Romero-Palazón, Ph.D., University of Alicante, Alicante, Spain, 1990.

70. E. Romero-Palazón, H. Marsh, F. Rodríguez-Reinoso, and P. Santana, Ext. Abst. 19th Biennial Conference on Carbon, American Carbon Society, Pennsylvania State University (1989), p. 178.

71. M.A. Forrest and H. Marsh, J. Mater. Sci. **18**, 991 (1983).

72. H. Marsh, F. Dachille, J. Melvin, and P.L. Walker, Jr., Carbon **9**, 159 (1971).

73. M. Makabe, H. Itoh, and K. Ouchi, Carbon **14**, 365 (1976).

74. H. Marsh and J. Griffiths, Ext. Abst. of International Symposium on Carbon, "New Processes and New Applications," Toyohashi, Japan, (1982), p. 81.

75. D. Crawford and H. Marsh, Fuel **55**, 751 (1976).

76. H. Marsh and E. Romero-Palazón, in *Proceedings of Third Australasian Aluminium Smelter Technology Course*, Sydney, Australia, 1989, p. 171.

77. H. Marsh and C. Cornford, in *Petroleum Derived Carbons*, Symposium 21 (ACS, Washington, DC, 1976), p. 266.

78. H. Marsh and D.E. Clarke, Erdol und Kohle **39**, 113 (1986).

79. H. Marsh and R. Menéndez, Fuel Proc. Technol. **20**, 269 (1988).

80. H. Marsh and R. Menéndez, in *Introduction to Carbon Science*, edited by H. Marsh (Butterworth, London, 1989), p. 37.

81. J.B. Donnet and R.C. Bansal, *Carbon Fibres* (Marcel Dekker, New York, 1990).

8

Mesophase State of Polyorganophosphazenes

V.G. Kulichikhin, E.M. Antipov, E.K. Borisenkova, and D.R. Tur

8.1 Introduction

Earlier it was believed [1] that an ordering at a level exceeding the short-range order of low molecular liquids cannot exist above the melting point in flexible-chain polymers which do not contain mesogenic groups. However, recently, a large number of polymers, for example, polyphosphazene, has been found, which cannot be described in the framework of the traditional classification of liquid crystals. An additional type of mesomorphic state, that of a conformationally disordered crystal, was suggested [2,3].

A more common approach is analyzing this type of mesophase as a three-dimensional ordered crystal having a different perfection extent along each of its crystallographic dimensions. In order to gain a better insight into this problem, it is necessary to accumulate experimental evidence.

Polyorganophosphazenes (PPh) with the following general formula:

$$\left[\begin{array}{c} OR \\ | \\ -P=N- \\ | \\ OR \end{array} \right]_n$$

are of interest from the viewpoint of the mesophase class of polymers [4], since many of the representatives of this class may have an ordering level intermediate between a crystalline and an amorphous state in a broad temperature range [5].

A comprehensive review of earlier works on the structure and properties of PPh was published in [6]. The number of publications on the description of the structure, morphology, and characteristics of the physical transitions and physical–mechanical properties of these polymers has considerably increased in recent years. However, there is no complete information on the structure of PPh, especially in the mesomorphous state.

The majority of publications is dedicated to the description of the meso-phase state of polyfluoroalkoxyphosphazenes [7–12]. Perhaps the presence of fluorine atoms at the side branches favors the occurrence of specific inter-action, leading to a stepwise melting of the crystalline structure. We believe that a detailed investigation of the structure and properties, particularly of this series of PPh, can help in creating general ideas about the formation of mesophases in polymers without mesogenic groups.

Thus, the items under investigation were poly-*bis*-fluoroalkoxyphospha-zenes with $R=OCH_2CF_3$, $—OCH_2CF_2CF_3$, and $—OCH_2CF_2CF_2CF_3$, sometimes with a noncomplete replacement of H atoms in the pendant groups. Special attention will be given to the first member of the investigated series poly-*bis*-trifluoroethoxyphosphazene (PFEPh) [13]. It is mainly for this polymer that specific synthetic conditions were developed [14,15], lead-ing to the preparation of high-molecular ($M \geq 10^6$), linear, and practically monodisperse products extending further to other poly-*bis*-fluoroalkoxypho-sphazenes.

8.2   Thermodynamics and Structure of Polyorganophosphazenes

The DSC heating thermograms for PFEPh after heating are shown in Fig. 8.1. The upper curve represents the oriented film prepared from solution; the lower curve relates to the sample obtained by extrusion. Both thermograms

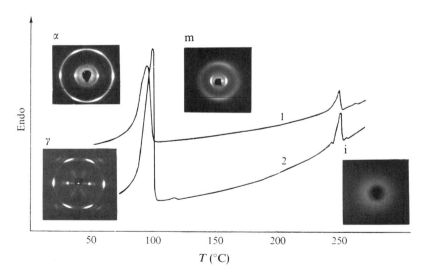

FIGURE 8.1. DSC curves of the PFEPh-oriented samples obtained from solution (1) and the mesophase melt at 500 K, (2), and x ray patterns of the α- and γ-crystalline modifications, mesophase (m), and isotropic melt (i).

260   V.G. Kulichikhin, E.M. Antipov, E.K. Borisenkova, and D.R. Tur

illustrate two phase transitions of the first order: the ratio of enthalpies of
low- and high-temperature transitions is approximately 8 to 10. The x ray
patterns also presented in Fig. 8.1 show the variety of the structural forms of
PFEPh in different temperature regions.

The equatorial diffractograms of PFEPh in the different phase states are
shown in Fig. 8.2. The quantitative treatment of the diffraction data has
shown that the film prepared from solution crystallized to form an orthor-

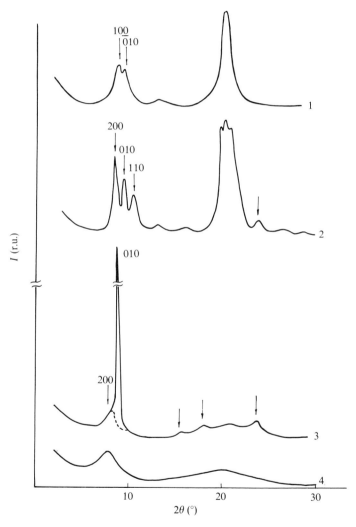

FIGURE 8.2. Equatorial x ray diffraction patterns of the various oriented PFEPh
samples: (1) α-orthorombic modification, 293 K; (2) γ-orthorombic structure, 293 K;
(3) mesophase, 373 K; and (4) isotropic melt, 523 K.

hombic lattice, with two monomers on a unit cell (α-modification). This sample, heated once above the transition point, does not form the initial type of structure even at high cooling rates, and crystallizes into an orthorhombic modification with the number of monomeric units on the cell being 4 (γ-lattice).

These data are consistent with the results obtained in [8–10], in which the structural state of the polymer between the two transitions is indentified as a mesophase. However, the structural organization of PFEPh in the meso-phase state is described ambiguously, possibly because of the different con-ditions for the transition to the mesophase of the α- and γ-crystalline modi-fications. In this connection, it is necessary to dwell in more detail on the formation processes of different crystalline modifications.

It should be noted preliminarily that two first-order transitions are only observable for those members of the polyfluoroalkoxyphosphazene series for which the increase in the number of fluorine atoms in the monomer unit leads to a decrease in the temperature of the crystal–mesophase transition $T(1)$, whilst the mesophase melting point T_m increases, resulting in a widening of the temperature range of the mesophase existence. The data presented in Table 8.1 provide an insight into the quantitative change in this range. It is important to point out that in the case of poly-*bis*-heptafluorobutoxypho-sphazene the mesophase state occupies an extensively broad temperature range of 300 °C.

Without discussing further the relaxation transitions of PFEPh, let us return to the conditions for the formation of its different crystalline modifica-tions. When PFEPh is crystallized from solution, the process goes on by-passing the preliminary mesophase formation stage to form a nonequili-brium α-modification. Subsequently, the temperature of the α-phase transi-tion to a mesomorphous state is ~ 20 °C lower than for the γ-phase, which is, on the contrary, a thermodynamically stable crystalline PPh form in this temperature range.

Extrusion through the capillary at 220 °C leads to a self-reinforcement system of well-oriented PFEPh fibrils [16] whose recovery conditions will be discussed below. The most important thing now is that, independent of the morphology, the extrudate structure is characterized by the γ-form. We failed to find a long periodicity over the whole range of temperatures under investi-

TABLE 8.1. Transition temperatures of poly-*bis*-fluoroalkoxyphosphazene.

Lateral group structure	T_g (°C)	$T(1)$ (°C)	T_m (°C)
—OCH$_2$CF$_3$	− 60	70–90	247–250
—OCH$_2$CF$_2$CF$_3$	− 50	48	300
—OCH$_2$(CF$_2$)$_2$CF$_3$	− 57	10	300

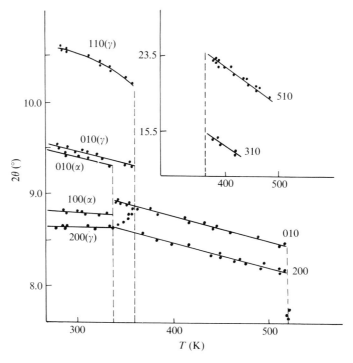

FIGURE 8.3. Temperature dependences of the equatorial reflexes positions for α- and γ-crystalline modifications and for the mesophase state. Dotted lines correspond to the phase transition points $T(1)$.

gation. The small-angle periodicity was also not observed for one PFEPh structure prepared from solution.

When PFEPh is heated above $T(1)$, irrespective of what type of packing was in the initial specimen, the system transforms into a mesomorphic state. The dependences of the positions of reflections $2\theta_{max}$, their intensities I, and the transverse dimensions of the coherent scattering domains L on the temperature are presented in Figs. 8.3 and 8.4, respectively.

The process proceeds in the following way. First, the Bragg reflection positions of both modifications shift monotonically toward the lower angles with temperature increasing, this corresponds to the ordinary heat expansion of the lattice and the intensity of the peaks is to some extent enhanced while the half-width is decreased. When the sample is further heated the α- and γ-phases evolution is substantially different. The positions of the α-phase reflections are drastically changed at 343 K. In this case, the intensity of the first peak increases sharply, the half-width decreases, the remaining crystalline reflections are damped out, and we observe only the diffraction maxima of the mesophase. The completion of the mesophase formation process occurs

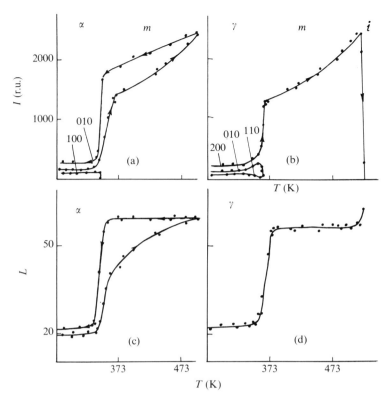

FIGURE 8.4. Temperature dependences of the main reflexes intensity (a, b) and the lateral crystallite dimensions (c, d) for the different PFEPh structures: (a, c) fiber and (b, d) extrudate.

at 365 K. As to the γ-phase, the transition to a mesomorphous state proceeds in two stages. Above 343 K we observe the coexistence of γ-crystals and the mesophase, and the crystalline phase disappears completely in only a relatively narrow temperature range of 360–365 K.

The x ray diffraction pattern of the mesophase (Fig. 8.2, curve 3) is characterized by a strong narrow reflection ($2\theta = 8.86°$) on the equator; at this small angle we can see a weak, relatively broad maximum as well as three peaks of low intensity at the large diffraction angles ($2\theta = 15.48°$; $17.92°$; $23.58°$). Strong diffuse reflections with a spacing of 0.43 nm are located on the first layer line. The indentity period determined from this reflection is 0.486 nm, i.e., it coincides with the interplanar spacing for (001) reflection of the crystalline α- and γ-modifications. The rapid decrease in the intensity is typical, with the diffraction angle rising as seen in the x ray diffraction pattern, which is simultaneously the result of the rotational (around the chain axis) and translational (along the chain axis) disordering in the system. A considerable broadening of the reflections on the first layer line indicates that the macro-molecules lose a conformational ordering.

In [8–10] the PFEPh mesophase structure was ascribed to be pseudo-hexagonal, while the diffuse meridional scattering is explained by the super-position of a number of reflections, broadened due to the rotation of the chain units and their random longitudinal shift relative to each other. How-ever, a comparison of the values calculated for the mesophase density, the crystallographic data obtained in this assumption compared with those ob-tained from the dilatometric experiments, shows an inadmissibly large dis-crepancy, $\Delta\rho = 140$ kg/m^3. The calculated value distinguishes from the true one toward lesser values, which render it possible to explain this fact by the presence of a notable fraction of the disordered amorphous component in the system. As for the crystalline modifications, the crystallographic and dilato-metric values coincide with experimental accuracy (± 5 kg/m^3).

Good agreement for the mesophase PFEPh can be attained if we suppose that the identity period along the macromolecule axis upon transition to the mesophase varies and becomes equal to 0.44 nm. However, the experimen-tally derived values, both in the present work and in those of other authors [9,10], are distinguished substantially, and as has been noted above is equal to 0.486 nm. It is appropriate to remember that for the polydiethylsiloxane mesophase, even at a small difference, $\Delta\rho = 80$ kg/m^3, researchers [17] had to give up describing the mesomorphous state structure within the frame-work of hexagonal modification. Thus, for PFEPh, it was necessary to sup-pose a packing different from the hexagonal one.

The correlation between the calculated and experimental data proved to be possible, if we suppose the existence of the disturbed γ-orthorhombic packing in the PFEPh mesomorphic structure. The corresponding Miller indices are given in Fig. 8.2. The strongly distinguishing character of the reflections (200) and (010) is interesting, in that the first of those reflections has a diffusive profile with a linear half-width of about 1.5°, whereas the second one reveals an exceptionally sharp shape of the line. Hence, it follows that, in the basal plane, the level of ordering in the two interperpendicular directions is essentially different. In turn, this circumstance indicates the dis-tinction of the chain symmetry in the mesophase from the cylindrical one. The large sizes of the cell and the lower density, in comparison with the crystalline packing, is obviously sufficient to resolve the partial rotational and translational disordering of the chains, but insufficient to get a hexagonal packing which is characterized by a free rotation of the chains relative to their axis.

In the case of PFEPh, such a free rotation requires a rather large volume because of the presence of the long side groups. Sterically and energetically this is not profitable as a result of which no pseudohexagonal mesomorphous structure with a full rotational disordering is realized, whereas the pseudo-orthorhombic one is characterized by a partial rotational disordering and a large density. The prefix "pseudo" means, in this case, the absence of a long-range positional order along the macromolecule axis, on the one hand, and the presence of a paracrystalline ordering along one of the basal plane directions on the other.

The following detail is also typical: a reverse change in the transverse dimensions of PFEPh crystallites takes place in the crystal-to-mesophase transition (Fig. 8.4). The sizes of the ordered mesophase domain exceed three to five times the corresponding values in the crystalline state. Under cooling, the large ordered formations disintegrate rapidly into small crystallites, probably due to the mechanism of casual nucleation. The reason for such a significant enlargement of those domains, in the direction perpendicular to the chain axis during the transition to mesophase, still remains, so far, unclear.

Indeed, the structure of the PFEPh mesophase is characterized by the one-dimensional monomolecular layer packing (Fig. 8.5). The conformational disordered chains are aligned along with the axis of orientation, and are so arranged as to put the backbones and the side groups in the layer plane. A short-range order is realized, both along the chain and normally thereto within the layer. The interlayer packing is characterized by a well-defined long-range order. So, the PFEPh mesophase structure has the following features:

(i) The PFEPh mesophase displays a well-pronounced, long-range, one-dimensional order for the monomolecular layers, within which a short-range order of paracrystalline type is realized along both other direc-

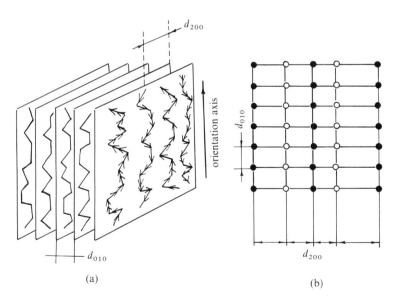

FIGURE 8.5. Schematic presentation of the PFEPh mesophase structure: (a) isometric sight and (b) basal plane projection.

tions. This type of packing is possible only for the partial rotational mobility of segments around the chain axis.

(ii) An orientational disordering is observed for the chain segments in the mesophase PFEPh with preservation of the uniaxial alignment of the macromolecules along the stream direction.

(iii) The lateral sizes of the coherent scattering areas at the mesophase transition are increased by a factor of three to five. Under cooling, these domains break up into small crystallites of the same dimensions as the original sample. A long periodicity has not been observed by the small-angle diffraction procedure. After the cooling of the mesophase-state sample, the longitudinal sizes of the crystallites determined by the half-width of the 001 reflection exceed 50 nm.

(iv) If the stiffness of the macromolecules is taken as a necessary reason [1] for a liquid crystalline (LC) order arising in a polymer system above the melting point, it should be concluded that an induced increase in the stiffness of the PFEPh chains occurs in the mesophase transition. This behavior, perhaps, may be due to the different interaction between inorganic backbones and regularly alternating side groups with three fluoroatoms at the ends. Independently of the nature of the interaction forces, such a contrast may be responsible for this mesophase state development in polyphosphazenes and polyorganosiloxanes in general.

The following examples can serve as confirmation of this fact. In a number of cases, no mesophase was found in polyphosphazenes containing side alkoxygroups of different length [18]. Polyethylene, being an organic flexible-chain polymer, does not form a mesophase spontaneously when heated. To induce LC ordering, it is necessary to have a mechanical field retaining the chain in an unfold state above the melting point [19,20]. However, a thermodynamically stable existence of the mesomorphous state of the condis-crystalline (or condis-mesophase) type in the temperature range 303–600 K was found in polytetrafluoroethylenes which, as distinct from polyethylene, have carbon atoms framed with fluorine instead of hydrogen atoms [21].

The transition of oriented PFEPh from mesophase to isotropic melt occurs very sharply at 523 K. Even at stepwise heating at 0.5 deg/min we failed to find the coexistence of a mesophase and isotropic melt. The diffractogram above 523 K (Fig. 8.2, curve 4) is characterized by the presence of two diffusion maxima, whose positions are 7.75° and 19.7°, respectively. At the moment of isotropization a jump-like change of the equatorial (Fig. 8.3) and meridional diffusion reflections takes place. The specific volume jump during the transition of the mesophase to isotropic melt is practically the same which, in the case of the transition of crystal to mesophase, is equal to 6% of the initial value.

Isotropization of the oriented materials is connected with the conformational folding of the chains and the formation of the liquid-like ordering in the system, which is indicated by the amorphous character of the scattering peaks in the diffractogram of the isotropic melt. It is possible to believe that in this state the full rotational disordering is realized, which should correspond, on average, to the pseudohexagonal packing at the level of two to three coordinational layers. The calculations, however, show that the inconsistency of the density values is too great for such a suggestion. Therefore, we have to make allowances for the fact that there is no full rotational freedom in this state also. The same suggestion for polydiethylsiloxane was made in [17].

Thus, it was recognized that the PFEPh structure in the crystalline state depending on the prehistory of preparing the specimen is characterized by two types of packing: metastable α-orthorhombic and equilibrium γ-orthorhombic modifications. The structure of PFEPh in the mesomorphous state is distinct from the pseudohexagonal structure typical for condis-type mesophases and possesses a one-dimensional (translational) layer packing of macromolecules in which the conformational disordering is realized in the chain axis direction; in the basal plane along one of the directions there is a long-range ordering but along another only a short-range ordering takes place. The PFEPh isotropic melt structure maintains the major features of the mesophase structure distinguished by the more fold macromolecule conformation in comparison with the mesophase state.

8.3   Relaxation Transitions of Poly-*bis*-Trifluoroethoxyphosphazenes

Although we give due respect to the DSC method for recording the phase and mesophase transitions, nevertheless, we wish to point out that for relaxation transitions, especially in the case of high-crystalline polymers, the method of dynamic mechanical analysis is in a number of cases more sensitive and informative. The above-mentioned fact is supported by the comparative analysis of the two PFEPh specimens prepared from the solution (PFEPH-1) and the mesophase melt state (PFEPh-2).

The apparatus unit used for investigation of solid polymer specimens is a reverse torsion pendulum operating in a self-generator regime at a constant small strain amplitude [22]. The modernized method of fixing the specimen led to obtaining a relative extension–compression strain at a level of 10^{-3}–10^{-4}. The temperature rise rate of the specimen was ~ 1.5 deg/min.

The temperature dependences of the loss modulus E'' for PFEPh-2, as well as for the mechanical loss tan δ for PFEPh-1 and PFEPh-2, are given in Fig. 8.6. Let us analyze those dependences starting from the low-temperature region. The first relaxation region is located at 83–133 K and is more intensive for PFEPh-1. It is likely that the relaxation mobility of the chain fragments, possibly the trifluorethoxy groups in the amorphous part of the poly-

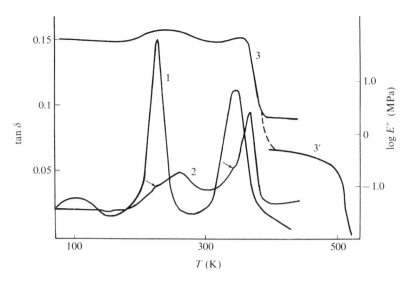

FIGURE 8.6. Dependences of tan δ for PFEPh-1 (1) and PFEPh-2 (2) and loss modulus for PFEPh-2 (3) on temperature. Curve 3' relates to data recalculated from the capillary viscometry results.

mer, is defrosted in this temperature range. The existence of the fragment mobility of the PFEPh macromolecules at such low temperatures allows us to expect the absence of a fragile destruction of the PFEPh samples under these conditions.

The main relaxation (glass transition) takes place at 189–272 K. The estimate of the glass transition temperature, T_g, from the $E' - T$ dependences infection, and the maxima of functions $E'' - T$ and tan $\sigma - T$, allows us to consider that T_g for PFEPh-1 is 234 K, which is in good accord with the earlier obtained data. As for as the PFEPh-2 specimen is concerned we can observe a shift of relaxation transition up to a temperature of about 266 K for this specimen provided a small modulus dispersion is preserved at 234 K (as indicated by the arrow). Such a singular "splitting" of the glass transition process in high-crystalline PFEPh can be due to the existence of two kinds of disordered regions. If the amorphous phase of PFEPh-2 is located at a distance from the crystallite, the large-scale chain mobility is decreased and that leads to an increase in the temperature of the main relaxation transition ranging from 234 K to 266 K. When discussing this question we should also not forget about the significant morphology distinction between PFEPh-1 and PFEPh-2. The PFEPh-1 film represents a monolithic system consisting of spherolites. At the same time the extrudate of PFEPh represents a fibrillized self-reinforced system with, very probably, thin layers, and with an intermediate level of structural organization between the highly oriented fibrils.

An interesting and, to some extent, an unusual experimental fact is the

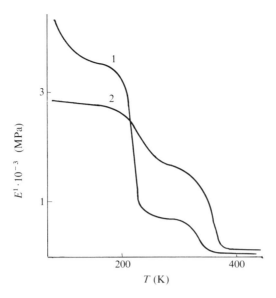

FIGURE 8.7. Dependences of the store modulus for PFEPh-1 (1) and PFEPh-2 (2) on temperature.

storage modulus inversion for the PFEPh-1 and PFEPh-2 specimens which is observed in the glass transition range (Fig. 8.7). In other words PFEPh-2, having an α-thorhombic structure and a low crystallinity degree up to T_g, is stiffer than the PFEPh-2 specimen having an γ-orthorhombic structure and a crystallinity level above 80%. This circumstance indicates that the glassy amorphous phase of PFEPh-1 has a higher Hook elasticity than the cooled crystalline phase prevailing in PFEPh-2. When T_g is attained, the situation changes reversely, i.e., at $T > T_g$ the determining contribution to the system rigidity is already given by the crystallinity degree which is higher in PFEPh-2.

When passing over to the high-temperature range, we should note one indisputable fact of the transition $T(1)$ temperature's difference: 348 K for PFEPh-1 and 372 K for PFEPh-2 (Figs. 8.6 and 8.7). The scale of difference in the values for $T(1)$ is near to the one obtained earlier [23], and supports the fact that the γ-phase structure is more in equilibrium and is refined, as compared with the α-phase. However, there also exists an additional new moment in the intepretation of the crystal–mesophase transition in that temperature range. The point is that for PFEPh-2 there exists a dissymmetry of the functions under analysis and at 348 K (indicated by the arrow) runs alongside the main modulus and the tan δ dispersion at 372 K. This fact can be due to the preservation of a small fraction of the α-phase, even after heating PFEPh-2 above $T(1)$ [24]. We failed to find any indication concerning the existence of the two types of orthorhombic cells at $T < T(1)$ in the literature. Now, if that is so, it is quite possible that there exist two sorts of macromolecules or their domains in the mesophase (one of the two is naturally prevailing) after cooling leading to the formation of either the γ- or α-phase.

At higher temperatures the specimens of the extrudate or film become plastic and lose their initial dimensions. Therefore, in order to have a complete general picture of the transitions involving the mechanical characteristics, the $E'' - T$ dependences were added by rheological experimental data up to a temperature of 523 K (Fig. 8.6, curves 3 and 3'). The noncoincidence of curves 3 and 3' is due to the difference in the deformation modes, as well as to the not entirely correct recalculation of the shear stress into the loss modulus.

The sharp drop in the shear stress connected with mesophase melting occurs at temperatures above 503 K. For this transition there is no longer a difference in the prehistory for preparing PFEPh samples which is removed by annealing the specimen in the mesophase state. So, viscoelastic characteristics "feel" the differences in the structure of PFEPh caused by the preparation and heat treatment conditions. This difference reflects on the glass point, the secondary (low-temperature) relaxation transition, and the phase transition $T(1)$ shifts along a temperature scale.

8.4   Properties of Diluted Solutions of Poly-*bis*-Fluoroalkoxyphosphazenes

The ability of poly-*bis*-fluoroalkoxyphosphazenes (PPh) to form thermotropic mesophases has aroused interest in respect of the equilibrium stiffness of their chains, and particularly to the magnitude of the statistic Kuhn segment as applied to its quantitative measure. This is especially important in order to formulate a general approach to the problem of mesophase creation by polymers which do not contain mesogenic groups.

This paragraph will present our data devoted to the investigation of the thermodynamic macromolecule's stiffness for three PPhs with branches of different length at PFEPh (I), $—[P(OCH_2CF_2CF_2H)_2{=}N]—$ (II), and $—[P(OCH_2CF_2CF_2CF_2H)_2{=}N]—$ (III).

PPhs were synthesized by the condensation of linear high-molecular polydichlorphosphazene ($[\eta] = 3.0–4.4$ dl/g, toluene, 25 °C) with appropriate sodium fluoralcoholates. The content of the unsubstituted PCl groups in PPhs (Table 8.2) was calculated from the residual quantity of chlorine in polymers, which was determined by the x ray fluorescent spectroscopy method (measurement accuracy ± 0.005 wt.%).

PPhs were fractionated by the method of distribution between two liquid phases in the following systems: DMAA-hexane- chloroform (I) and N-methylpirrolidone-hexane-chloroform (II and III). The intrinsic viscosities $[\eta]$, the sedimentation constants S_0, $M_{s\eta}$ calculated from the Flory–

TABLE 8.2. PPh characteristics.

Polymer	[PCl], (mol.%)	$[\eta]$ (THF, 25 °C) (dl/g)	$\bar{V}$ (THF, 25 °C) (cm^3/g)	$S_0 \cdot 10^{13}$ (cm/s·dyne)	$\bar{M}_{s\eta} \cdot 10^{-6}$	$\bar{M}_w/\bar{M}_n$
I	≤ 0.03	4.58	0.817	97.0	22.4	—
II	0.09	6.35	0.789	55.6	9.9	1.39
III	0.10	6.40	0.731	61.0	9.1	1.09

Mandelkern equation, and the specific partial volume $\bar{V}$ are presented in Table 8.2.

The polydispersity coefficients $\bar{M}_w/\bar{M}_n$ and the dependences of $[\eta]$ and S_0 on M were calculated from the MWD, constructed from the fractionation data and the measurements of $[\eta]$, S_0 and $M_{s\eta}$ for fractions (Tables 8.2 and 8.3). The exponent α in the Mark–Houwink equation for the PPh series increases from 0.5 to 1.00 with an increase in the number of carbon atoms in the side fluoroalkoxy substituents from two to five. When the polymerization degree, $X_{s\eta}$, is the same, for example, at $X_{s\eta} = 2.4 \times 10^4$, the values of $[\eta]$ in THF for I, II, and III are 1.38, 4.19, and 8.50 dl/g, respectively, i.e., they also increase significantly with an increase in the length of the side fluoroalkoxy-groups.

In order to estimate the parameters of the unperturbed PPh macromolecule sizes, we considered the following models of their behavior in solution: the Gaussian impermeable coil (GIC) [24], a chain of finite length (CFL) [25], and a persistent chain (PCh) [26]. The length of the projection of the repeated link —P(OR)=N— on the molecule axis is $l = 2.8$ Å. The geometric parameters (link length, PN 1.60 Å, valence angles, NPN 130° and PNP 120°) used to calculate l were estimated from the averaged structural data for polyphosphazenes as described in the literature [27].

The mean square distance between the chain ends $(\bar{h}^2/M)^{1/2}$ and the Kuhn segment lengths A, calculated on the basis of the $[\eta]$ and $M_{s\eta}$ measurements according to the models under consideration, are given in Table 8.4. The most plausible are the values of those parameters obtained in describing the behavior of the PPh molecules in solution by the PCh model. The values of $(\bar{h}^2/M)^{1/2}$ and A, calculated from the PCh model, increase by increasing

TABLE 8.3. Parameters of equations $[\eta] = K_\eta M^\alpha$ and $S_0 = K_s M^{1-b}$.

Polymer	K_η	α	K_s	$1-b$	Range $X_{s\eta} \cdot 10^{-3}$
I	$1.77 \cdot 10^{-4}$	0.59	$3.52 \cdot 10^{-15}$	0.47	45.0–122.1
II	$6.10 \cdot 10^{-6}$	0.85	$1.17 \cdot 10^{-14}$	0.39	3.6–63.1
III	$6.74 \cdot 10^{-7}$	1.00	$2.95 \cdot 10^{-14}$	0.33	7.1–23.7

TABLE 8.4. Parameters of PPh macromolecule equilibrium stiffness.

Polymer	$(\bar{h}^2/M)^{1/2}$ (Å)	A (Å)	$(\bar{h}^2/M)^{1/2}$ (Å)	A (Å)	$(\bar{h}^2/M)^{1/2}$ (Å)	A (Å)
	GIC [24]		CFL [25]		PCh [26]	
I	0.632	35	0.707	44	0.714	44
II	0.520	30	1.200	160	1.052	122
III	0.141	4	*	*	1.304	300

* This is impossible to calculate due to the negative slope of the $M/[\eta] = f(M^{1/2})$ dependence.

the side fluoroalkoxy substituents length as in the case of an increase in the above-considered a and $[\eta]$ values. Figure 8.8 gives the corresponding plots from which $(\bar{h}^2/M)^{1/2}$ and A were calculated for the PCh model in THF. Besides, for a fraction of I, the values of those parameters for the PCh model were confirmed in dioxane: $(\bar{h}^2/M)^{1/2} = 1.693$ Å and $A = 120$ Å.

The fact that the unreal values for A were obtained when the GIC model was used for II and III proves the impossibility of referring them to the flexible-chain polymers. At the same time, the molecular characteristics of PFEPh (I) can be described within the GIC-model framework, i.e., it is a flexible-chain polymer [28].

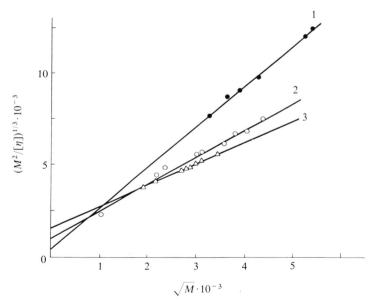

FIGURE 8.8. Dependence $(M^2/[\eta])^{1/3}$ versus $M^{1/3}$ for solutions in THF in accordance with the PCh model for PPh-1 (1), PPh-II, (2) and PPh-III (3).

8.5   Rheological Properties of Concentrated Poly-*bis*-Trifluoroethoxyphosphazene Solutions

The data presented in the previous section proves that the enhanced PPh micromolecule stiffness regularly increases with an increase in the fluorine atoms content in the side chains and their lengths. If we notice that the PFEPh can yet be a flexible-chain polymer, the PPh II and III are typical semistiff chain polymers. The stiffness of their macromolecules is similar to that of the derivatives of cellulose, therefore, in principle, we might expect for them the appearance of a lyotropic LC-phase. As for PFEPh, in spite of the indication [29] concerning the transition of a concentrated solution to an LC-state, we failed to confirm this observation knowing the Kuhn segment value of this polymer is quite natural. The addition to PFEPh of only 5% dimethylsulfoxide leads to a loss in its ability to form a mesophase [30]. A more complicated case is connected with PFEPh solutions in ethylacetate, for which an ordering of solutions with a concentration $> 25\%$ is observed in the low-temperature range (< 230 K). However, the nature of the low-temperature transition and novel phase structure are not so far identified.

The properties of moderately concentrated PFEPh solutions are also very specific. From the literature it was found [31] that they do not satisfy the Cox–Merz rule when comparing the steady-state and dynamic shear experiments. It was shown that the character of the polymer–polymer interaction is unusual, and that with an increase in concentration the fluorolkoxyphosphazene copolymer solutions display a behavior similar to that in the high-elasticity plateau region of the flexible carbonchain polymers. We investigated the rheological properties of PFEPh solutions in acetone, ethylacetate (EA), and dimethylacetamide (DMAA).

The investigation of the rheological properties of diluted PFEPh solutions in all of the solvents has revealed their anomalously low viscosity, which indicates an extraordinarily low hydrodynamic resistance of the macromolecules under flow. The dependences of the inherent viscosity on concentration for PFEPh solutions are characterized by the low slope and, consequently, by the small Huggins constant values (0.15–0.35). The intrinsic viscosity values for acetone, EA, and DMAA are 4.1, 3.42, and 2.52 dl/g, respectively. These values indicate a deterioration in the thermodynamic quality in the series of solvents. The change of $[\eta]$ under a variety of solvent quality is natural, nevertheless, the small absolute values of $[\eta]$ in comparison with the flexible-chain polymers of similar M have attracted attention. Thus, at $M = 1.2 \times 10^7$, the $[\eta]$ value for PS in toluene is 13.60 dl/g ($[\eta] = 2.89 \times 10^{-1} \times M^{0.66}$ [12]), and in decaline it is 34.5 dl/g ($[\eta] = 7.26 \times 10^{-4} \times M^{0.35}$ [32]). Such $[\eta]$ values, typical for ultrahigh molecular weight polymers, are once more indicative of the specificity of the isolated PFEPh macromolecules behavior in a hydrodynamic field.

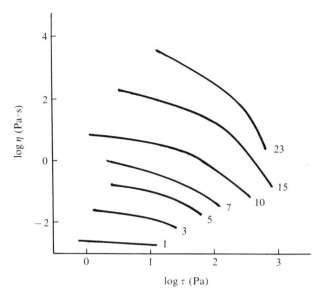

FIGURE 8.9. Flow curves of PFEPh solutions in EA at 20 °C. Figures indicate PFEPh contents, wt.%.

The situation is the same for the effective viscosity of solutions in a wide range of concentrations. The dependences of PFEPh solutions in the EA viscosity η on the shear stress τ are given in Fig. 8.9. We can single out three concentration regions on the flow curves:

(i) 0.5–1.0% (the solutions behave like Newtonian fluids);
(ii) 3–10% (the solutions have weakly expressed non-Newtonian effects); and
(iii) 15–23% (the solutions exhibit an essentially nonlinear behavior).

In the case of concentrated solutions, the Newtonian region practically disappears on the flow curves. The difficulties in achieving a region of zero shear viscosity is most likely due to the existence of a developed system of intermolecular contacts. An especially notable effect of the decrease in viscosity with increasing τ manifests itself after transition to the concentrated solution $(C \geq 15\%)$.

The PFEPh solution in DMAA (Fig. 8.10) is also characterized by small absolute viscosity values and the well-pronounced dependence of non-Newtonian behavior, not only on the concentration but also on the temperature. Taking into account that DMAA, as distinct from EA, is not a highly volatile solvent, the viscous properties of the PFEPh solutions in DMAA were investigated in the 10°–95° range. At 95°, the 3% solution represents a Newtonian fluid whose viscosity is constant in all of the investigated τ values. An increase in polymer concentration and a decrease in temperature of the

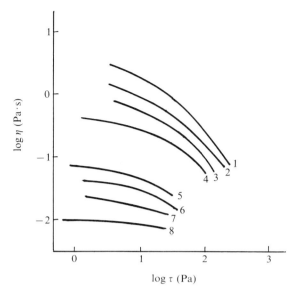

FIGURE 8.10. Flow curves of the solutions with 7% (1–4) and 3% (5–8) content of PFEPh in DMAA at 10 °C (1, 5), 30°C (2, 6), 60 °C (3, 7), and 95 °C (4, 8).

solution leads to the appearance of η on τ dependence. The presented data can indirectly witness for the existence of an upper critical solution temperature for PFEPh solutions (for liquid–liquid equilibria) or a liquidus line (in a case of crystalline equilibria), since with increasing temperature the rheological behavior of the solutions tends more linearity, i.e., typical for non-structurized systems. Parallel with this, we may suppose that 3% solutions are located fairly close to the phase boundary.

When investigating the temperature dependence of the PFEPh solutions viscosity in EA, the steel ball filling time was measured in a cylindrical sealed ampule. In which manner the solutions viscosities were measured from $-10°$ up to a temperature higher than the EA boiling point. The analysis of the data rendered possible the determination of the flow activation energy values, E, in a sufficiently broad temperature range. It appeared to be 5 kJ/mole for solutions with $C < 3–5\%$ and 20 kJ/mole for solutions with $C > 3–5\%$ (Fig. 8.11). Such small E values at $C < 5\%$ are more typical for low molecular weight fluids than for polymers, even in terms of enough high chain flexibility and weak intermolecular interaction. However, values of E for high-concentrated solutions in terms of absolute values are four times higher than E for diluted solutions, which indicates a change in the mechanism of the PFEPh solution's flow with increasing concentration. In this connection, the 5% solutions of the polymer display a threshold character. The E values for PFEPh solutions in DMAA in the investigated concentration range is

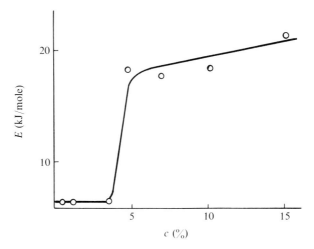

FIGURE 8.11. Flow activation energy versus the concentration of PFEPh in EA solutions.

about the same as in the case of EA, i.e., no specific influence of the solvent on the temperature coefficient of viscosity was observed.

The concentration dependence of the PFEPh solution's viscosity in EA, represented in log–log coordinates in Fig. 8.12, can be approximated by three linear sections with different slopes, each of which corresponds to different hydrodynamic behavior of the polymer–solvent system. Such a representation allows us to disclose two regions of variation of the exponent β in the power dependence of viscosity on concentration $\eta \approx C^{\beta}$. The first bend is observed at $C = 0.4\%–0.5\%$ (it is clearly seen in Fig. 8.12(a) in which the concentration dependence is given in $\eta_{\text{inh}}/C - C$ coordinates). The second bend is seen in the $3.5\%–7\%$ concentration interval.

Let us now consider in more detail the three linear sections of the viscosity concentration dependence. The first one is described by the formula $\eta \approx C^{0.4}$ and, in a sense, it corresponds to the diluted PFEPh solution. Note that such a low value of the exponent for the dilute solution regions was, earlier, practically never discussed. It is generally recognized that in polymer solutions there exists a concentration C_c corresponding to the interpenetrating of coils (in other words, the formation of a continuous entanglement network) below which $\eta \approx C$, and above which $\eta \approx C^{3-6}$ [34–35] (both the flexible-chain and stiff-chain polymers are ascribed by the given exponent values). In this connection, the first critical concentration for PFEPh solutions, $C'_c = 0.5\%$, is not quite clear, as it separates the regions with low-growth rates of hydrodynamic interaction with rising concentration. If we assume, after all, that this is responsible for the transition from isolated to interpenetrating coils it

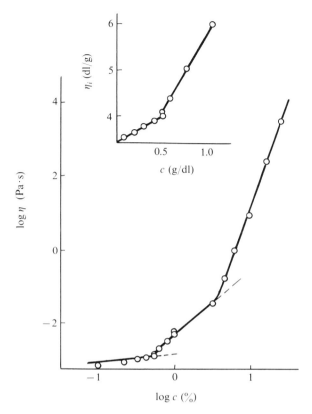

FIGURE 8.12. Concentration dependence of the viscosity of PFEPh solutions in EA at 20 °C. The low-concentration region is presented at the top section of the drawing.

is necessary to find the cause, leading to a drastic decrease (four to five times) of β in comparison with the universal values for different polymer classes.

Let us analyze an alternative hypothesis stating that the overlapping is accounted for by the second critical concentration ($C'' \approx 5\%$), while in the first case there occurs a change in the macromolecule conformation. Then the question arises as to the reasons leading to such high β values at $C > C_c''$ (≈ 8). Earlier, similar values were obtained only for the solutions of the aromatic copolyamides [36] in sulfuric acid in a quite special concentration range adjacent to the transition to the liquid crystalline state. In this region, the homo- and heterophase fluctuations are essential. In all the rest of the cases β did not exceed six.

It should be noted that at $C > C_c''$ the optical properties and high elasticity of the PFEPh solutions are varied. As far as the optical characteristics are concerned their change is associated with the appearance of turbidity of the solution which is quite visible. As for the elasticity, it is necessary to underline

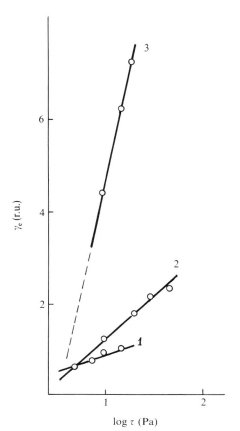

FIGURE 8.13. Dependence of the recovery strain on shear stress for 9% (1), 12% (2), and 15% (3) solutions of PFEPh in EA.

that we can measure the recoverable strain γ_e only for the solutions with $C > C_c''$. Figure 8.13 displays the dependence $\gamma_e(\tau)$ for 9%, 12% and 15% PFEPh solutions in EA. Considering that the limit of sensitivity of the used device is a γ_e-value 0.2, we can see that at $C \ll 9\%$ it is difficult to expect a notable elasticity. At the same time, using the values of the first difference of the normal stresses N_1 calculated according to the relationship $N_1 = 2\gamma_e\tau$ and the analysis of the $N_1(\dot\gamma)$ dependence, we can consider that the concentrated PFEPh solutions at small $\dot\gamma$ is soon in the linear region ($N_1 \sim \dot\gamma^2$) than in the region of the plateau, as was supported by [31] for solutions of a similar polyphosphazene. The rheological nonlinearity of the viscous behavior and linearity of the elastic characteristics are one more specific feature of PFEPh solutions.

Thus, at C_c'' we observe a change in the physical–chemical properties and the structural characteristics of the PFEPh solutions in EA. Similar behavior was observed for DMAA solutions. The difference in the absolute viscosity values of either solution is due to natural reasons, namely, the difference in

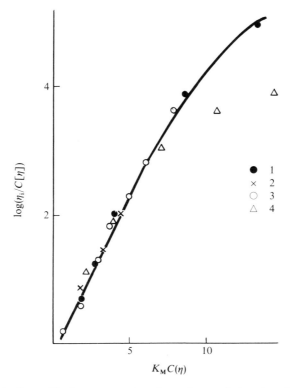

FIGURE 8.14. Generalized concentration dependence of the viscosity for PFEPh solutions in EA (1), DMAA (2), and polyfluoroalkoxyphosphazene in THF (3), and polyaryloxyphosphazene in THF (4) [31].

the thermodynamic quality of the solvents and their viscosities. Taking into account these reasons, through normalizing the specific viscosity by the $C[\eta]$ product and using the $K_M C[\eta]$ product as the concentration factor (K_M is the Martin constant), we can reduce the data on the concentration dependence of the solution viscosity of PFEPh in EA and DMAA to a general curve (Fig. 8.14). We have in mind not only the diluted, but also the concentrated, solutions, i.e., under $C > C_c''$ conditions. The experimental points also lie satisfactorily on the curve for polyfluoroalkoxyphosphazene solutions in THF. At the same time, in the case of polyaryloxyphosphazene copolymers, they are distinctly different from the generalized dependence at high concentrations (the experimental data are taken from [31]). Such distinctions may be related to another kind of chemical structure of the side groups in the same polyphosphazene main chain. Particularly, the latter copolymer does not contain fluorine atoms. The concentration reduction for such various systems should be realized in terms of differences in the critical concentrations [35].

In order to answer the question as to the legitimacy of one or another hypothesis accounting for the dramatic change in the viscous properties of the PFEPh solutions along the C scale, let us calculate the critical concentration C^* responsible for the occurrence of an intermacromolecular contacts network in the solutions. For this purpose we will use the approach as described in [37], according to which C^* is associated with the coil volumes per unit of M ($\langle S^2 \rangle^{3/2}/M$):

$$C^* = AM/N_A \langle S^2 \rangle^{3/2},$$

where N_A is the Avogadro number, $\langle S^2 \rangle$ is the mean square gyration radius of the macromolecular coil, and A is a constant depending on the way in which the equivalent spheres with radius R are packed and on the polymer density. For hexagonal packing $R = 0.735 \langle S^2 \rangle^{1/2}$, $A = 0.154$ (PFEPh density equals 1.7 g/cm^3). On the basis of the preceding data, we may consider that in diluted solutions PFEPh molecules exist in the form of isolated spherical particles whose diameter (for $M \sim 10^7$) is 600 Å. Close $\langle S^2 \rangle^{1/2}$ values are given in the literature for polychlorphenoxyphosphazene [38,39]. Taking into account the monodispersity of the synthesized polymer we may assume that $R = 300$ Å. Then $C^* \cong 4.6\%$, which is fairly close to C_c'', i.e., to the second critical concentration determined from the viscosity–concentration dependence (Fig. 8.11).

For PFEPh solutions we may also obtain the same results from simpler considerations on the basis of the specific behavior of polyphosphazenes molecules in solution, viz., the attraction between the chains is inconsiderable because of the three electronegative fluorine atoms in each side chain, which is confirmed by the small values of the flow activation energy. So, the macromolecules in diluted solution are of homogenous spheres. In this case, there arises the possibility of estimating the distance between the centers of the neighboring spheres at the concentrations C_c' and C_c''.

Without taking into account the character of the macromolecule packing in the system, we can calculate this distance from the formula, $l = 1/n$ [40], where n is the number of spheres per cubic centimeter of solution, determined as $n = N_A C_c/M \times 100$. By inserting the corresponding values in the formula, we obtain at C_c'', $l \cong 600-800$ Å, and at C_c', $l \cong 1500$ Å, i.e., by this reverse way we have arrived at the same result as earlier: the coils contact at C_c'', and are at a distance of ≈ 5 radii between the centers at C_c'.

Thus, me may suppose that in the diluted solution region ($C < C_c'$) the PFEPh macromolecules are coils which are more compact than in the case of other flexible chain polymers of similar M, and do not practically interact with one another. It is this reason that leads to low $[\eta]$ values, Huggins and Martin constants, and the K coefficient in the Mark–Houwink equation.

On exceeding the C_c' value, some kind of conformation transition takes place as a result of which the coil sizes increase somewhat (the scale of this growth is not large, which is shown by the coincidence of the above-calculated data), though the interaction between them still remains low. We concede that even in the case of the PFEPh coils collision, the presence of a large number of electronegative fluorine atoms can generate the "detachment" effect consisting of the dynamic change in the chain conformation while passing the adjacent chain through it. This effect also takes place to some extent for the overlapping coils in the concentrated solutions ($C > C_c''$), as the viscosity in this case is also not high. However, under these conditions, the growth rate of the viscosity with concentration increases essentially which, in cooperation with earlier discussed changes in the optical and elastic properties of the solutions, can prove the formation of a continuous network (at $C \sim 5$–8%), but with stronger contacts than in the case of the dispersion fluctuation interaction networks.

It is quite obvious that they have a crystalline nature, since, as a result of the regularity of the PFEPh structure, they are prone to crystallization, and in concentrated solutions the fixation in the arrangement of elements of different chains can be caused by the coil interpenetration effect. Earlier, the local crystallization phenomenon was observed in solutions of aromatic polysulphoneamide [41]. The distinction in the structure formation of PFEPh and polysulphoneamide solutions is the different degrees of completion of the crystallization process. If, in the case of PFEPh, the process is practically stopped at the stage of local crystallization (the viscosity of the structured solutions remains invariable for several months) for polysulphoneamide, the local crystalline contacts represent the nuclei of total crystallization, and the solutions become completely crystallized (solidified) for several days. The reasons for these differences is undoubtedly due to the chemical nature of the compared polymers.

8.6 Rheological Properties of Poly-*bis*-Fluoro-alkoxyphosphazenes in the Mesophase State

The specificity of the PFEPh mesophase structure does not give a priori an affirmative answer about its flow ability. Factually, it is a labile layer structure, but it is unknown whether such a lability for an authentic flow with a nonreversible displacement in a stream of molecular fragments or their aggregates (domains) is sufficient.

The flow curves of PFEPh are given in Fig. 8.15 at different temperatures. The evolution of the flow curves shape, with a decline in temperature, practically reflects the transition of the Newtonian fluid to a viscous–plastic state with a yield point τ_m, and then onto a solid-like state. At $T > T_m$ the isotropic melt has an unusual low viscosity for a flexible-chain polymer with $M \sim 10^6$; the viscosity being equal to only 3×10^3 Pa·s at $\tau = 10^3$ Pa. Note that at the molecular weights similar to PFEPh, the melt viscosity of the linear PE is two to three orders of magnitude higher [42]. The reasons for such a PFEPh isotropic melt behavior are not sufficiently clear, although there exists a hypothesis [43] about the high compactness of the PFEPh macromolecule coils

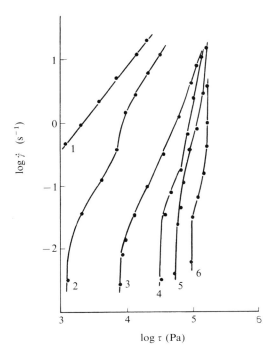

FIGURE 8.15. Flow curves of PFEPh melts at 250 °C (1), 240 °C (2), 225 °C (3), 190 °C (4), 162 °C (5), and 120 °C (6).

and the possible effect of the dynamic change in the conformation ("repulsion") as the monomer units of the adjacent macromolecules approach each other. Both factors facilitate the decrease in hydrodynamic resistance in the flow.

At temperatures below T_m, the plastic flow region appears and develops as the temperature drops. At low shear stresses, the flow curves transform into practically vertical lines with parallel ordinate axes. At $\tau > \tau_m$, PFEPh behaves like a Newtonian fluid. Deviation from linearity (viscosity anomaly) increases sharply with decreasing temperature. The same rheological behavior is also displayed by polypentafluoropropoxy- and polyheptafluorobutoxyphosphazenes.

Thus, polyfluoroalkoxyphosphazenes in the mesophase state behave as strongly structurized systems, but prone to plastic deformation. Definite conceptions about the flow mechanism can be obtained from the similarity curves' shape for PFEPh and LC polymers [44]. The flow of such systems in a specific range of shear rates and shear stresses is realized at domain level. The structure features of the PPh mesophase also allow us to support the domain character of the flow in this case. Moreover, for such a peculiar layer structure, the difference in the defectness level in each crystallographic direction is quite natural. Therefore, we can expect the existence of extended areas maintaining an interior structure under strain action. In this case, the yield point corresponds in all likelihood to the strength of the mesophase matrix in the least ordering direction. Overcoming the yield stress means violation of the system continuity with the order preservation in the rest directions. When the temperature is raised, the distortion of such a structure becomes easier as proved by the decrease in the viscous and viscoelastic characteristics as well as by the yield point values. On the whole, the approach of the experimental temperature to T_m, in combination with the high values of the shear stress, leads to "shattering" of the domain structure and transition to the molecular flow. In the rheological sense, the liquid crystalline smectics show the same behavior whereas, in the case of nematics, the viscosity in the isotropic state is higher than in anisotropic state [44].

The specific features of the structure and the rheological properties of the PFEPh mesophase lead to a phenomenon unusual in flexible-chain polymers, and which consists of the occurrence of continuous fibers (macrofibrils) of several microns in diameter in the extrudates. It is impossible to determine

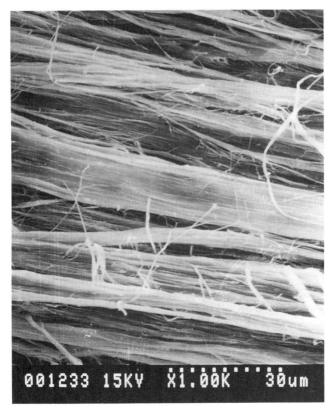

FIGURE 8.16. SEM micrograph of PFEPh fibrillized extrudite. Magnification × 2000.

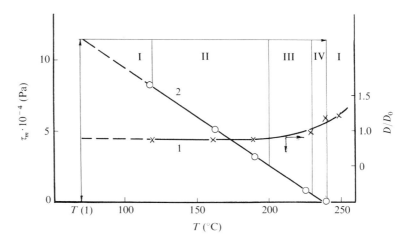

FIGURE 8.17. Diagram of different morphological forms of PFEPh extrudates and dependences of die-swell (1) and yield point (2) on temperature. I—brittle nonoriented extrudate; II—layer morphology; III—fiber morphology; IV—fiber skin and brittle core.

precisely their diameter, as the macrofibrils are complex in morphological respect, i.e., they contain micro- and submicrofibrils (Fig. 8.16).

The macrofibrils fraction in the cross section of the extrudate is determined by the temperature and the applied shear stress, to be more exact, its removal from the yield point. Thus, at 240° and stress 1×10^3 Pa, i.e., close to τ_m, the fibrous structure is formed only in the peripheral extrudate layer. The thin envelope and the core remain monolithic. With an increase in τ, up to 4×10^3 Pa, the thickness of the fibrous layer increases, and at higher values of τ, the fibers already penetrate over all the extrudate cross section.

The orginal diagram of the different morphological PFEPh structures' existence is given in Fig. 8.17. The same figure illustrates the dependence of the yield stress and die swell (D/D_0) values on temperature. The appearance of fibers is observed in the temperature range 160°–240° (zones II–IV). At lower and higher temperatures (I zone), fragile unoriented extrudates were obtained. In general, the perfection of the fibrous structure became worse with a decrease in temperature. From the x ray diffraction patterns, recorded at room temperature after fast cooling from the mesophase state (Fig. 8.18), we can see that PFEPh in the fibers is in a highly oriented crystalline state.

Thus, PFEPh, under certain conditions, can form a highly oriented self-reinforced system during extrusion. The maintainance of orientation in such a system at the channel exit explains the unusual data on the temperature dependence of the extrudate diameter (Fig. 8.17). The ratio D/D_0, where D is the extrudate diameter and D_0 is the channel diameter, characterizing the elastic recovery of PFEPh at the capillary exit, which in a broad temperature

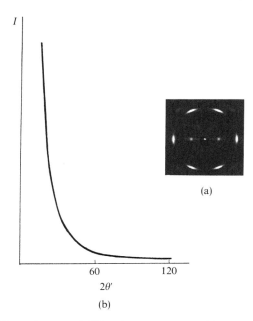

FIGURE 8.18. X ray photo and diffraction patterns at wide (a) and low (b) diffraction angles.

range appears to be close to or even less than one unit, whilst in the case of isotropic flexible chain polymer melts it approaches several units. Such low-swelling factors were noted for certain LC polymers and filled systems [45,46] and, in general, were related either to the existence of a yield point, which prevents stress relaxation and recovery of the elastic strain, or to the presence of a negative first difference of the normal stresses.

The formation of fibrous structures in the extrudates was observed earlier in individual LC thermotropic stiff-chain polymers [45,47,48], as well as in incompatible polymer blends containing [49], or not containing, LC polymers [50]. Further, the problem of fiber-formation polymer blends based on PFEPh will be considered in detail. Here we will dwell on the fibrillation of individual mesophase polymers [51]. Apparently, the fibrillation during the extrusion of PFEPh (as well as LC polymers) is based on the orientation of the domains at the channel entrance zone where there exist two types of flow: shear and uniaxial extension. The extension is a more oriented kind of deformation than the shear, the former is realized largely in the channel axis zone, while the latter is mostly intensive at the walls. The yield point in extension $\sigma_m = (3\tau_m)^{1/2}$. Therefore, we may suppose that in the case of a low pressure difference at the ends of the capillary only τ_m overcomes at the prewall region, while in the case of a high pressure difference we may also expect an overcoming in σ_m and an expansion in the zone of action of the high-shear

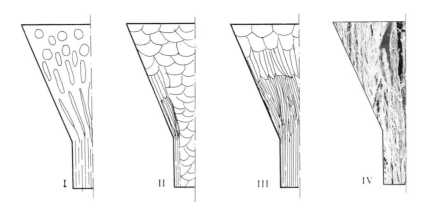

FIGURE 8.19. Scheme of fibrillation in the entrance zone: I—blend of incompatible polymers; II and III—PFEPh at different shear stresses; and IV—micrograph of PFEPh morphology in channel. Magnification × 600.

stresses in the direction of the channel central zone. Depending on the contribution of the kinds of flow and the level of stresses, we may obtain either a fibrous envelope and a monolithic unoriented core (Fig. 8.19) or a thin monolithic envelope filled with macrofibrils. The above-cited supposition was supported experimentally while performing the demountable capillary and PFEPh morphology analysis at the channel entrance zone by the SEM method. As was assumed, the fibrillation appeared to occur at the entrance cone of the channel.

The reason for the occurrence, during simple forcing through the capillary of the discrete system, is due to the instability of the ideally oriented ensemble of macromolecules and the necessity of its disintegration into a set of micro-regions separated by interfaces during cooling and crystallization. However, this problem needs further theoretical and experimental analysis.

The data under consideration reveal the complicated character of the PFEPh rheological behavior. It is characterized by the presence of a yield point in the mesophase state, by exclusively low viscosity in the isotropic state, and by a self-reinforcing phenomenon with the formation of oriented fibrous structures when forcing the mesophase polymer through the capillary. The range of the above features nears PFEPh with the filled compositions of the isotropic polymers containing highly structurizing fillers and with mixtures of incompatible polymers. Besides, the rheological behavior of PFEPh in the mesophase state is in many respect similar to the stiff-chain LC polymer behavior.

With respect to both systems, the flow in the mesophase state occurs at the level of the displacement, strain, and disintegration of the domains. The process of the rheological domains formation considered as flow units is responsible for the yield point appearance. The lability of the LC and mesomorphous structures gives rise to the sensitivity of the rheological properties to the thermal and mechanical prehistory. However, a principal distinction in the structure of the macromolecules (in contrast to LC stiff-chain systems, PFEPh does not contain mesogenic groups), and hence is the mechanism responsible for changing the stiffness along the temperature scale, does not allow us, as yet, to put the equality sign between the stiff-chain and flexible-chain mesomorphous systems. In the latter case, we have to bear in mind the circumstance that the size of the Kuhn segment, determined in the diluted solutions, does not reflect precisely the natural stiffness of the PFEPh macro-molecules in the mesophase state. It might be that we are dealing with a system whose macromolecular stiffness can increase with rising temperature due to the specificity of the macromolecular interaction.

8.7   Polymer Blends on a Base of Poly-*bis*-Trifluoroethoxyphosphazene

The application of polymer blends and alloys is very enlightening for mod-ifying the complex of individual polymer properties. From the point of view of the physical–mechanical characteristics increase, of special interest are the blends of LC and isotropic polymers. Examples of LC polyester and com-mercial polyethyleneterephtalate (PET) blends show that it is possible to raise essentially the strength and elasticity modulus of PET by introducing a thermotropic component into it [52]. It appears that the processing of the blend compositions is improved as a consequence of decreasing the blend melt viscosity, in comparison with the commercial thermoplast viscosity [53].

In this section, the PFEPh was used as an additive to polyethylene (PE) of different M. As was established above, the PFEPh mesophase has another kind of structure distinct from the typical LC systems, and is characterized by different ordering levels in each of the three crystallographic directions, and only one of them has a good translation order leading to a layer mesophase structure. It was of interest to recognize the behavior of such a structure in a blend with PE under extrusion conditions.

The blend with PFEPh ($M = 2.2 \times 10^7$) and PE with $M = 10^7$ (Bl-1), 1.4×10^6 (Bl-2), 0.6×10^6 (Bl-3) and 0.3×10^6 (Bl-4) as well as PFEPh with $M = 9.7 \times 10^6$ and PE with $M = 1.4 \times 10^6$ (Bl-5) were investigated. The values of the initial components and blend viscosities at $\tau = 3.1 \times 10^4$ were used to construct the viscosity composition plots.

Figure 8.20 illustrates the dependences of the melt viscosity on M for the samples under investigation: PE, isotropic and mesophase PFEPh. For PE melts, the slope of the dependence lg η(lg M) is close to the traditional value 3.4. For isotropic PFEPh melts it is equal to 2.6. Surprising was the absence of the viscosity dependence on the molecular weight of PFEPh in the meso-

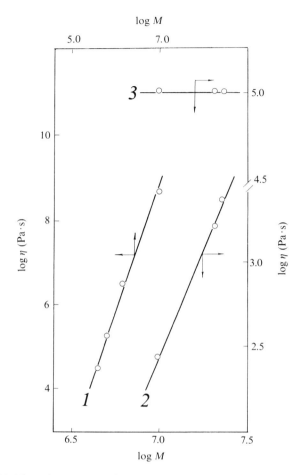

FIGURE 8.20. Viscosity versus M for PE (1), isotropic (2), and mesophase PFEPh (3) at 225 °C (1, 3) and 250 °C (2).

phase state. It is likely that this circumstance is a confirmation of the domain flow mechanism of the mesophase PFEPh, while the size of the domains are independent of M. In this connection, we cannot apparently consider the PFEPh melt at 250 °C to be completely isotropic. The elements of the layer order were mentioned earlier as being preserved in it. Possibly, it is because of this reason that the power index for the dependence $\log \eta = f(\log M)$ is substantially lower than 3.4.

The sharp distinction in the PE and PFEPh viscosities in isotropic melts is attractive by itself. This discrimination reaches the six orders (at similar M's). Probably lower viscosities of PFEPh in the isotropic state are due to

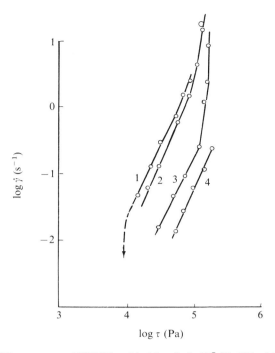

FIGURE 8.21. Flow curves of PFEPh with $M = 2 \cdot 2 \cdot 10^7$ (1), PE with $M = 1 \cdot 4 \cdot 10^6$ (4), and their blends with PFEPh content 30% (2), and 10% (3) at 225 °C.

the above-mentioned unusual hydrodynamic interaction between the macromolecules. Earlier in this chapter a supposition was put forward concerning the possible effect of the dynamic change in the conformation of the PFEPh molecules in a solution as a result of "repulsion" of the lateral groups. Such an effect can also take place in melts which, together with the residues of a layer structure, leads to a low strength of the entanglements in the pseudoisotropic melts.

The PE and PFEPh flow curves, as well as the compositions curves (10–30%), are given in Fig. 8.21. The addition of 10% PFEPh to PE brings about a 3.5 times decrease in viscosity. At $\tau \geq 1.5 \times 10^5$ Pa, we can observe a viscosity anomaly appearance, which is already typical for 100% PFEPh [54]. The viscosity anomaly effect is more pronounced for a blend with 30% PFEPh. Besides, this composition has a viscosity close to that of PFEPh. The electron microscopic patterns have shown that PFEPh appears on the surface and plays the role of a low-viscous lubricant at the blend composition flow. The redistribution of the incompatible blend components, with migration of the low-viscous component to the stream periphery, is, in principal, a known fact [55]. However, such a strong influence on the rheological prop-

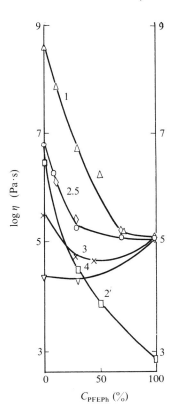

FIGURE 8.22. Viscosity versus PFEPh content at 225 °C (1–5) and 250 °C (2′). Curve numbers correspond to blend numbers (see text).

erties of the compositions is typical, namely, for the investigated and LC systems and is likely related to the unusual mechanism of the mesophase PFEPh flow. We can observe a monotonous change in the viscosity during the transition from PE to PFEPh occurring in the isotropic zone.

The viscosity–composition dependences for different blends are shown in Fig. 8.22. It is seen that for the mesophase PFEPh the character of the viscosity–composition curves is determined, in many cases, by the ratio of the components viscosities. Thus for (Bl-1), where $\eta_{PE}/\eta_{PFEPh} \cong 4 \times 10^3$, the viscosity decreases systematically with an increase in the low-viscous component content, and only at 70% PFEPh it becomes practically equal to its viscosity. We may suppose that with such a PFEPh content in the indicated system, this polymer becomes responsible for blend viscosity.

When the PE/PFEPh viscosity ratio is of several tens (Bl-2), the blend viscosity becomes practically independent on the PFEPh concentration, already at 30% of its content. Namely, for this system, we illustrated the flow curves in the previous figure, and for it the considerations are related concerning the formation of the PFEPh envelope around the composition extrudates. The hypothesis covering the domain mesophase PFEPh flow is supported by the universal character of the $\eta(C_{\text{PFEPh}})$ dependences for various PFEPh (Fig. 8.22, curves 2 and 5).

Of the $\eta_{\text{PE}}/\eta_{\text{PFEPh}}$ values, several units (Bl-3), the $\eta(C_{\text{PFEPh}})$ dependence passes through minimum (at $\sim 50\%$ PFEPh). The same picture was observed in [56] for amorphous nylon 6 and LC polyester blends at similar initial viscosity values. This can be accounted for by the circumstance that the anisotropy viscosity effect (or molecular orientation along the extrusion direction) is more pronounced in the prewall layers where the shear stresses are higher.

The blend with $\eta_{\text{PE}} < \eta_{\text{PFEPh}}$ (Bl-4) behaves like a typical filled system whose viscosity increases with an increasing content of highly viscous component particles. This case is reversed for the behavior of blend (Bl-1).

The determining influence of the components' viscosity ratio on the shape of the viscosity–composition curve is confirmed by the experiment made for (Bl-2) at 250°, i.e., at a temperature at which PFEPh is practically isotropic. The transition from 225° results in a drastic drop in η_{PFEPh} (almost 100-fold), while this in turn increases the $\eta_{\text{PE}}/\eta_{\text{PFEPh}}$ ratio and leads to a transformation of the $\eta(C_{\text{PFEPh}})$ curves from a case typical of (Bl-2) to a case typical of (Bl-1). The question concerning the PFEPh migration in the prewall layers for the above case still remains to be clarified.

The interaction effect between the PE and PFEPh phases was studied using DSC, x ray diffraction, and electron microsocopy techniques. The evolution of the thermograms is typical for the incompatible or partially compatible systems as the content of one of the components is increased. We can observe a systematic decrease in the temperatures of the PE and PFEPh transitions when 70% of the second is added which, however, does not exceed $1°-3°$, and therefore it need not be taken into consideration. The low declination from the additivity of the melting enthalpies with a change in the declination sign at the $\sim 60-70\%$ PFEPh content does not allow us to have an appropriate judgment on the interactions of the components too.

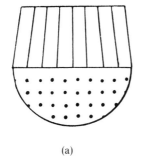

(a)   (b)

FIGURE 8.23. Sketch of reinforced sample with envelope (a) and SEM pattern (b) of blend extrudate (Bl-2).

X ray diffraction investigation of the blend extrudates shows the increase
of the PFEPh orientation beginning with its content in blend $\sim 30\%$. It is this
concentration range that leads the envelope forms for (Bl-2) and (Bl-3), while
the anisodiametric particles (in the limiting case, fibers) penetrate the extrudate cross section. The SEM picture of blend extrudates, after removing
PFEPh by selective solvent (the dark holes, PFEPh traces), and a sketch
explaining its constitution is shown in Fig. 8.23. We have to stress that this
effect takes place at a definite component viscosities ratio.

The corresponding rheological behavior diagram for 30% PFEPh additives in the coordinates of "blend viscosity normalized on component viscosity–component viscosities ratio" is illustrated in Fig. 8.24 (curve 1 is normalized on PFEPh viscosity, curve 2 on PE viscosity). In the diagram there are
three typical points, A, B, and C. At point A, $\log(\eta_{Bl}/\eta_{PE}) = 0$, i.e., $\eta_{Bl} = \eta_{PE}$
and in this case the flow occurs according to PE. Here $\eta_{PE}/\eta_{PFEPh} < 1$ and
PFEPh plays the role of a filler (see the election microscopic patterns in Fig.
8.25(a)). At point B, $\eta_{Bl} = \eta_{PFEPh}$ and $\eta_{PE} > \eta_{PFEPh}$. Such a situation is most
interesting, and it is in this case that the PFEPh envelope is formed around
the blend extrudate penetrated by the PFEPh fibers (Fig. 8.23).

At point C, $\eta_{PE} = \eta_{PFEPh}$, the blend viscosity appears to be less than each
component's viscosity. In other words, we have here a viscosity–composition
curve with a minimum ((Bl-3) in picture 23). In such a viscosity ratio the best

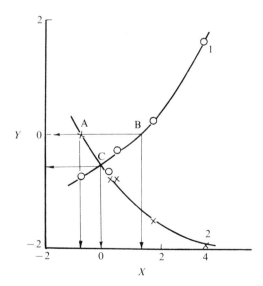

FIGURE 8.24. Diagram of the rheological behavior of blends. Y is the blend viscosity
normalized on component viscosity (curve 1 is normalized on PFEPh viscosity, curve
2 is normalized on PE viscosity); X is the viscosity ratio of the initial components.

(a)
(b)

FIGURE 8.25. SEM patterns of longitudinal and cross sections of (Bl-4) (a) and (Bl-3) (b) extrudates.

distribution of PFEPh and PE and their mutual orientation under extrusion takes place (Fig. 8.26(b)).

The unusual rheological properties of isotropic and mesophase PFEPh render it possible to modify the technological properties of flexible-chain ultrahigh molecular weight PE. Thus, comparatively small PFEPh additives decrease dramatically the PE viscosity during extrusion and allow us to obtain extrudites of a high-surface quality. Besides, the simple method of producing laminated extrudates and fibers (under specific conditions) offers promise for modifying the surface properties of polymers.

8.8   Conclusions

Polyfluoroalkoxyphosphazenes have a labile structure and, depending on the condition of polymer obtaining, the thermal prehistory of the different crystalline modifications is formed. Most interesting is the fact that their existence in the mesophase state, which in contrast to the well-known LC state structure, is characterized by the maintainance of a one-dimensional positional order between the planes accommodating the macromolecules. Their packing within the planes is characterized by the presence of an oriented order relative to the long macromolecule axis, by the short-range order in the lateral direction, and practically conformational disordering along the chain. It is likely that the such a structure can be attributed to the conformationally disordered state [2,3].

The main reason for the formation of the mesophase can be the specificity or, more exactly, the interaction contrast of the main and side chains containing a large number of electronegative fluorine atoms. Also it is important that an alternation of the P and N atoms in the main chain is substantially different from the internal structure (remember that another popular class of mesophase polymers–polyorganosiloxanes also contain different Si and O atoms in the main chain).

The specificity of the chemical PPh structure predetermines a fairly high thermodynamic stiffness of molecules which is, by the way, insufficient to form a lyotropic LC phase, particulary when dealing with PFEPh. Nevertheless, the extraordinarily low viscosities of the diluted and concentrated solutions are evidence either for very compact PFEPh coils, which is unlikely taking into account the necessity of a dense packing of six atoms of the same kind in each monomer link or for an effect of the "dynamic repulsion" under the coils interpenetration conditions. However, in the concentrated solutions range, this effect is changed by another, due to the regularity of the PFEPh macromolecular structure and leading to local contacts of a crystalline or pseudocrystalline nature. The microscopic manifestation of the presence of such a contact network is the turbidity and rubber-like behavior of the solutions.

The PPh mesophases behave like viscoplastic bodies with a yield point and strong viscosity anomaly. They practically do not show any die swelling, which is most likely due to the domain flows, i.e., the absence of unnatural molecular conformations which are realized during shear. Inside the domains, the macromolecules maintain their equilibrium conformation and are practically truly oriented along the extrusion direction. When samples with such an orientation are cooled and crystallized, the system becomes unstable and disintegrates into the microfibrils.

Using PFEPh in a blend with PE allows us to produce the same microfibers in the flexible-chain polyolefine matrix. Simultaneously with this process, a PFEPh envelope is formed which plays the role of lubricant and leads to a sharp decrease in the blend viscosity. This effect is so strong that it allows the ultrahigh-molecular polyethylene with PFEPh additives to be extrudated. The above-mentioned modifying ability of PFEPh expands substantially the region of their practical application.

References

1. P.J. Flory, Adv. Polymer Sci. **59**, 1 (1984).

2. B. Wunderlich, M. Müller, J. Grebovicz, and H. Baur, Adv. Polymer Sci. **87**, 1 (1988).

3. B. Wunderlich and J. Grebovicz, Adv. Polymer Sci. **60/61**, 2 (1984).

4. *Liquid Crystalline Order in Polymer*, edited by N.S. Schneider, G.R. Desper, and J.J. Beres (Academic Press, New York, 1978).

5. S.V. Vinogradova, D.R. Tur, and I.I. Minosiants, Uspekhi Khim. **53**, 87 (1984).

6. G. Allen, C.J. Levis, and S.M. Todd, Polymer. J. **11**, 44 (1970).

7. T.P. Russel, D.P. Anderson, R.S. Stein, G.R. Desper, J.J. Beres, and N.S. Schneider, Macromolecules **17**, 1795 (1984).

8. T. Masuko, R.L. Simeone, J.H. Magill, and D.J. Plazek, Macromolecules **17**, 2857 (1984).

9. M. Kojima and F.H. Magill, Polymer. J. **26**, 1971 (1985).

10. M. Kojima and F.H. Magill, Makromol. Chem. **187**, 649 (1985).

11. J.H. Magill, J. Petermann, and U. Rieck. Colloid. Polymer. Sci. **264**, 570 (1986).

12. V.S. Papkov, V.M. Litvinov, N.I. Dubovik, G.L. Slonimsky, D.R. Tur, S.V. Vinogradova, and V.V. Korshak, Dokl. Akad. Nauk SSSR **284**, 1423 (1985).

13. H.R. Allcock and W.J. Cook, Macromolecules **7**, 284 (1974).

14. S.V. Vinogradova, D.R. Tur, I.I. Minosiants, L.I. Komarova, and V.V. Korshak, Acta Polymerica **33**, 331 (1982).

15. D.R. Tur, V.V. Korshak, S.V. Vinogradova, S.A. Pavlova, G.I. Timofeeva, Ts.A. Goguadse, and N.O. Alikhanova, Dokl. Akad. Nauk SSSR **291**, 364 (1986).

16. V.G. Kulichikhin, E.K. Borisenkova, E.M. Antipov, D.R. Tur, S.V. Vinogradova, and N.A. Platé, Vysokomolek. Soedin. B **29**, 484 (1987).

17. D.Ya. Tsvankin, V.S. Papkov, V.P. Shukov, Yu.K. Godovsky, V.S. Svistunov, and A.A. Zhdanov, J. Polymer Sci., Polymer Chem. Ed. **23**, 1043 (1985).

18. N.B. Sokolskaya, Ya.S. Freidzon, V.V. Kochervinsky, and V.P. Shibaev, Vysokomolek. Soedin. A **28**, 300 (1986).

19. V.P. Popov, E.M. Antipov, S.A. Kuptsov, N.N. Kuzmin, I.I. Bezruk, and S.Ya. Frenkel', Acta Polymerica **36**, 131 (1985).

20. P.J. Lemstra, N.A.Y.M. Aerle, and C.V.M. Vastiaansen, Polymer J. **19**, 8598 (1987).

21. N.G. Shirina, Ph.D. Thesis, Moscow, 1986.

22. A.V. Semakov, E.K. Borisenkova, B.S. Hodyrev, D.R. Tur, and V.G. Kulichikhin, Vysokomolek. Soedin. B **31**, 830 (1989).

23. E.M. Antipov, V.G. Kulichikhin, E.K. Borisenkova, D.R. Tur, and N.A. Platé, Vysokomolek. Soedin. A **31**, 2385 (1989).

24. W.H. Stockmayer and M. Fixman, J. Polymer Sci. C **12**, 137 (1963).

25. J.E. Hearst, J. Chem. Phys. **40**, 1506 (1964).

26. M. Bohdanecky, Macromolecules **16**, 1483 (1983).

27. Y. Chatani and K. Yatsuyanagi, Macromolecules **20**, 1042 (1987).

28. D.R. Tur, G.I. Timofeeva, Z. Tuzar, and S.V. Vinogradova, Vysokomolek. Soedin. B **31**, 712 (1989).

29. S.M. Aharoni, Polymer Preprints **22**, 116 (1981).

30. L.K. Golova, G.Ya. Rudinskaya, S.A. Kuptsov, and N.V. Vasil'eva, Vysokomolek. Soedin. B **32**, 605 (1990).

31. P.K. Ho and M.C. Williams, Polymer Engrg. Sci. **21**, 233 (1981).

32. L. Sotles, D. Mikulasova, and J. Hudec, Chem. Zwesti. **35**, 543 (1981).

33. C.A. Simonescu, B.C. Simonescu, J. Neamtu, and S. Joan, Polymer J. **28**, 165 (1987).

34. *Rheology of Polymers*, edited by G.V. Vinogradov and A.Ya. Malkin (Mir, Moscow, 1980).

35. V.G. Kulichikhin, A.Ya. Malkin, E.G. Kogan, and A.V. Volokhina, Khim. Volokna. **6**, 26 (1978).

36. V.G. Kulichikhin, V.A. Platonov, L.P. Braverman, T.A. Rozdestvenskaya, E.S. Kogan, N.V. Vasil'eva, and A.V. Volokhina, Vysokomolek. Soedin. **29**, 2537 (1987).

37. G.C. Berry, H. Nakayasu, and T.G. Fox, J. Polymer Sci., Polymer Phys. Ed. **17**, 1825 (1979).

38. G. Allen, C.J. Lewis, and S.M. Todd, Polymer J. **11**, 3144 (1970).

39. B. Chu and C. Gulari, Macromolecules **12**, 445 (1979).

40. S.P. Papkov, Vysokomolek. Soedin. **24**, 869 (1982).

41. A.Ya. Malkin, L.P. Braverman, E.P. Plotnikova, and V.G. Kulichikhin, Vysokomolek. Soedin. **18**, 2596 (1976).

42. N.P. Krasnikova, D.E. Dreval, E.V. Kotova, E.P. Plotnikova, G.V. Vinogradov, G.P. Belov, and Z. Pelczbauer, Vysokomolek. Soedin. **24**, 1423 (1982).

43. N.V. Vasil'eva, V.G. Kulichikhin, L.K. Golova, D.R. Tur, S.V. Vinogradova, and S.P. Papkov, Vysokomolek. Soedin. A **31**, 852 (1989).

44. V.G. Kulichikhin, A.Ya. Malkin, and S.P. Papkov, Vysokomolek. Soedin. A **26**, 451 (1984).

45. D.G. Baird, in *Polymer Liquid Crystals,* Proceedings of the 2nd Symposium Division Polymer Chemistry, New York (1985).

46. G.V. Vinogradov, E.K. Borisenkova, and M.P. Zabugina, Dokl. Akad. Nauk SSSR **277**, 614 (1984).

47. E. Baer, A. Hiltner, and H.D. Keith, Science **235**, 1015 (1987).

48. L.C. Sawyer and M. Jaffe, J. Mater. Sci. **21**, 1897 (1986).

49. V.G. Kulichikhin, O.V. Vasil'eva, I.A. Litvinov, E.M. Antipov, I.L. Parsamyan, and N.A. Platé, J. Appl. Polymer Sci., **42**, 363 (1991).

50. M.V. Tsebrenko, G.V. Vinogradov, T.I. Ablazova, and A.V. Yudin, Kolloid. J. **38**, 200 (1976).

51. V.G. Kulichikhin, E.K. Borisenkova, D.R. Tur. V.V. Barancheeva, I.I. Konstantinov, E.M. Antipov, V.E. Dreval, and N.A. Platé, Vysokomolek. Soedin. A **31**, 1636 (1989).

52. M. Amano and K. Nakagawa, Polymer J. **28**, 263 (1987).

53. A. Siegman, A. Dagan, and S. Kenig, Polymer J. **26**, 1325 (1985).

54. E.K. Borisenkova, D.R. Tur. I.A. Litvinov, V.G. Kulichikhin, and N.A. Plate, Vysokomolek. Soedin. A **32**, 1505 (1990).

55. R.A. Weiss, H. Wansoo, and L. Nicolais, Polymer Engrg. Sci. **27**, 684 (1987).

56. K.G. Blizard and D.G. Baird, Polymer Engrg. Sci. **27**, 653 (1987).

9

Chiral Nematic Mesophases of Lyotropic and Thermotropic Cellulose Derivatives

D.G. Gray and B.R. Harkness

9.1   Mesophase Formation

Polymer molecules are capable of forming lyotropic and thermotropic liquid crystalline phases. These can be grouped into three general categories; *side-chain liquid crystalline polymers*, *main-chain liquid crystalline polymers*, and *rigid-rod liquid crystalline polymers*. Side-chain liquid crystalline polymers have small mesogenic groups attached to the flexible polymer backbone. Main-chain liquid crystalline polymers contain mesogenic groups within the polymer backbone, joined by flexible segments. It is the mesogenic groups that are responsible for the liquid crystalline behavior of these macromolecules. Rigid-rod liquid crystalline polymers have a rigid or semirigid backbone. The polymer may adopt a helical secondary structure that results in the molecule having a rod-like shape. This is the case for certain polypeptides which possess an α-helical conformation that arises as the result of intramolecular hydrogen bonding and steric effects. Liquid crystalline rigid-rod polymers may also be composed of conformationally rigid monomer units which restrict the flexibility of the chain. These rigid monomers may be para-substituted aromatic groups as in polyamides or polyesters, or the β-linked glucopyranosic units found in cellulose and its derivatives.

Several theories of ordered phase formation relevant to this last group of polymers have been presented. Onsager, using a virial approach, showed that a solution of hard, asymmetric, rod-like molecules should phase separate into two phases at a critical concentration that depends on the axial ratio of the molecule [1]. Flory proposed a lattice theory to account for the phase separation of rigid rod-like macromolecules in solution. Ignoring intermolecular interactions with the exception of the interactions of infinite magnitude that occur when two molecules occupy the same position in the lattice, the critical concentration at which phase separation occurs can be expressed as

$$V_p^* \approx \frac{8}{X}\left(1 - \frac{2}{X}\right),$$ (9.1)

where V_p^* is the critical volume fraction of the polymer and X is its axial ratio [2]. The phase separation of the solution can be attributed to a favorable entropy of forming a liquid crystalline phase at high polymer concentrations. Maier and Saupe showed that the formation of a nematic phase may arise from energy differences arising from orientation-dependent interactions [3,4]. It appears that both entropy and enthalpy contribute to the stability of nematic mesophases, although for liquid crystalline polymers the dominant factor is the entropy term which is controlled by the size and shape of the molecule. Flory and coworkers [5,6] have modified the lattice theory to account for asymmetric attractions between polymer chains.

There are several problems associated with the application of the Flory and Onsager theories to real liquid crystalline polymers. The problem that has received most attention recently concerns the actual rigidity of the polymers. The original theoretical treatments assumed that the chains were completely rigid and rod-like, but even the most rigid polymers have some degree of flexibility at large molecular weights. To account for the semiflexible nature of liquid crystalline polymers, Khokhlov and Semenov have extended the Onsager approach from rods to worm-like chains [7,8]. It has been shown by Odjik that the theory of Khokhlov and Semenov can accurately predict the phase behavior for such polymers as schizophyllan, poly(γ-benzyl L-glutamate) and DNA [9], although in these cases the polymers are relatively stiff with large persistence lengths. Sato and Teramoto extended Cotter's scaled particle theory and found good agreement with phase boundaries and osmotic pressures of schizophyllan and poly(hexyl isocyanate) in dichloromethane and toluene [10]. Recently, Hentschke has modified the Khokhlov and Semenov theory to obtain an equation of state for persistent-flexible main-chain polymers that is valid over a wide range of polymer concentrations [11]. Experimental data on poly(γ-benzyl L-glutamate) in dimethylformamide was in good agreement with the theoretical predictions, but the theory has not been applied to polymers with short persistent lengths.

The lattice theory of Flory has also been modified to account for chain flexibility in liquid crystalline polymers [12,13]. Flory suggested that the critical concentration for mesophase formation from chains of freely jointed rods depended on the axial ratio of the individual rods, and not on the contour length of the chain. It has since been assumed that semiflexible polymers that behave as Kuhn chains can be treated in the same way as stiff polymers if the Kuhn segment length, k, rather than the contour length of the polymer is used to evaluate the axial ratio. Recently, the Flory theory has been extended to account for the phase behavior of semiflexible polymers with flexible side chains [14].

9.2   Mesophases of Cellulose Derivatives

Cellulose is a naturally occurring polymer consisting of a chain of 1–4 linked β-D-glucose units. The pyranose rings of the repeat units are in the chair conformation with the glycosidic bonds between carbons 1 and 4 in an equatorial position. A variety of cellulose derivatives have been prepared by substituting side-groups onto the hydroxyl groups at positions 2, 3, and 6 of the anhydroglucose units. The nature of the side-group as well as the number of side-groups (the degree of substitution, DS, is defined as the average number of substituted hydroxyl positions per anhydroglucose unit) and their distribution along the chain can dramatically affect the physical properties of cellulose derivatives.

The conformation of cellulose and its derivatives has been investigated by means of conformational energy calculations, x ray diffraction, electron microscopy, and NMR spectroscopy. For cellulose-based polymers, the conformation of the chain depends on the angles of rotation about the two bonds to each glycosidic oxygen atom. The angles of rotation ϕ and ψ are defined as the dihedral angles between C1–H1 and O4–C4 and between O4–C1 and C4–H4, respectively. The results of energy minimizing calculations indicate that the cellulose chain adopts an extended helical conformation which results from a limited number of energetically favorable conformations for the angles ϕ and ψ [15–17]. Leung et al. have analyzed the crystal structure of the model compound β-D-acetyl cellobiose and determined that the glycosidic torsion angles ϕ and ψ correspond to the maximum extension of the molecule [18]. The crystal structures of several cellulose derivatives indicate that the cellulose backbones of these derivatives have a helical conformation in the crystalline state [19]. Perlin et al. have measured the vicinal $^{13}C–^1H$ coupling across the glycosidic linkage of cellobiose and obtained evidence that this disaccharide adopts a conformation in solution which is similar to that in the solid state [20].

The molecular dimensions of cellulose and cellulose derivatives have also been examined by their hydrodynamic behavior [21]. These studies have indicated, in general, that many cellulose derivatives have an extended conformation in solution.

In light of the apparent stiffness and extended conformation of cellulose-based polymers, it is in retrospect not surprising that many form liquid crystalline phases. The first published report of the formation of a liquid crystalline phase by a cellulose derivative [22] showed that concentrated aqueous solutions of hydroxypropyl cellulose form lyotropic mesophases that display iridescent colors over a specific concentration range. Since this initial report in 1976, many cellulose derivatives have been found to form lyotropic and thermotropic liquid crystalline phases.

The semiflexible nature of cellulose derivatives results in these polymers having a critical concentration for mesophase formation which is greater than that of rigid rods of the same molecular weight. The critical concentration varies with the nature of the solvent, indicating that the solvent plays a role in determining the stiffness of the cellulose backbone, but the critical concentration was found to be unaffected by the degree of polymerization of HPC as expected for a semiflexible chain [23,24]. Similar results were obtained for HPC in dimethylacetamide [25] and (acetoxypropyl)cellulose in dibutyl phthalate [26]. However, these results were for polydisperse samples with chain lengths much greater than their persistence lengths. The critical concentration for cellulose acetate in trifluoroacetic acid [27,28] and cellulose in N-methylmorpholine N-oxide [29] was found to depend on the degree of polymerization. Evidently, these two polymers assume a more rigid conformation in the given solvents than HPC and its derivatives. Recent calorimetric and turbidimetric measurements on carefully fractionated and characterized HPC solutions in water [30,31] also showed a marked dependence of critical concentration on molar mass. This work also demonstrated that the "gel phase" that separates when a *dilute* aqueous solution of HPC is heated is, in fact, a stable emulsion of the concentrated anisotropic phase dispersed in dilute isotropic solution. In addition to the usual wide and narrow biphasic regions, the phase diagram shows a curious intermediate region (Fig. 9.1). On heating a biphasic solution initially in the narrow region (about

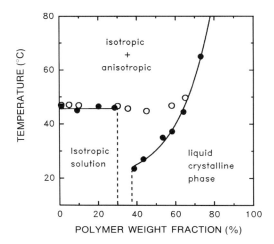

FIGURE 9.1. Phase diagram for the fraction of hydroxypropylcellulose (molar substitution = 5.0, $M_w = 209,000$ g·mol^{-1}) in water [31]. The onset of phase separation was detected by turbidimetry (●). Transition temperatures measured with the scanning Calvet calorimeter are also shown (○).

30–37 wt.% polymer for the fraction shown in Fig. 9.1), not one but two turbidimetric transitions are observed. The transitions are not due to changes in the number or nature of the phases, but rather to sharp changes in their composition [30]. Precise measurements of the temperature and enthalpy of phase separation with a scanning Calvet calorimeter suggest that the observed thermal effects occur in the isotropic phase, and the isotropic–anisotropic transition is driven by entropic factors alone [31]. This provides an explanation for the absence of discontinuities in temperature (Fig. 9.1) and enthalpy at the critical concentration for phase separation. This phase behavior of the HPC–water system stems in part from the tendency of many aqueous solutions to demix on heating [30], and is thus rather different from that observed for most lyotropic cellulose derivatives.

The rigidity of cellulose derivatives in solution may be modeled by the Kratky–Porod worm-like chain, in which the chain stiffness is measured in terms of the persistence length q, where $q = k/2$ in the limit of infinitely long chains. The persistence length can be determined experimentally by light scattering [32] and viscosity measurements [33]. It has been shown that the persistence lengths for cellulose derivatives are generally 10 nm or shorter [25,26,34,35] although there are some exceptions [36]. The predicted results of the Khokhlov–Semenov theory for some cellulose derivatives deviate significantly from the experimental data [26], but the theory may be more accurate for derivatives with larger persistence lengths. The Flory lattice theory has been applied quite extensively to the phase behavior of cellulose derivatives. It has been found that the critical concentration for mesophase formation is in reasonable agreement with equation (9.1) when x is taken as the Kuhn segment axial ratio of the polymer [23,26,34,35,36], although it has been suggested that soft asymmetric attractions may also be involved [34,35].

Many cellulosics also form thermotropic liquid crystalline phases with chiral nematic properties. (Acetoxypropyl)cellulose, prepared by acetylating HPC, was the first reported thermotropic [37], and a large number of other heavily substituted esters and ethers of (hydroxyalkyl)celluloses were subsequently found to form equilibrium chiral nematic phases, even at ambient temperatures [38,39,40,41]. There is a dearth of careful studies on the thermotropic phase behavior of well-characterized fractions of cellulose derivatives; the chiroptical properties of some of these polymers are discussed in the following sections.

9.3   Chiral Nematic Properties

9.3.1   Theories of the Chiral Nematic State

As discussed above, cellulose derivatives form liquid crystalline phases primarily due to the higher combinatorial entropy of mixing in the ordered phase. However, cellulose derivatives are chiral, and as a result there is also a chiral contribution to interactions between the rods in the mesophase, leading to a chiral nematic structure. Several theories have been proposed to account for the twisting force between chiral rod-like mesogens in the liquid crystalline phase. Goossens was the first to propose that the chiral nematic structure is the result of an anisotropic dispersion energy between chiral mesogens [42]. Samulski and Samulski calculated the intermolecular forces between two rod-like chiral mesogens in a dielectric medium [43]. They determined that the introduction of an asymmetric dispersion energy results in adjacent mesogens having a slight twist relative to each other. An increase in the temperature of the mesophase would be expected to increase the frequency of molecular rotation about the long molecular axis, resulting in a decrease in the dispersion energy and an increase in the pitch. An important contribution of the Samulski and Samulski theory was the prediction that the macroscopic twist sense should also depend on the dielectric properties of the medium in addition to the chirality of the constituent molecules.

The observed handedness and pitch of the helicoidal structure depends on the concentration and structure of the components, and is also sensitive to temperature. The observed temperature dependence of the pitch for chiral nematic phases can be classified into four categories, according to Lin-Liu et al. [44]:

(i) The pitch increases to a finite value with a decrease in temperature. This is the most common observation for chiral nematic phases composed of pure small molecule chiral mesogens and is observed for some chiral rod-like polymers.

(ii) The pitch increases with a decrease in temperature, reaching infinity at the nematic temperature T_N, and then decreases with the opposite twist sense. This has been observed for mixtures of left- and right-handed small molecule chiral nematic phases.

(iii) The pitch increases to a finite value with an increase in the temperature. This has been observed for chiral nematic phases composed of chiral rod-like polymers.

(iv) The pitch increases with an increase in temperature, reaching infinity at the nematic temperature T_N, and then decreases with the opposite twist sense. This has been observed for chiral nematic phases composed of rod-like chiral polymers.

Keating was the first to propose a theory describing the temperature dependence of the pitch for small molecule chiral nematic mesophases [45]. Modeling the temperature dependence of the pitch as a rotational analogue of thermal expansion, Keating obtained the following expression:

$$P = \frac{4\pi a I \omega_0^4}{AkT},$$ (9.2)

where A is a constant, ω_0 is the angular frequency of rotation of the rod-like molecule about its long axis, I is the moment of inertia of the molecule, a is the interplanar distance, and k is the Boltzmann constant. The observed temperature effects are more complex than predicted by this equation. Subsequently, Lin-Liu et al. developed a theory to account for all of the above temperature effects [44,46]. According to their model, the temperature dependence of the pitch is determined by the shape and position of the intermolecular potential as a function of the intermolecular twist angle.

Kimura et al. have proposed a statistical theory that describes the role of attractive and repulsive asymmetric intermolecular forces in determining the pitch of a lyotropic mesophase of long rod-like macromolecules such as synthetic polypeptides [47,48]. The relationship between the pitch and temperature, concentration, and polymer molecular weight is expressed as

$$\frac{1}{P} = \frac{12\lambda\Delta}{\pi^2 LD} cf(c)\left(\frac{T_N}{T} - 1\right),\qquad(9.3)$$

where T_N is the temperature at which the mesophase is nematic, c is the polymer volume fraction, L is the length of the rod, D is the width of the rod, λ is a numerical factor, and Δ is the height of the ridge of the coil. Δ corresponds to the length of the side-group measured from the core of the rod. The concentration function $f(c)$ is defined as

$$f(c) = \frac{(1 - \frac{1}{3}c)}{(1 - c)^2}.\qquad(9.4)$$

The values of λ and Δ in (9.3) depend on the shape of the molecule with Δ depending on the conformation of the side-groups. Kimura et al. have concluded that a change in the twist sense of poly(γ-benzyl L-glutamate) mesophases can occur thermally without a change in the apparent helical sense of the polypeptide secondary structure.

As described in detail in Chapter 1, Osipov has developed a statistical theory taking into account steric and chiral interactions between polypeptide molecules in solution to predict the influence of temperature and solvent dielectric constant on the pitch and twist sense of poly(γ-benzyl L-glutamate) liquid crystals [49]. The pitch is given by

$$\frac{1}{P} = \frac{\rho^2}{12\pi}(J_0(\varepsilon_m) - \lambda k_B T)S(S + 2)/K_{22},\qquad(9.5)$$

where ρ is the number density of peptide molecules, λ is a pseudoscalar, k_B is a constant, S is the orientational order parameter, K_{22} is the twist elastic constant of the liquid crystal, and $J_0(\varepsilon_m)$ is the coupling constant which is dependent on the molecular polarizability, molecular optical activity, and the solvent dielectric constant. The steric contribution to the pitch is given by the term $-\lambda k_B T$. The steric and coupling terms are of opposite sign, so a thermally induced change in the chiral nematic twist sense will occur when $T = J_0(\varepsilon_m)/\lambda k_B$. The behavior thus depends on the dielectric constant of the solvent ε_m. Osipov concluded that the magnitude of the pitch and the sign of the chiral nematic twist sense are determined by the influence of the polymer-solvent interaction on the optical activity of the polypeptide. He has extended his treatment of polypeptide mesophases to those of cellulose derivatives [50]. Assuming that the cellulose chain adopts a "twisted belt" conformation in solution, a relationship similar to equation (9.5) was obtained

$$2\pi P^{-1} = -\tfrac{1}{2}\rho^2(\chi - \lambda kT)K_{22}^{-1}, \tag{9.6}$$

where the parameter λ depends on the steric repulsion between the chains, χ is related to the attraction interaction, P is the pitch, k is a constant, ρ is the number density of rigid segments, and K_{22} is the twist elastic constant of the liquid crystal.

The temperature dependence of the pitch for polypeptide liquid crystalline phases has also been expressed by the empirical relationship [51]

$$P^{-1} = a(1 - T/T_N), \tag{9.7}$$

where the value of a depends on the concentration of the polymer, and is given by

$$a \propto c^n, \tag{9.8}$$

where the exponent n varies in the range of 1 to 2 for polypeptides. The value of T_N is also dependent on the concentration of the polymer.

9.3.2   Chiral Nematic Properties of Cellulose Derivatives

Mesophases of cellulose derivatives form chiral nematic structures as a result of the chirality of the cellulose backbone. They exhibit optical properties analogous to those observed from chiral nematic mesophases of polypeptides and chiral small molecule mesogens. These optical properties have been found to be sensitive to several factors such as the nature of the side-groups, the degree of substitution, the molecular weight of the polymer, temperature, the nature of the solvent, and the polymer concentration. The optical properties of hydroxypropyl cellulose mesophases were among the first to be investigated.

Based on the form of the wavelength dependence of the optical rotation close to the chiral nematic reflection bands [22], Werbowyj and Gray observed that lyotropic mesophases of hydroxypropyl cellulose in water, acetic acid, and methanol have a right-handed chiral nematic twist sense [52]. Several ethers and esters of hydroxypropyl cellulose prepared by etherifying or esterifying the hydroxyl groups on the HPC polymer also form right-handed chiral nematic phases, as indicated by the shape of the optical rotation (ORD) or apparent circular dichroism (CD) curves close to the reflection bands [48–52]. The form of such curves is illustrated in Fig. 9.2, which shows chiroptical spectra for a thin (20–30 mm) layer of (ethoxypropyl)cellulose thermotropic mesophase. Curve (a) shows the optical rotatory dispersion (ORD) at 140 °C; the shape is similar to the Cotton effect observed for an

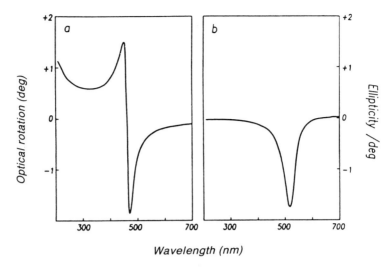

Wavelength (nm)

FIGURE 9.2. Chiroptical spectra for a thin (20–30 mm) layer of ethoxypropyl-cellulose [41]. Curve (a) shows the optical rotatory dispersion (ORD) at 140 °C, and curve (b) shows the apparent circular dichroism (CD) or circular reflectivity at 152 °C. The sample forms a thermotropic right-handed chiral nematic phase with a reflection band close to 500 nm at these temperatures.

isolated chromophore in a chiral molecular environment in dilute solution [53], but in this case the magnitude of the optical rotation due to the helicoidal supramolecular structure [54] is much greater. Curve (b) shows the apparent circular dichroism (CD) at 152 °C; again this resembles the circular dichroism observed for an isolated chromophore in dilute solution, but the apparent absorption band is much stronger, and is due to the reflection of light by the mesophase. The shapes of the curves indicate a right-handed chiral helicoidal structure with a reflection band close to 500 nm at these temperatures. (For a left-handed structure, the peak shapes would be inverted, with the ORD peak changing sharply from negative to positive with increasing wavelength, and a positive apparent CD peak.) The ORD and apparent CD signals from a reflection band are of course related; in practice, the measurement of apparent CD with a modern CD spectrometer is more convenient and sensitive for weakly reflecting samples, while the ORD method is more suitable for measuring the reflection wavelength of strong reflection bands.

Measurement of the apparent absorption spectra may also be used to characterize the magnitude and wavelength of the reflection band. A perfect chiral nematic with monodomain planar texture should reflect all of one hand of circularly polarized light, and transmit all of the other hand, thus giving a transmittance at the peak maximum of 50% incident unpolarized

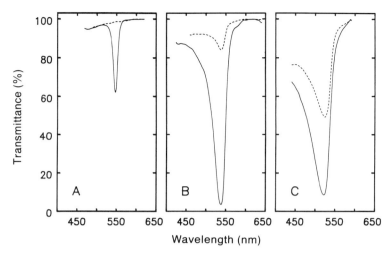

FIGURE 9.3. The effect of sample thickness on the UV-visible transmittance of circularly polarized light through a HPC–water mesophase with a reflection band at ≈ 540 nm [55]. The thickness of the mesophase was (A) = 160 μm, (B) = 660 μm, and (C) = 860 μm. The solid and broken lines are for right- and left-handed circularly polarized light, respectively.

light. This reflection requires that the sample be about an order of magnitude thicker than the cholesteric pitch [54]. Curiously, on increasing the thickness of the sample, the decrease in transmittance has been observed to be greater than 50% [55]. Thus, thicker samples must reflect both components of circularly polarized light. This is demonstrated in Fig. 9.3, which shows that a thin sample (A) of carefully prepared HPC–water mesophase reflects only one hand of circularly polarized light, whereas the polarization of light reflected by a thick sample (C) is much lower. Sample (B) approaches the theoretical reflectivity for a planar chiral nematic. The reason for the loss of selectivity in thick samples is not obvious, but it should be noted that thick samples are not planar, but show more complex focal conic textures, so oblique and multiple reflections of light must occur within the sample.

9.3.3   Reversal of Handedness of Chiral Nematic Cellulosics

HPC and its ether and ester derivatives generally form chiral nematic mesophases that have a right-handed chiral nematic twist sense, with a pitch that increases with increasing temperature and decreasing polymer concentration. This simple state of affairs might have been expected for a family of polymers that share the same chiral cellulose backbone. (The side-chain substituents are normally achiral.) However, recent observations indicate more complex patterns of behavior. The first left-handed chiral nematic mesophase of a cellulose derivative was observed by Vogt and Zugenmaier for a solution of ethyl cellulose in glacial acetic acid [56]. This was the first indication that the nature of the side-groups attached to the cellulose backbone can influence the chiral nematic twist sense. There have been several subsequent reports of a change in the chiral nematic twist sense that occurs with a change in the side-group substituents. Cellulose acetate in trifluoroacetic acid forms a left-handed chiral nematic mesophase which inverts to a right-handed structure when trifluoroacetate groups are added to the polymer backbone [57]. The substitution of trifluoroacetate groups also influenced the magnitude of the concentration dependence of the pitch. The twist sense of ethyl cellulose/chloroform mesophases is also dependent upon the nature of the side-group substitution. The acetylation of the free hydroxyl groups on ethyl cellulose can change the chiral nematic twist sense from being right-handed to left-handed, in chloroform, when a critical degree of acetylation has been reached [58], the pitch of the left-handed mesophases decreased with an increase in temperature whereas the right-handed mesophases showed an increase in the pitch with increasing temperature (Fig. 9.4). Other modifications that have been observed to change the twist sense are the acetoacetylation of hydroxy-propyl cellulose [59] and the substitution of a chlorine atom at the para-position of the cellulose tricarbanilate phenyl groups [60,61]. Further exam-

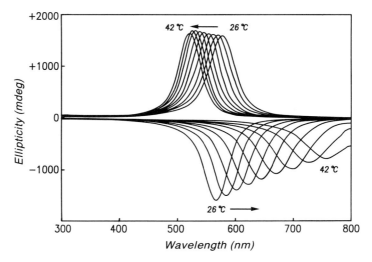

FIGURE 9.4. Reversal of handedness and temperature dependence of chiral nematic pitch with degree of substitution [58]. The positive CD curves are for a left-handed mesophase of (acetyl)(ethyl)cellulose with acetyl DS = 0.06 and ethyl DS = 2.5. The negative curves are for a right-handed mesophase of (acetyl)(ethyl)cellulose with acetyl DS = 0.5 and ethyl DS = 2.5. Concentration $\approx 50\%$ by weight in chloroform.

ples for (trityl)cellulose derivatives are given below. Evidently, the sign and magnitude of the pitch are quite sensitive to chain structure.

There are cases where the chiral nematic twist sense is determined by the solvent. The mesophase of ethyl cellulose in acetic acid and dichloroacetic acid changes from a left-handed to right-handed structure with an increase in the concentration of dichloroacetic acid [62]. In this case, it is possible that the solvent may be reacting with the polymer and therefore the change in the twist sense may be the result of a modification of ethyl cellulose. In another example, cellulose tricarbanilate was found to form a right-handed chiral nematic structure in methylpropyl ketone and a left-handed structure in diethyleneglycol monoethyl ether [60].

As mentioned above, the pitch of chiral nematic mesophases of cellulose derivatives is sensitive to temperature. The usual behavior is an increase in the pitch with an increase in temperature, although there are several examples of a decrease in the pitch with an increase in the temperature. An example of a thermally induced inversion of the twist sense analogous to that observed for thermotropic polypeptides has been reported for oligomers of tri-O-2-(2-methoxyethoxy)ethyl cellulose (TMEC) and tri-O-heptyl cellulose [63]. The pitch of TMEC increased with temperature from 80 °C to 106 °C and then decreased with the opposite twist sense as the temperature was increased above 106 °C.

The concentration of the polymer also influences the magnitude of the pitch. The usual behavior for lyotropic mesophases of cellulose derivatives is a decrease in the pitch with an increase in the polymer concentration. However, an increase in the polymer concentration can also give an increase in the pitch [56,61]. The pitch also varies as a function of the molar mass of the polymer. For thermotropic (acetoxypropyl)cellulose fractions, the short chain fraction showed the longest pitch [40], and a similar trend was observed for HPC in water [30]. However, Siekmeyer and Zugenmaier have reported that the pitch of a cellulose tricarbanilate/diethylene glycol monoethyl ether mesophase increases with an increase in the molar mass of the polymer [60].

The dependence of the chiral nematic pitch on concentration and temperature is, in principle, predicted by equations (9.3) and (9.4), developed for a helically grooved rod model [47,48]. Attempts to rationalize the experimental concentration dependence of the pitch for hydroxypropyl cellulose in water [52] and the temperature dependence of the pitch for (acetoxypropyl)-cellulose in dibutylphthalate [40] according to this model are unconvincing, because the compensation temperature, T_N (the temperature where the pitch becomes infinite and changes handedness) could not be experimentally determined, and is almost certainly a function of polymer concentration in these systems.

The change in the twist sense of chiral nematic mesophases can be accommodated in the theories of Kimura et al. [47,48] and Osipov [49,50]. For mesophases of helical molecules such as polypeptides, the twist sense of the mesophases depends *inter alia* on the sense of the helicity of the individual polymer chains [51]. The twist sense of the mesophases of cellulose derivatives may also be influenced by a helical contribution from the cellulose backbone secondary structure. However, the evidence for chain helicity in solutions or melts of cellulose derivatives is currently much more tenuous than in the case of polypeptides, and there has been no clear correlation between the conformation or possible helicity of cellulose derivatives in solution and the chiral nematic twist sense of their liquid crystalline phase. Measurements of circular dichroism and induced circular dichroism should provide information relevant to this problem.

9.4 Circular Dichroism and Induced Circular Dichroism

Chiroptical measurements can provide information concerning macromolecular conformation. Molecules that exhibit chiroptical activity can be divided into two groups [64]. The first group consists of molecules that are intrinsically chiral and exhibit chiroptical activity when in the gas, liquid, or solid phase. The second group consists of achiral molecules that are extrinsically chiral, exhibiting chiroptical behavior when in the solid phase or in a chiral environment.

Intrinsically chiral molecules can be subdivided into two categories [65]. The first includes molecules that contain inherently dissymmetric chromophores such as twisted biphenyls and skewed dienes. The second category includes molecules possessing inherently symmetric chromophores that are asymmetrically perturbed. In this case, the optical activity of the chromophore is the result of a dissymmetric molecular environment. Phenyl and carbonyl groups are examples of this type of chromophore.

The sign and magnitude of the chiral activity for an electronic transition in an intrinsically chiral molecule is related to the electronic and structural properties of the molecules in the ground and excited states. These properties may be inferred from spectroscopic measurements of the circular dichroism (CD) displayed by the molecule. In its simplest form, the CD band for an electronic transition has the same general shape as the absorption band, with the sign of the ellipticity determined by the asymmetry of the molecular environment. It is still not possible to predict accurately the sign and magnitude of the chiroptical activity of the chiral molecules such as carbohydrates. Several empirical or semiempirical approaches have been used to interpret the optical rotation of these molecules in terms of their stereochemistry [66–69]. These methods involve the summation of the partial rotatory contributions of the various conformation elements and have led to rules that allow for the estimation of the optical rotations for several cyclic sugars. Until recently, there has been a limited number of studies into the circular dichroism of carbohydrates due to the absence of chromophores in accessible regions of the spectrum. For this reason, most of the CD studies of carbohydrates have involved derivatives containing carbonyl or aromatic groups. Stevens and coworkers have recently extended the study of carbohydrate circular dichroism into the vacuum ultraviolet region of the spectrum to measure the electronic transitions responsible for the Na_D rotation [70]. It was found that an electronic transition near 150 nm accounts for most if not all of the observed Na_D rotation. Stevens and coworkers have also proposed a semiempirical theory for saccharide rotation that includes the coupling of chiral contributions from the electronic transitions of all CC, CO, and CH groups in the carbohydrate [71,72].

When two or more chromophores are in close proximity, the electronic transition dipole moments may interact by exciton coupling if they are held in a fixed or restricted chiral orientation relative to each other [53]. This is often observed in polymers with a helical array of chromophores along the polymer backbone [53] and in small molecules with a chiral arrangement of chromophores [73]. The chiral orientation of the chromophores results in chiroptical activity that is dependent on both the chirality of the single perturbed chromophore as well as the optical activity resulting from exciton coupling. Exciton coupling of strong electronic transitions results in a CD band with two components separated in wavelength by $\Delta\lambda$, symmetrically disposed at higher and lower frequencies about the chromophore absorption maximum λ_0. The chirality of the exciton interaction depends on the skew sense of the interacting chromophores and the polarization directions of their transition moments. The magnitude of the exciton coupling depends on the strength and orientation of the interacting electronic transitions as well as on the distance between the two chromophores. Theoretically, the interaction between the transition dipoles decreases as the inverse third power of the interchromophore distance [53]. Electric–dipole forbidden transitions are usually too weak to give rise to exciton interactions.

Exciton chirality has been useful in determining the absolute stereochemistry of natural products such as chiral diols containing dibenzoate moieties. The hydroxyl configuration of sugars has been determined by this method [74]. The analysis of exciton effects in biological polymers, particularly polypeptides, has been used to provide evidence for the presence of a helical secondary structure in solution [53]. Exciton effects have been observed in vacuum–UV CD measurements on cellulose acetate [75]. Some experimental evidence for the splitting of induced CD peaks, for dilute solutions of cellulose ologomers and methylcellulose complexed with Congo red [76] and of carboxymethylcellulose complexed to acridine orange [77], have been attributed to exciton coupling resulting from a helical secondary structure.

Optical activity in polymers has been shown to arise from chromophores perturbed by a dissymmetric environment and from exciton coupling that results from a helical arrangement of chromophores. Optical activity may also result from a dissymmetric array of aggregated molecules. This is referred to as induced optical activity. Chiral nematic mesophases have been observed to induce optical activity in electronic transitions as a result of the imposed tertiary helicoidal arrangement of chromophores [64]. Circular dichroism, induced by the chirality of a liquid crystalline phase, is called "liquid crystal-induced circular dichroism" (LCICD). Achiral molecules such as anthracene and pyrene show detectable CD signals in mesophases of poly-(γ-benzyl L-glutamate), but no detectable signals in isotropic solutions of the same polymer [78]. LCICD has also been observed for acridine orange in a chiral nematic mesophase of cellulose acetate and trifluoroacetic acid [79] and for Congo red in cellulose films with a chiral nematic structure [80].

LCICD has been observed in chromophores intrinsic to the polymer. There are no detectable CD signals for the phenyl side-groups of poly(γ-benzyl L-glutamate) in the isotropic state, but at concentrations above the critical concentration for mesophase formation strong CD signals are observed [51,81]. The observation of LCICD is therefore associated with the chiral nematic mesophase and disappears when that order is lost.

There have been many examples of LCICD in both polymer and small molecule liquid crystalline phases [64]. The signs and magnitudes of the LCICD from the ordered chromophores has been found to depend on the chiral nematic twist sense, the linear birefringence of the mesophase, the position of the pitch band, λ_0, relative to the absorption band of the chromophore, λ_{ab}, and the linear dichroism of the chromophore. The twist sense of a chiral nematic mesophase may be determined from the sign of a LCICD, if a correlation between the LCICD and the chiral nematic twist sense has been found. For example, it has been shown that the sign of the LCICD for the phenyl side-groups of PBLG are of opposite sign for chiral nematic mesophases with opposite twist sense [51].

The coupling of an anisotropic medium with the imperfect optics and electronics inherent in CD spectrometers may lead to artifacts [82,83], but the observation and analysis of LCICD spectra in a variety of chiral media lend credence to the method.

9.5 Chiroptical Properties of Specifically Substituted Cellulose Derivatives

The pitch and chiral nematic twist sense of the mesophases of cellulose derivatives are strongly influenced by the nature of the side-group substitution and the solvent. It is not known how or why different achiral substituents on the cellulose chain, or changes in the solvent, can influence the chirality of these tertiary helicoidal structures. One possibility is that solvent–polymer interactions influence the conformation of the cellulose backbone and this determines the chirality of the chiral nematic structure. Differences in the conformations of cellulose derivatives in solution may be detected by circular dichroism spectroscopy. The problem with cellulose derivatives is that, in many cases, there are no chromophores with absorption bands above 200 nm, and measurements below 200 nm are more difficult because of restrictions on solvents and instrumentation. Most of the work to date has been conducted on polysaccharide derivatives in which chromophore-containing groups have been attached to the carbohydrate backbone or as polysaccharide-dye complexes, but no attempt has been made to correlate the optical activity of these polymers with the chirality of their mesophases. Chiroptical studies on derivatives where the cellulose side-group chromophores are attached to a specific hydroxyl group along the chain are obviously required. The chiroptical activity of chromophores, arranged in a specific regular array, would be much easier to interpret than the more complicated case where there is no specific substitution pattern, particularly when exciton effects are involved. In order to determine if an exciton effect is the result of a helical conformation of the cellulose derivative in solution, it is necessary to ensure that the exciton interaction is between chromophores on adjacent monomer units and not between chromophores within the same monomer unit. Ideally, the cellulose derivative should have chromophores attached to one specific position on each anhydroglucose unit along the chain.

In the many studies of the liquid crystalline properties of cellulose derivatives discussed above, there has been little work on derivatives that have a specific substitution pattern along the cellulose backbone. It is known that the degree of side-group substitution can change the twist sense of chiral nematic mesophases of cellulose derivatives, but it is not known if the distribution of these substituents has a similar effect.

(Triphenylmethyl)cellulose or trityl cellulose (2) is a good starting material for the preparation of specifically substituted cellulose derivatives. The heterogeneous preparation of trityl cellulose has been known for many years and extensive research has shown that the trityl protecting group shows high selectivity for the primary hydroxyl group at position 6 of the repeating anhydroglucose units [84]. Hearon et al. have shown that, under moderate reaction conditions, trityl chloride reacts 13.8 times as fast with the primary hydroxyl at position 6 compared to the secondary hydroxyls at positions 2 and 3 [85]. This results in degrees of tritylation of approximately 1.0, with 90% substitution at the primary hydroxyl group position

(1)   (2)

These trityl cellulose derivatives contain specifically substituted bulky aromatic side-groups that are also chromophores, with electronic transitions located in the ultraviolet region of the spectrum. The close proximity of these bulky groups to the cellulose backbone enhances the possibility for the observation of optical activity resulting from the perturbation of the phenyl electronic transitions by the chiral cellulose chain.

Phenyl chromophores usually show three absorption bands in the spectral region between 180 nm and 300 nm [86]. A weak band, called the ^1L_b transition, appears at a wavelength around 260 nm and originates from an electronically forbidden transition that usually exhibits considerable fine structure. In spite of the forbidden nature of this transition, it can become optically active when the phenyl group occurs in an optically active molecule that has a rigid or conformationally restricted structure. The phenyl chromophore also shows two strong absorption bands in the spectral regions 200–220 nm and 180–190 nm. The most generally accepted assignments for these two bands are the forbidden ^1L_a and allowed ^1B_a transitions, respectively [53]. These bands also become optically active when perturbed by a dissymmetric molecular environment.

Trityl cellulose itself has not been found to produce a lyotropic mesophase; it is not readily soluble in common solvents, and tends to form gels when dissolved in high concentrations in polar solvents. In order to increase the solubility of tritylated cellulose in organic solvents, the free hydroxyl functionalities at positions 2 and 3 were substituted with benzyl or alkyl substituents to generate 6-O-trityl-2,3-O-benzyl cellulose or 6-O-trityl-2,3-O-alkyl cellulose derivatives

1 . NaOH
2 . RI
DMSO

(2)   (3)

The polymers were prepared by dissolving trityl cellulose in DMSO and adding sodium hydroxide and an alkyl iodide or benzyl chloride to the solution [87,88,89]. The resultant polymers dissolved quite readily in organic solvents and formed lyotropic liquid crystalline phases above a critical polymer concentration. Decomposition occurred at the temperatures required to form a thermotropic mesophase (> 230 °C).

The optical properties of the mesophases of trityl–alkyl cellulose derivatives are sensitive to the nature of the alkyl-group substitution. Several of the

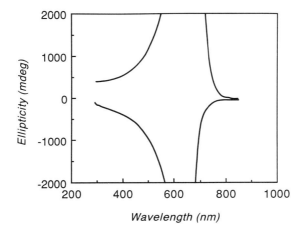

FIGURE 9.5. Apparent circular dichroism spectra for lyotropic mesophases of (trityl)-(hexyl)cellulose (positive band) and (trityl)(pentyl)cellulose (negative band); polymer concentrations of 66 and 46 mass %, respectively, in tetrahydrofuran [89]. The intense reflection bands produce apparent CD signals that are off-scale.

mesophases exhibited chiral nematic reflection bands in the visible region of the spectrum, and both left- and right-handed chiral nematic structures were observed [89]. For example, the polymer 6-O-trityl-2,3-O-pentyl cellulose was found to form a right-handed chiral nematic structure in tetrahydrofuran whereas 6-O-trityl-2,3-O-hexyl cellulose formed a left-handed chiral nematic structure (Fig. 9.5). One possible reason for the difference in the twist sense of these two mesophases is that the two polymers have different conformations in solution. All of the 6-O-trityl-2,3-O-alkyl cellulose mesophases that displayed visible reflection bands showed an increase in the pitch with increasing temperature and with decreasing polymer concentration.

6-O-trityl-2,3-O-alkyl cellulose derivatives show dilute solution chiroptical activity from the phenyl chromophores along the polymer backbone [88,89]. This presented an opportunity to examine the chiroptical activity of these chromophores in the mesophase and isotropic solution, and to try and correlate this behavior with the chiral nematic twist sense. The signs of the CD bands for the ^1L_a and ^1L_b phenyl electronic transitions in the mesophase depend on the twist sense. The CD spectra of mesophases with a right-handed chiral nematic structure showed positive bands for the ^1L_a and ^1L_b transitions, whereas these bands were negative in a left-handed mesophase (Fig. 9.6). These results indicate that the chiral nematic twist sense correlates with the chiroptical activity of the phenyl chromophores in the mesophase. The CD signals from the phenyl chromophores in the mesophase were found to have different signs and magnitudes compared to the signals observed in dilute solution (Fig. 9.7), reflecting the additional chiral contribution of the

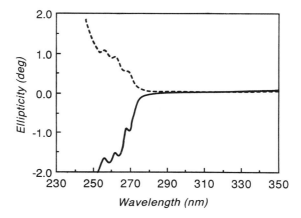

FIGURE 9.6. CD spectra for lyotropic mesophases of (trityl)(hexyl)cellulose (solid line) and (trityl)(pentyl)cellulose (dashed line); polymer concentrations of 42 and 49 mass %, respectively, in tetrahydrofuran [89]. The weak ^1L_b band around 260 nm appears as shoulders on the side of the stronger ^1L_a band centered around 220 nm.

chiral nematic structure to the chiroptical activity. In dilute solution (no chiral nematic structure), the CD signals arise from the perturbation of the phenyl chromophores by the chiral cellulose backbone. If differences in the conformation of the trityl–alkyl cellulose polymers were responsible for the left- and right-handed chiral nematic structures, then this should be reflected in the dilute solution chiroptical activity of the phenyl chromophores. However, the CD signals from these chromophores in dilute solutions were

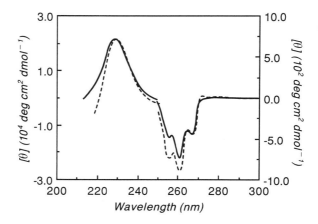

FIGURE 9.7. CD spectra for isotropic solutions of (trityl)(hexyl)cellulose (dashed line) and (trityl)(pentyl)cellulose (solid line) in tetrahydrofuran [89]. The ^1L_b bands (right) and ^1L_a bands (left) are of the same signs and magnitudes for the two polymers.

virtually identical for all of the (trityl)cellulose derivatives, suggesting that differences in the chiral environment surrounding the trityl group are not responsible for the differences in the twist sense. Furthermore, the absence of exciton splitting for the phenyl ^1L_a and ^1L_b transitions suggests that there is no helical secondary structure for these polymers in solution that could influence the chirality of the mesophase.

Replacement of the trityl group with a naphthyl chromophore allows examination of the ^1B_b transition, located around 225 nm. It has already been shown that naphthyl groups are useful chromophores for the determination of a helical secondary structure in polypeptides by the exciton chirality method [90,91]. The trityl group was removed by treating trityl–pentyl cellulose in chloroform with gaseous hydrochloric acid [92]. NMR results confirmed that all of the trityl groups were removed by this process and also showed that most of the primary hydroxyl groups at position 6 were unsubstituted

The polymer 6-*O*-α-(1-methylnaphthalene)-2,3-*O*-pentyl cellulose (5) was prepared by reacting 2-3-*O*-pentyl cellulose in dimethylacetamide with potassium *t*-butoxide and 1-(chloromethyl)naphthalene

The polymer forms a thermotropic right-handed chiral nematic mesophase that showed an anisotropic–isotropic phase transition in the temperature range of 90–110 °C [92]. This is in contrast to the trityl–alkyl cellulose derivatives that formed only lyotropic liquid crystalline phases. The dilute

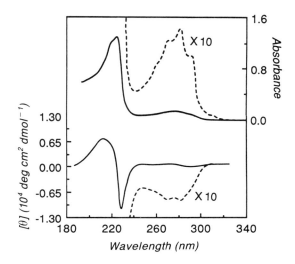

FIGURE 9.8. CD (upper) and absorbance (lower) spectra for dilute solutions of (naphthylmethyl)(pentyl)cellulose in cyclohexane [92]. The CD peak for the naphthyl ^1B_b transition at 220 nm shows exciton splitting.

solution CD spectrum of 6-O-α-(1-methylnaphthalene)-2,3-O-pentyl cellulose shows a weak negative signal for the ^1L_a transition with the signal for the ^1B_b transition having the anomalous peak shape expected for an exciton split band (Fig. 9.8). This suggests that the polymer may have a helical secondary structure in solution, although it is not known if or how such a secondary structure influences the chiral nematic twist sense of the mesophase.

The formation of a chiral nematic structure changes the appearance of the CD signals from the naphthyl chromophores, in a manner similar to that observed for the (trityl)(alkyl)cellulose derivatives described above. In the mesophase, the ^1L_a transition for the naphthyl group showed positive ellipticity whereas in dilute solution this band was negative [92]. The positive signal in the mesophase is of the same sign as the CD peak for the right-handed (trityl)(alkyl)cellulose mesophases. In principle, the sign and magnitude of the induced circular dichroism depends on the orientation of the transition moment of the chromophore relative to the chiral nematic structure, but a quantitative interpretation will require measurements on well-oriented planar samples.

9.6   Concluding Remarks

Despite past efforts, described in several reviews [23,93,94], there remains much to be learned about the lyotropic and thermotropic liquid crystalline phases of cellulosic polymers. Phase separation of lyotropic systems may be qualitatively predicted from dilute solution properties, as described above. However, there are some cellulose derivatives that appear to form liquid crystalline solutions readily in certain solvents and not in others. (Ethyl)-cellulose is one example; it forms liquid crystalline solutions in chloroform, dioxane, cresol, and pyridine, but apparently not in ethyl acetate, n-butyl acetate, or methylethyl ketone, even though there is no striking trend in dilute solution viscosities that would indicate significantly greater stiffness in the former group of solvents [95]. Possibly, a change in polymer–solvent interactions with concentration in some solvents may lead to significant changes in chain dimensions with concentration.

Chain length polydispersity, virtually unavoidable with cellulosics, must also affect the phase behavior, especially for chains whose lengths are of the order of the persistence length or less. An unexpected biphasic region observed for a thermotropic cellulose derivative [96] was ascribed to chain length polydispersity. However, the observations might more plausibly result from the polydispersity in chain flexibility that has been demonstrated by Stupp et al. [97] to explain subsequent observations on other thermotropic polymer systems. Certainly, most cellulosic polymers are chemically disordered, in that there is a distribution of substituted units of different chemical compositions along the chain, resulting in a "polydispersity of persistence lengths" [25]. The problem can be avoided experimentally by preparing either completely tri-substituted derivatives, or uniform, specifically substituted cellulose derivatives, as described above. The increase in chain regularity for tri-substituted derivatives has had some interesting consequences; oligomers of cellulose trialkanoates have recently been found to form columnar thermotropic liquid crystalline phases, with properties quite different from those of the usual chiral nematic phases [98].

A better understanding of the chiroptical properties of chiral nematic cellulose derivatives also requires measurements on polymers with regular substitution patterns. The use of monodisperse oligomeric derivatives may prove advantageous; this is illustrated by the recent observation of a temperature-induced reversal of handedness for a thermotropic mesophase of an oligomeric trisubstituted cellulose ether [63]. Regularly substituted derivatives are known to form helical molecular conformations in the crystalline state, often as solvates [99], but whether these conformations are maintained in the liquid crystalline solutions or melts has not been confirmed. Thus, the interpretation of experimental results on the magnitude and handedness of the helicoidal chiral nematic structure in terms of the chain substituents is not yet possible. Perhaps a clue to future directions is the recent unexpected observation [100] that aqueous suspensions of unsubstituted cellulose crystallites form *chiral* nematic phases at relatively low concentrations. Here, the interacting species are rodlet (typically 200 nm long and 7 nm wide) of crystalline cellulose, stabilized by surface charge; phase separation occurs at concentrations of around 5 vol.%, and is sensitive to electrolyte content (as is the chiral nematic pitch).

References

1. L. Onsager, Ann. N.Y. Acad. Sci. **51**, 627 (1949).

2. P.J. Flory, Proc. Roy. Soc. London Ser. A **234**, 73 (1956).

3. W. Maier and A. Saupe, Z. Naturforsch. **149**, 882 (1959).

4. W. Maier and A. Saupe, Z. Naturforsch. **159**, 187 (1960).

5. P.J. Flory and G. Ronca, Mol. Cryst. Liq. Cryst. **54**, 311 (1979).

6. M. Warner and P.J. Flory, J. Chem. Phys. **73**, 6327 (1980).

7. A.R. Khokhlov and A.N. Semenov, Physica **108A**, 546 (1981).

8. A.R. Khokhlov and A.N. Semenov, Physica **112A**, 605 (1982).

9. T. Odjik, Macromolecules **19**, 2313 (1986).

10. T. Sato and A. Teramoto, Mol. Cryst. Liq. Cryst. **178**, 143 (1990).

11. R. Hentschke, Macromolecules **23**, 1192 (1990).

12. P.J. Flory, Macromolecules **11**, 1141 (1978).

13. P.J. Flory, Adv. Polymer Sci. **59**, 1 (1984).

14. B. Jung and R. Stern, Macromolecules **22**, 3628 (1989).

15. I. Simon, H. Scheraga, and R. St. J. Manley, Macromolecules **21**, 983 (1988).

16. D.A. Rees and R.J. Skerret, Carbohydr. Res. **7**, 334 (1968).

17. A. Pizzi and N. Eaton, J. Macromol. Sci.-Chem. **A22**, 105 (1985).

18. F. Leung, H.D. Chanzy, S. Perez, and R.H. Marchessault, Can. J. Chem. **54**, 1365 (1976).

19. P. Zugenmaier, J. Appl. Polymer Sci., Appl. Polymer Symp. **37**, 223 (1983).

20. A.S. Perlin, N. Cyr. R.G.S. Ritchie, and A. Parfondry, Carbohydr. Res. **37**, C-1 (1974).

21. W. Brown, *Cellulose and Cellulose Derivatives*, Part IV, edited by N.M. Bikales and L. Segal (Wiley, New York, 1971), p. 557.

22. R.S. Werbowyj and D.G. Gray, Mol. Cryst. Liq. Cryst. Lett. **34**, 97 (1976).

23. D.G. Gray, J. Appl. Polymer Sci., Appl. Polymer Symp. **37**, 179 (1983).

24. R.S. Werbowyj and D.G. Gray, Macromolecules **13**, 69 (1980).

25. G. Conio, E. Bianchi, A. Ciferri, A. Tealdi, and M.A. Aden, Macromolecules **16**, 1264 (1983).

26. G.V. Laivins and D.G. Gray, Macromolecules **18**, 1753 (1985).

27. P. Navard, J.M. Haudin, S. Dayan, and P. Sixou, J. Polymer Sci., Polymer Lett. Ed. **19**, 379 (1981).

28. D.L. Patel and R.D. Gilbert, J. Polymer Sci., Polymer Phys. Ed. **21**, 1079 (1983).

29. H. Chanzy and A. Peguy, J. Polymer Sci., Polymer Phys. Ed. **18**, 1137 (1980).

30. S. Fortin and G. Charlet, Macromolecules **22**, 2286 (1989).

31. L. Robitaille, N. Turcotte, S. Fortin, and G. Charlet, Macromolecules **24**, 2413 (1991).

32. H. Benoit and D. Doty, J. Phys. Chem. **57**, 958 (1953).

33. H. Yamakawa and J. Fuji, Macromolecules **7**, 128 (1974).

34. E. Bianchi, A. Ciferri, G. Conio, L. Lanzavecchia, and M. Terbojevich, Macromolecules **19**, 630 (1986).

35. W.R. Krigbaum, H. Hakemi, A. Ciferri, and G. Conio, Macromolecules **18**, 973 (1985).

36. J.M. Mays, Macromolecules **21**, 3179 (1988).

37. S.L. Tseng, A. Valente, and D.G. Gray, Macromolecules **14**, 715 (1981).

38. S.L. Tseng, G.V. Laivins, and D.G. Gray, Macromolecules **15**, 1262 (1982).

39. S. Bhadani and D.G. Gray, Mol. Cryst. Liq. Cryst. **99**, 29 (1983).

40. G.V. Laivins and D.G. Gray, Polymer **26**, 1435 (1985).

41. A.M. Ritcey and D.G. Gray, Macromolecules **21**, 1251 (1988).

42. W.J.A. Goossens, Mol. Cryst. Liq. Cryst. **12**, 237 (1971).

43. T.V. Samulski and E.T. Samulski, J. Chem. Phys. **67**, 824 (1977).

44. Y.R. Lin-Liu, Y.M. Shih, and C.W. Woo, Phys. Rev. A **15**, 2550 (1977).

45. P.N. Keating, Mol. Cryst. Liq. Cryst. **8**, 315 (1969).

46. Y.R. Lin-Liu, Y.M. Shih, and C.W. Woo, Phys. Rev. A **14**, 445 (1976).

47. H. Kimura, M. Hoshino, and H. Nakano, J. Phys. (Paris) **40**, C3-174 (1979).

48. H. Kimura, M. Hoshino, and H. Nakano, J. Phys. Soc. Jpn. **51**, 1584 (1982).

49. M.A. Osipov, Chem. Phys. **96**, 259 (1985).

50. M.A. Osipov, Nuovo Cimento, Soc. Ital. Fis. **10D**, 1249 (1988).

51. See, for example, I. Uematsu and Y. Uematsu, Adv. Polymer Sci. **59**, 37 (1984).

52. R.S. Werbowyj and D.G. Gray, Macromolecules **17**, 1512 (1984).

53. E. Charney, *The Molecular Basis of Optical Activity* (R.E. Krieger, Malabar, FL, 1985), Chapters 2, 4, and 5.

54. S. Chandrasekhar, *Liquid Crystals* (Cambridge University Press, Cambridge, 1977), Chapter 4.

55. G. Charlet and D.G. Gray, J. Appl. Polymer Sci. **37**, 2517 (1989).

56. U. Vogt and P. Zugenmaier, Ber. Bunsenges. Phys. Chem. **89**, 1217 (1985).

57. A.M. Ritcey and D.G. Gray, Macromolecules **21**, 2914 (1988).

58. J.X. Guo and D.G. Gray, Macromolecules **22**, 2086 (1989).

59. W.P. Pawlowski, R.D. Gilbert, R.E. Fornes, and S.T. Purrington, J. Polymer Sci., Part B, Polymer Phys. **25**, 2293 (1987).

60. M. Siekmeyer and P. Zugenmaier, Macromol. Chem., Rapid Commun. **8**, 511 (1987).

61. H. Steinmeier and P. Zugenmaier, Carbohydr. Res. **173**, 75 (1988).

62. P. Zugenmaier and P. Haurand, Carbohydr. Res. **160**, 369 (1987).

63. T. Yamagishi, T. Fukuda, T. Miyamoto, T. Ichizuka, and J. Watanabe, Liq. Cryst. **7**, 155 (1990).

64. F.D. Saeva, in *Liquid Crystals: The Fourth State of Matter*, edited by F.D. Saeva (Marcel Dekker, New York, 1979), Chapter 6.

65. A. Moscowitz, in *Advances in Chemical Physics*, edited by I. Prigogine (Interscience, New York, 1960), Vol. IV, p. 67.

66. Reference [53], pp. 191–201.

67. C.S. Hudson, J. Am. Chem. Soc. **47**, 268 (1925).

68. D.H. Whiffen, Chem. Ind. (London) 964 (1956).

69. J.H. Brewster, J. Am. Chem. Soc. **81**, 5483 (1959).

70. E.S. Stevens and B.K. Sathyanarayana, Biopolymers **27**, 415 (1988).

71. E.S. Stevens and B.K. Sathyanarayana, J. Am. Chem. Soc. **111**, 4149 (1989).

72. E.S. Stevens and B.K. Sathyanarayana, and E.R. Morris, J. Phys. Chem. **93**, 3434 (1989).

73. N. Harada and K. Nakanishi, *Circular Dichroic Spectroscopy, Exciton Coupling in Organic Stereochemistry* (University Science, Mill Valley, CA, 1983), Chapter 1.

74. N. Harada and K. Nakanishi, Chem. Commun. 1691 (1970).

75. A.J. Stipanovic and E.S. Stevens, J. Appl. Polymer Sci., Appl. Polymer Symp. **37**, 277 (1983).

76. A.M. Ritcey and D.G. Gray, Biopolymers **27**, 479 (1988).

77. K. Nishida and A. Iwasaki, Z. Polymer **251**, 136 (1973).

78. F.D. Saeva and G.R. Olin, J. Am. Chem. Soc. **95**, 7882 (1973).

79. J. Lematre, S. Dayan, and P. Sixou, Mol. Cryst. Liq. Cryst. **84**, 267 (1982).

80. A.M. Ritcey and D.G. Gray, Biopolymers **27**, 1363 (1988).

81. E. Iizuka and J.T. Yang, Mol. Cryst. Liq. Cryst. **29**, 27 (1974).

82. Y. Shindo, M. Nakagawa, and Y. Ohmi, Appl. Spectrosc. **39**, 860 (1985).

83. Y. Shindo and Y. Ohmi, J. Am. Chem. Soc. **107**, 91 (1985).

84. B. Helferich and K. Koester, Ber. **57**, 587 (1924).

85. W.M. Hearon, D.D. Hiatt, and C.R. Fordyce, J. Am. Chem. Soc. **65**, 2449 (1943).

86. P. Crabbe and W. Klyne, Tetrahedron **23**, 3449 (1967).

87. B.R. Harkness and D.G. Gray, Macromolecules **23**, 1452 (1990).

88. B.R. Harkness and D.G. Gray, Liq. Cryst. **8**, 237 (1990).

89. B.R. Harkness and D.G. Gray, Can. J. Chem. **68**, 1135 (1990).

90. M. Sisido, S. Egusa, and Y. Imanishi, J. Am. Chem. Soc. **105**, 1041 (1983).

91. M. Sisido and Y. Imanishi, Macromolecules **19**, 2187 (1986).

92. B.R. Harkness and D.G. Gray, Macromolecules **24**, 1800 (1991).

93. R.D. Gilbert and P.A. Patton, Progr. Polymer Sci. **9**, 115 (1983).

94. P. Sixou and A. Ten Bosch, in *Cellulose Structure, Modification and Hydrolysis*, edited by R.A. Young and R.M. Rowell (Wiley, New York, 1986), Chapter 12.

95. D. Budgell, Ph.D. Thesis, McGill University, Montreal, 1989.

96. G.V. Laivins, D.G. Gray, and P. Sixou, J. Polymer Sci., Polymer Phys. Ed. **24**, 2779 (1986).

97. S.I. Stupp, J.S. Moore, and P.G. Martin, Macromolecules **21**, 1228 (1988).

98. T. Yamagishi, T. Fukuda, T. Miyamoto, Y. Yakoh, Y. Takashima, and J. Watanabe, Liq. Cryst. **10**, 467 (1991).

99. P. Zugenmaier, in *Cellulose Structure, Modification and Hydrolysis*, edited by R.A. Young and R.M. Rowell (Wiley, New York, 1986), Chapter 13.

100. J.-F. Revol, H. Bradford, J. Giasson, R.H. Marchessault, and D.G. Gray, Int. J. Biol. Macromol. **14**, 170 (1992).

10

Bowlics

L. Lam

10.1   Introduction

December 4, 1979, 6:30 p.m., Hotel Ashok, Bangalore, India. I was from Beijing, invited by S. Chandrasekhar of the Raman Research Institute to attend this wonderful conference on liquid crystals. Since India is a country with a long history, unlike most other international conferences, there was a cultural event almost every evening during this particular conference. And they were free! There was still half-an-hour to go before the performance by the very beautiful and renowned classical dancer, Sonal Mansingh (Fig. 10.1). (Sonal and I were born in the same year, but this is inconsequential to what was going to happen.)

I was sitting in a room, waiting for the performance to begin. Incidentally, this was the first international liquid crystal conference I ever attended in my life and there were no old acquaintances in that room. So I was sitting by myself, and having nothing more meaningful to do, I looked up to the ceiling. It was a ceiling decorated by some regular hexagons connected to each other in an irregular pattern, or so it seemed. Each one of these hexagons was like the convex cover of a fruit plate except that it was now attached upside down to the ceiling. In other words, it was a three-dimensional object.

Let me go back a little bit to that conference. As every liquid crystalist knows these days, the first discotic was hexagonal and was created in the laboratory by Chandrasekhar's group in 1977 [1], and many different varieties were immediately synthesized in other parts of the world. In this particular conference, organized by Chandrasekhar himself, a mere two years after the discovery, it was only natural that paper after paper on discotics was presented day after day. Discotics was hot! Working in Beijing in those two years was quite isolated and handicapped. (There was only one Chinese-made copying machine in the whole Institute of Physics where I worked, which broke down more often than not and was off limits to any scientist, Ph.D. or not. The machine was attended by two "specialists." In addition to that, the main

FIGURE 10.1. Sonal Mansingh in
an almost C_2 pose.

library at the Academy of Sciences to which my Institute belonged, was *not*
open-shelf.) Obviously, there was no way that I could play catch up in
discotics. I got to leapfrog this hot topic and found something new to do.

When I was admiring these hexagons and their pattern of arrangement in
the ceiling, the discotics came to mind. But wait a minute, these hexagons
were different from the discotics! A discotic molecule is flat like a pancake
(Fig. 10.2)—a two-dimensional object, while these hexagons on the ceiling
were three dimensional, like a rice bowl. (Being an oriental, I use rice bowls
three times a day. It is not surprising that rice bowls pop into the mind even
when one is not actually hungry.) If pancakes can stack up in columns to
form columnar discotics, shouldn't rice bowls be able to do the same? I saw
them in the kitchens, no doubt about that! As they say, the rest is history.

I could not shake the thought of bowl-like liquid crystals out of my mind.

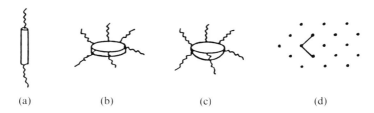

(a)   (b)   (c)   (d)

FIGURE 10.2. Shapes of molecules that form liquid crystal phases: (a) rodic (1D); (b) discotic (2D); (c) bowlic (3D); and (d) mass point (0D)—hexatic phase.

During one of the coffee hours after that fateful night I worked up my courage to tell it to the experts, in this case, Adriaan de Vries of Kent and Christian Destrade of Boudeaux. (By the way, Sonal's dance was splendid, and I promised myself that I would come back to every conference that Chandrasekhar organized in Bangalore. Which I did, almost. But that is another story.)

Back to Beijing. Even though I wanted very much to write up my thoughts on the bowl-like liquid crystals, it would have been very difficult, if not impossible, to get it published in China, or elsewhere. In China, you did not pay anything to the journal when your paper was accepted for publication; instead, they paid you! That was one of the two and only two ways for a Chinese scientist to supplement his or her low income; the other was to referee a paper. In other words, to get a paper published was not a purely academic matter. Since there was no experimental evidence to support the existence of bowl-like liquid crystals at that time, I did not think I stood a chance of getting my speculations published. Besides, I was busy with the nematic–isotropic phase transition [2] and other problems [3]. Yet, I did talk to a few chemists in Beijing and asked them to try the synthesis. They were not interested; they were interested in discotics!

The chance of getting my beloved speculations into print finally came in 1981. I was writing a review on the liquid crystal phases for the journal *Wuli*, which is the Chinese word for physics. (This journal is the Chinese equivalent of *Physics Today* minus the advertisements, except that it also has a section for original papers.) I was trying to give a systematic survey of the rodics (liquid crystals with long molecules), the discotics, and the less familiar hexatic phases [4], and I decided on a scheme of introducing them according to the "dimension" of their constituent molecules (Fig. 10.2). It then followed that we could have liquid crystal phases from molecules of dimension zero (hexatics), one (rodics), or two (discotics). Following this reasoning, the conjecture of the existence of liquid crystals with molecules of three dimensions (bowlics) was natural and almost irresistable. I was right. The paper [5], "Liquid Crystal Phases and the 'Dimensionality' of Molecules," got through the referee and the editor without a hitch.

Knowing that *Wuli* was not the most popular journal in the world, after its publication in 1982, I sent out an English translation (of the Introduction and Section IV) of this paper to the major liquid crystal laboratories outside China. A year later, in an article [6] summarizing the development of liquid crystal research in China from 1970 to 1982, I devoted a whole paragraph to my predictions of the bowlics.

Nothing happened, not until 1985. In Halle (Saale), of the then German Democratic Republic, during the 6th Liquid Crystal Conference of Socialist Countries, August 26–30, 1985, I was informed by Christian Destrade that bowlics had been successfully synthesized, and Anne Levelut of Orsay had just finished doing an x ray study of them. I was more than excited. On my way home to New York after the conference, I stopped over at Orsay and talked to Levelut. She kindly provided me with a preprint of her work [7] and copies of the two papers announcing the synthesis [8,9].

Of course, none of these authors called these compounds bowlics. Neither did I. Zimmermann et al. [8] called them "pyramidics," and Malthête and Collet [9] called them "cone-shaped." After all, Egypt is closer to Europe than China is. In fact, I myself had been calling them bowl-likes, even though "pyramid or hemisphere shapes" were also mentioned earlier in my paper [5]. The word bowlic was formally introduced in my invited talk [10] at the 11th International Liquid Crystal Conference at Berkeley, July 2, 1986, ten years after America's Bicentennial and three years before the Tiananmen massacre in Beijing.

Since then, there have been more studies [11–27] of bowlics. The monomers do form the usual columnar liquid crystal phases as expected [5], and sometimes do possess some unique properties [11] not shared by the discotics. However, the predicted ferroelectricity and thermotropic nematic phases [5] have not yet been found. Bowlic Langmuir films have been studied [22,23]. Bowlic polymers have been discussed theoretically [24–27] and predicted to be capable of forming ultrahigh T_c superconductors [24,25]. These polymers remain to be synthesized. All these developments will be summarized in the rest of this chapter.

As will be emphasized below, there exist no fundamental obstacles for these exotic properties of bowlics to be realized in the laboratory, and hence their novel applications in industry. What is needed is more effort on the part of chemists and materials scientists.

10.2   The First Paper

The first paper predicting the existence of bowlics [5] contains an introductory section, outlining the systematics of the then existing liquid crystal phases and the motivation for the bowlics, and five other sections. The bowlics and their unique properties are discussed in Section IV, which is followed by the conclusions in Section V. In the following, the English translation of the relevant parts of this paper is presented.

Liquid Crystal Phases and the "Dimensionality" of Molecules

Lin Lei

Institute of Physics, Chinese Academy of Sciences

Liquid crystal was first observed in cholesteryl benzoate by the Austrian botanist Reinitzer in 1888 and named by the German crystallographer Lehmann one year later. The two words "liquid crystal" represent a new state of matter intermediate between liquid and crystal, which is now very important in both basic research and applications.[1] The history of liquid crystal research in our country goes for 11 years already.[2]

In the 89 years span between the discovery of liquid crytals and 1977, the only known liquid crystals are found in compounds made up of long organic molecules. In these molecules there is usually a central part that is long and rigid of lathy shape with flexible end chains fixed at one or both ends.[3] Liquid crystal states (phases) of long organic molecules can be chiefly divided into three types, viz., nematics, cholesterics, and smectics (see Sec. I). The basic properties of these liquid crystal phases may be explained by treating the molecules as volumeless rod-like (1-dimensional) entities.

In 1977, Chandrasekhar et al. of India first observed liquid crystal phases in organic compounds of circular disc-like molecules. Later, liquid crystals were consecutively found in other compounds of rectangular or asymmetric disc-like molecules. All these molecules may be considered to be 2-dimensional in shape. In these mesophases of disc-like molecules there exist also nematics and cholesterics, but no smectics; on the other hand, there are columnar phases[1,4,5] not found in the case of long molecules.

One of the characteristics of liquid crystals is its optical anisotropy (with the exception of the D phase, see Sec. I below). In the cases of long and disc-like molecules mentioned above the anisotropy of the liquid crystal phases comes from the (1- or 2-dimensional) asymmetric shape of the molecules.

Halperin and Nelson[6,7] in 1978 considered a 2-dimensional lattice of mass-point molecules. They suggested that a hexatic liquid crystal phase may appear before the 2-dimensional crystal melts into liquid. Recently, this idea was generalized to a 3-dimensional lattice by Nelson and Toner[8] who proposed the possible existence of a cubic liquid crystal phase in this case. According to what we said above, the mass-point molecules forming hexatic and cubic liquid crystals are molecules of 0 dimension. Although there are no decisive experiments so far, the question of the existence of hexatic and cubic mesophases has already aroused wide attention and many research activities. From the standpoint of liquid crystal research, mesophases of 0-dimensional molecules break away from the traditional concept that the liquid crystal phases must come from organic molecules of asymmetric shape. This development is very important.

Rigorously speaking, all molecules in nature are of course 3-dimensional. The "dimension" of a molecule discussed here really means the dimension of "molecules" in the physical models used in the description of the liquid crystal phases. In other words, it means the minimal (or important) character of the molecular shape when mesophases are formed.

Since 1- and 2-dimensional moleclues have been proved to have liquid crystal phases and 0-dimensional molecules may probably have liquid crystal phases, it is then natural to ask: Do 3-dimensional molecules also possess liquid crystal phases? In our opinion, the answer is yes!

In recent years, there has been tremendous progress on the study of liquid crystal phases. The aim of this paper is to introduce and summarize, from the viewpoint of the "dimensionality" of molecules, what is known about: (1) the different structures of (thermotropic) liquid crystal phases, especially those of smectics and disc-like molecules; (2) the sequences of phase transitions between the different mesophases including reentrant phenomena; and (3) to discuss the possible existence and properties of mesophases made up of 3-dimensional molecules. The last part is obviously speculative in nature. It is our hope that there will be more investigations by theorists and experimentalists in this direction.

The headings of Sections I–III are: I. One-Dimensional Molecules. 1. Nematics. 2. Cholesterics. 3. Smectics. 4. D phase and blue phases. 5. Sequences of phase transitions. 6. Reentrant phenomena. II. Two-Dimensional Molecules. 1. Nematics and cholesterics. 2. Columnar phases. 3. Sequences of phase transitions. III. Zero-Dimensional Molecules. Here Sections I–III will be skipped, and the translation is continued with Sections IV and V.

IV. Three-Dimensional Molecules

Although many long molecules do carry with them strong electric dipoles[3] but the arrangement of these molecules in the mesophases is such that they always compensate each other (Fig. 13) resulting in vanishing net polarization. Consequently, in nematics, cholesterics, and most of smectics there is not ferro-electricity (see Sec. I). In the case of 3-dimensional molecules, e.g., the bowl-like molecules shown in Fig. 14, due to the special shape of the molecules, the cancellation of the dipoles of the molecules will be more difficult (the probability of the molecules lying toward a common direction is larger). It is then possible to observe ferroelectric nematic phases. The corresponding phase for chiral molecules will be cholesterics.

In this case, the physical quantities of the mesophase do not necessarily have

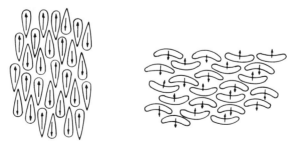

FIG. 13. Two cases in which the electric dipoles of rodic molecules cancel each other.

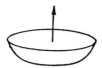

FIG. 14. A bowlic molecule. The arrow represents the electric dipole.

the $\mathbf{n} \to -\mathbf{n}$ symmetry, where $\mathbf{n}$ is the unit vector of the director (representing the average molecular orientation). For liquid crystal of bowl-like molecules, there is still the local symmetry of uniaxial rotations. The elastic energy of nematics is identical to the Frank free energy of 1-dimensional molecules (since the derivation[19] of the Frank free energy does not depend on the symmetry $\mathbf{n} \to -\mathbf{n}$). However, there are extra terms in the dissipation function in comparison with the case of long molecules [see (5.6) of Ref. 20], which lead to some new phenomena not found in nematics of 1-dimensional molecules. For example, thermal gradients can change the molecular orientations and vice versa.

It can be imagined that in the case of bowl-like molecules there will also be columnar phases, but no smectics. In addition, since it is easier for bowl-like molecules (in comparison with disc-like molecules) to pile up on each other, there may exist a new type of phase which is intermediate between nematic and columnar phases, viz., among the spatially separate individual molecules as in nematics there exist short columns of pile-up molecules.

Apart from bowl-like molecules, other 3-dimensional molecules of, for example, pyramid or hemisphere shapes may also be considered. But in these latter cases, the compensation of the dipoles is easier to achieve and the occurrence of ferroelectric mesophases are less likely. It should be pointed out that the term "bowl-like" used above is symbolic. In experiments, one only has to select those molecules which are slightly flat but asymmetric in the up and down directions. There should be some, but not too strong, piling capacity of the molecules; otherwise, once piled up the molecules may form crystals rather than liquid crystals.

Since mesophases of 3-dimensional molecules are ferroelectrics there should be many practical applications. Their properties are quite different from those of known mesophases.[21] Further investigations are worthwhile and most desirable.

V. Conclusions

These days the study of liquid crystal phases and their transitions is a hot topic[1] in condensed matter physics. It was only in the last few years that reentrant phenomena, ferroelectric liquid crystals, discotics, 2-dimensional hexatic and cubic phases, etc., were discovered or proposed. In this article, presented from the viewpoint of the "dimensionality" of molecules for the first time, these new developments are introduced; the existence and the unique properties of ferroelectric liquid crystals formed by 3-dimensional molecules are predicted.

10.3   Bowlic Monomers

All that is known about bowlic monomers will be summarized here. Theoretical studies of bowlic polymers will be presented in Section 10.4.

10.3.1   The Molecules

The three types of bowlic molecules synthesized so far are sketched in Fig. 10.3 and summarized in Table 10.1. Compounds **I** consist of a rigid bowl structure with three benzene rings, and six identical flexible side chains. In particular, **I.1** is hexaalkyloxytribenzocylononene (*n*HETB), **I.2** is hexaalkanoyloxytribenzocylononene (*n*HATB), and **I.4** is hexaalkoxybenzoyloxytribenzocylononene (*n*HBTB). Compound **II** has a similar core but only three tails attached to the three benzene rings in the core; they are prepared in racemic and optically active forms. Both **I** and **II** are derivatives of the cyclotriveratrylenes [28], and the core is in the crown form. In contrast, **III** is formed from a core of four benzene rings and eight flexible tails; in the columnar phase the core is in the sofa form with C_{2h} symmetry [14,15].

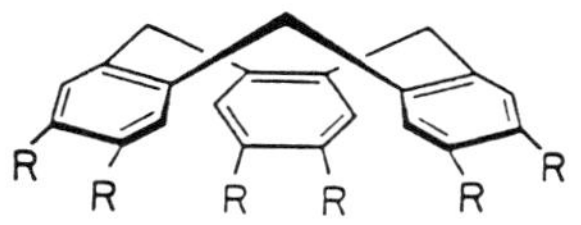

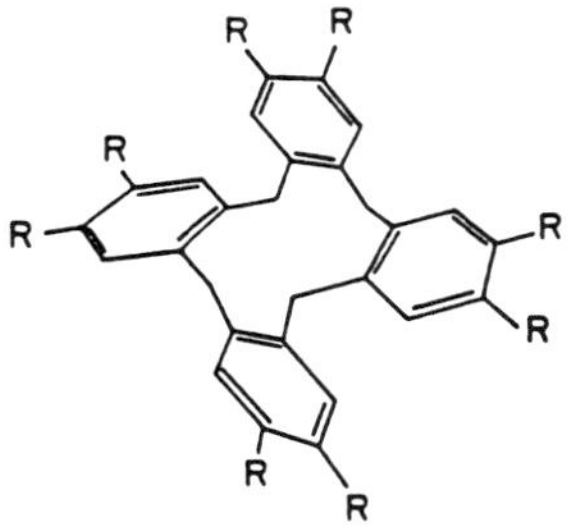

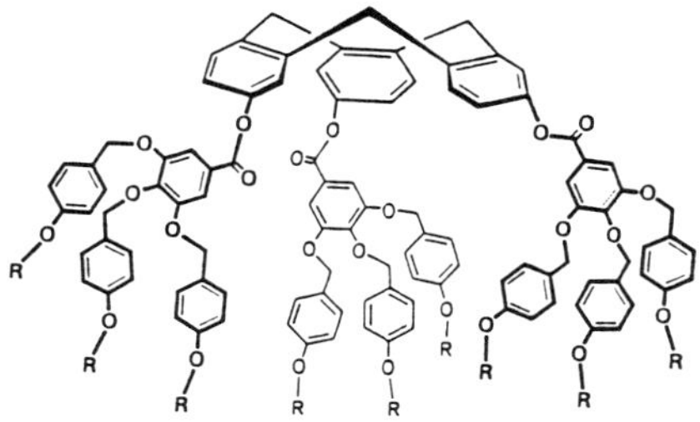

FIGURE 10.3. The three types of bowlic monomeric molecules synthesized: hexasubstituted cyclotricatechylene (**I**); a triester of cyclotriphenolene with a 3,4,5-trisubstituted benzoic acid (**II**); and octasubstituted tetrabenzocyclododecatetraene (**III**).

TABLE 10.1. Summary of bowlic monomers synthesized. Molecules **I**, **II**, and **III** refer to those in Fig. 10.3.

Compound	Tail R	n	Year	Reference	Remarks
I.1	$C_nH_{2n+1}O$	4–12	1985	Zimmermann et al. [8]	Pressure study (n = 7, 9, 10, 11) [16]; x ray diffraction (n = 8, 9, 10) [17]
I.2	$C_nH_{2n+1}CO_2$	7–14	1985	Zimmermann et al. [8]	x ray diffraction (n = 9, 11, 12, 14) [17–19]
		9, 11, 15	1985	Malthête and Collet [9]	
I.3	$C_nH_{2n+1}ØCO_2$	7, 8, 10	1986	Zimmermann et al. [11]	
I.4	$C_nH_{2n+1}OØCO_2$	12	1985	Malthête and Collet [9]	x ray diffraction [7]
		10	1986	Zimmermann et al. [11]	
		8	1988	Wang and Pei [12]	See also [18]
I.5	$C_nH_{2n+1}CH{=}CHC_nH_{2n+1}CO_2$	8	1988	Wang and Pei [12]	See also [18]
II	C_nH_{2n+1}	12	1987	Malthête and Collet [13]	
III.1	$C_nH_{2n+1}O$	8, 10	1988	Zimmermann et al. [14]	
III.2	$C_nH_{2n+1}CO_2$	11, 13	1988	Zimmermann et al. [14]	
III.3	$C_nH_{2n+1}ØCO_2$	10	1988	Zimmermann et al. [14]	
III.4	$CH_3O(CH_2CH_2O)_n$	1, 2, 3	1989	Zimmermann et al. [15]	

10.3.2 Mesophases and Phase Transitions

All bowlic compounds synthesized so far exhibit thermotropic columnar mesophases. In addition, methoxydiethyleneoxide and methoxytriethylene-oxide, **III.4** with $n = 2$ and 3, respectively, are also lyotropic and can sustain (at room temperature) up to 40 wt.% water. At a higher water content methoxydiethyleneoxide exhibits another lyomesophase M_F which is more fluid and *nematic-like* [15].

All the thermotropic bowlic mesophases discovered are columnar. There are at least five of them, denoted by B_1-B_5 here (corresponding to $P_A, \ldots, P_D$ in [8]). Tables or diagrams of phase transitions for the homologues and sometimes mixed bowlic compounds can be found in [8–11,14,15,18]. Compared with the tribenzos **I**, the tetrabenzos **III** have:

(i)   much less tendency for polymorphic mesomorphism;
(ii) an increase in the melting temperature; and, in most cases; and
(iii) a significant increase in the clearing temperatures and clearing enthalpies [14].

The tetrabenzos are therefore more highly ordered than the tribenzos.

Some typical transition sequences are listed here, where K (I) represents the crystal (isotropic liquid) phase:

I.2, $n = 7$ (7HATB) [8]:

$$K \xrightleftharpoons[16.1\ \text{kJ}]{5.2\ °C} B_4 \xrightleftharpoons[31.0\ \text{kJ}]{153.1\ °C} I.$$

I.2, $n = 11$ (11HATB) [8]:

$$K \xrightleftharpoons[48.4\ \text{kJ}]{58.1\ °C} B_4 \xrightleftharpoons[3.1\ \text{kJ}]{118.8\ °C} B_3 \xrightleftharpoons[20.2\ \text{kJ}]{140.6\ °C} I.$$

III.2, $n = 11$ [14]:

$$K \xrightleftharpoons[75.9\ \text{kJ}]{81.9\ °C} B \xrightleftharpoons[27.3\ \text{kJ}]{246.3\ °C} I.$$

All the phase transitions are first order; some of the transition enthalpies are extremely small. Note that 7HATB and 8HATB are both bowlic columnars at room temperature. When the tail R in **I** is too short, as is the case in the rodics, mesophases are not formed. In fact, 5HATB (6HATB) was found [18] to melt directly from the isotropic liquid phase at 152 °C (153 °C) to a "disordered crystalline phase" [20], in which the bowlic molecules stack in a triclinic system with space group $P\,\bar{1}$.

For the tetrabenzos **III.1**, **III.2**, and **III.3**, it was suggested [14] that the rapid interconversion between the two symmetry-related sofa conformations may result in an average planar four-fold symmetry for the molecules and

hence nonpolar columns in the columnar phases. This remains to be confirmed by x ray diffraction studies.

The predicted thermotropic bowlic polar or nonpolar nematics [10] have
not yet been found in the thermotropic bowlic compounds already synthesized. In principle, we do not see any basic difficulty for them to be realized.
It seems that more effort in chemical synthesis is needed. In this regard, it
should be emphasized that nematic-like lyotropic bowlics (the M_F mentioned
above) already exist, which can even be readily and rapidly ($\sim 0.1 \ \mu s$, in
contrast to ~ 1 ms in thermotropic rodic nematics) aligned by a magnetic
field [15]. The potential industrial applications of these lyotropics, as fast
magnetooptic switches or displays, should be explored. No parallel results of
electrical realignment on these compounds are available yet.

When the two R tails attached to each of the three benzene rings of **I** differ
from each other, the molecule is chiral and bowlic cholesterics may be formed
[10]. Compound **II** consists of exactly such molecules, in which a columnar
but not cholesteric phase was found [13]. Of course, as in the case of the
bowlic nematics, this result does not imply that bowlic cholesterics cannot
exist.

In [10], other bowlic mesophases such as the "stringbean," the "donut,"
and the "onion" have been proposed (Fig. 10.4). It seems that the two-dimensional version of the donut or onion mesophase may have been observed in
the bowlic Langmuir films (see Section 10.3.5).

The structures of the columnar phases have been found to be B_{ho}, B_{ro},
and B_{to} [7,17,18]. Here the subscripts "h" and "r" refer to vertical columns
with two-dimensional hexagonal and rectangular structures, respectively; "t"
refers to tilted columns; and "o" refers to ordered periodic arrangement of the
molecules within a single column. There is no correlation among the molecules from different columns. (Of course, the opposite is true in the crystalline
phases of 5HATB and 6HATB [20].) For example, the high (low) temperature B_3 (B_4) phase of 11HATB is B_{ho} (B_{to}) with a period of 4.82 Å along
a column and a lattice constant of 49.38 Å [7]. A spiral structure exists
for the low temperature columnar phase of 12HBTB in which the paraffinic
medium, the external shell of each column, forms a helix [7]. This spiral
structure is also found in discotics [29], and is not that uncommon in
polymers [30].

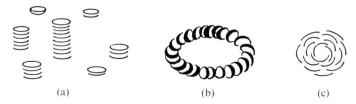

FIGURE 10.4. Three types of bowlic mesophases proposed in [10]: (a) stringbean;
(b) donut; and (c) onion.

10.3.3   Physical Properties

The director $\mathbf{n} \to -\mathbf{n}$ symmetry is a mesoscopic or macroscopic symmetry which exists in the rodics and discotics but, because of the steric effect, may break down in the bowlics. Consequently, bowlics should be true ferro-electrics, or antiferroelectrics in other cases [5,10]. In the bowlic columnars it has not yet been found, for the obvious reason: Experimentally, the sample is not aligned before the columnars are formed. Once the columnar is formed the bowl shape of the molecular core leads to the lock-in of the bowls within the columns [9] with both up and down orientations coexisting in each column—the column is not polar and there is no ferroelectricity (or antiferroelectricity).

In this regard, it is encouraging to note that the half-life of the crown conformer is only ~ 4 min at $3°$ below the clearing temperature in **II** so that, in principle, the columns can be made polar by the crown inversion process under an electric field and the resulting ferroelectricity can be preserved at a lower temperature by quenching [13]. By the same reasoning, it should be easier to align electrically the bowlic molecules in the isotropic liquid phase where the crown half-life is even shorter (~ 0.1 s at 200 °C [9]) provided that the molecular polarization is large enough. This isotropic liquid is paraelectric and the lower mesophases obtained from cooling should be ferroelectric [10].

All previous experiments on bowlics were done with unaligned samples. However, some alignment methods have been tried. For example, 9HATB at 60 °C was put between two glass plates, with one coated with DMOAP and the other untreated. It seems that the coated surface was able to align the bowlic columns and some hexagonal structures were observed in the textures [31]. This approach of alignment is worth further pursuit. Also, some degree of orientation is possible when the mesophase is spread out on a glass plate with a spatula [18]—a method used before for the discotics.

For **I.1** and **I.2** the columnar mesophases B_1 and B_3 are optically uniaxial, and B_2 and B_4 are apparently biaxial [8]. The compounds **III.1** and **III.2** are optically uniaxial while **III.4** could be optically biaxial; all these three com-pounds have a negative optical anisotropy [14]. A very interesting optical phenomenon was observed in **I.3** and **I.4**, viz., the optical anisotropy reverses sign at some well-defined temperature within the mesophase region. This is interpreted as due to the conformational changes involving the side-chain benzene rings [11].

Many textures of the bowlics look like those in the discotics since both exhibit columnar phases. In particular, hexagonal dendrites are formed when the bowlics are slowly cooled from the isotropic liquid phase (see Plate 2 in [14] for bowlics and [32] for discotics). The growth of these dendrites can be easily controlled and could serve as very good candidates in the study of pattern formation [33]. Second harmonic generation studies of bowlics on a water film have been carried out. The effect is similar to those found in the case of the rodics [34].

Pressure studies for nHATB reveals that the transition temperatures increase monotonically with pressure, and there exists no reentrant phenomenon. Volume changes for the transitions and the ratio of the expansion and isothermal compressibility coefficients for the mesophases are reported [16].

Discussions on other physical properties such as elasticity, flexoelectricity, defects and textures, and hydrodynamics unique to bowlics are given in [5,10]. In particular, coupling between the molecular orientation and temperature gradient (like those in the Lehmann rotation phenomenon in the rodic cholesterics) is expected in bowlic nematics.

10.3.4   Bowlic Nematics in Electric Fields

Bowlic nematics may exist in the polar (i.e., ferroelectric) or nonpolar form [10,21]. The static and dynamic response of polar nematics to external electric fields [35] will be presented in this section. The results are relevant to a new type of fast-switching liquid crystal displays using these polar nematics [10].

Statics

The free energy density describing a polar nematic in the presence of an external electric field $\mathbf{E}$ is given by [10]

$$F = F_0 + K_0 \nabla \cdot \mathbf{n} - a\mathbf{n} \cdot \mathbf{E} - (\varepsilon_a/8\pi)(\mathbf{n} \cdot \mathbf{E})^2, \tag{10.1}$$

where F_0 is the Frank free energy given by $F_0 = \frac{1}{2}K_1(\nabla \cdot \mathbf{n})^2 + \frac{1}{2}K_2(\mathbf{n} \cdot \nabla \times \mathbf{n})^2 + \frac{1}{2}K_3(\mathbf{n} \times \nabla \times \mathbf{n})^2$.

We now consider a planar polar nematic in which the molecules are parallel to the two glass plates at $z = \pm d/2$. Here the z axis is along the cell normal, and d is the cell thickness. Under the action of a field $\mathbf{E}$ normal to the cell, the molecules will tend to align parallel to $\mathbf{E}$ (since $a > 0$). Let us first consider the case that the dielectric interaction is weak, i.e., $\varepsilon_a = 0$. F becomes

$$F = \frac{1}{2}(K_1 \cos^2 \theta + K_3 \sin^2 \theta)\left(\frac{d\theta}{dz}\right)^2 + K_0 \nabla \cdot \mathbf{n} - aE \sin \theta, \tag{10.2}$$

where θ is the angle between the director $\mathbf{n}$ and the horizontal x axis. The Lagrange equation of motion is given by

$$(K_1 \cos^2 \theta + K_3 \sin^2 \theta)\frac{d^2\theta}{dz^2} + (K_3 - K_1) \sin \theta \cos \theta \left(\frac{d\theta}{dz}\right)^2 + aE \cos \theta = 0, \tag{10.3}$$

resulting in

$$\frac{1}{2}(K_1 \cos^2 \theta + K_3 \sin^2 \theta)\left(\frac{d\theta}{dz}\right)^2 = -a \sin \theta + aE \sin \theta_{\mathrm{m}}, \tag{10.4}$$

where θ_m is the maximum tilt angle at $z = 0$. Setting $\sin \theta_m \equiv k$ and $\sin \theta \equiv k \sin \lambda$, and following the usual procedure [36], we obtain

$$\sqrt{2aE} \left(\frac{d}{2} \right) = \int_0^{\pi/2} \left[\frac{K_1(1 - k^2 \sin^2 \lambda) + K_3 k^2 \sin^2 \lambda}{k - k \sin \lambda} \right]^{1/2} \frac{k \cos \lambda \, d\lambda}{(1 - k^2 \sin^2 \lambda)^{1/2}}.$$

(10.5)

The threshold field E_c is obtained by taking the limit $k \to 0$ on the right-hand side of (10.5). We then have

$$\sqrt{2aE_c} \left(\frac{d}{2} \right) = \lim_{k \to 0} \int_0^{\pi/2} \frac{\sqrt{K_1} k \cos \lambda \, d\lambda}{\sqrt{1 - \sin \lambda}}$$

$$= \lim_{k \to 0} \int_0^1 \frac{\sqrt{K_1} k \, dt}{\sqrt{1 - t}} = 0,$$

(10.6)

i.e., $E_c = 0$. In other words, there is no Freedericksz transition for polar nematics. The molecules reorient as soon as the electric field is turned on. This conclusion remains true even when the dielectric interaction is included and is independent of the magnitude of ε_a. Similar results are obtained for the other two kinds of Freedericksz transitions (i.e., E parallel to the glass plates for either planar or homeotropic cells).

For the case considered above, the variation of θ_m as a function of E obtained numerically is plotted in Fig. 10.5. Note that E_c remains zero for a twisted polar nematic cell with arbitrary twist angle. Consequently, many grey scales are possible in a liquid crystal display using polar nematics.

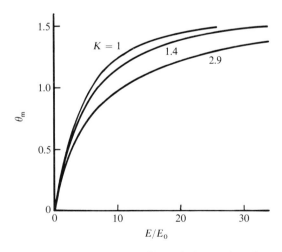

FIGURE 10.5. Variation of the maximum tilt angle θ_m as a function of electric field E in a planar polar nematic cell. $K \equiv K_3/K_1$ and $E_0 \equiv 2K_1/(ad)$.

Dynamics

The orientational motion of molecules in polar nematics may be described by

$$\gamma \frac{\partial \theta}{\partial t} = K \nabla^2 \theta + aE \cos \theta, \tag{10.7}$$

where γ is a viscosity coefficient, the one-elastic-constant approximation is assumed for simplicity, and the dielectric term is ignored. Equation (10.7) is the overdamped sine–Gordon equation, or the double sine–Gordon equation if the dielectric term is included [37]. Similar equations appear in the ferroelectric smectic C* case [38] and can be handled accordingly. In particular, soliton solutions are possible. Response time could be in the microsecond range and we have a very fast liquid crystal switch device [10].

10.3.5 Bowlic Langmuir Films

In a bowlic Langmuir film a monolayer of bowlic molecules can rest on a film of liquid. There are at least four possibilities as far as the orientation of the molecules with respect to the air–liquid interface is concerned, viz., the bowl-shaped core can be up, down, vertical, or tilted (Fig. 10.6).

Experiments on Langmuir films with 8HATB (**I.2**) and 9HETB (**I.1**), respectively, on purified water or a NaOH solution were carried out recently by El Abed et al. [22,23]. In the bulk, these two bowlic compounds form columnars at room temperature. The surface pressure π versus molecular area A isotherms, with purified water as the subphase, show the existence of plateaus (Fig. 10.7). The location and height of the plateau change when water is replaced by the NaOH solution [22]. The nonplateau parts at high A are understood to correspond to a monolayer arrangement, even though it is not entirely clear whether the molecules are in the up or down orientations [39]. The nonplateau parts at low A are interpreted to be due to the formation of an upper layer by a "roll-over" collapse. The difference between the two isotherms is attributed to the difference of the polar groups (COO— and

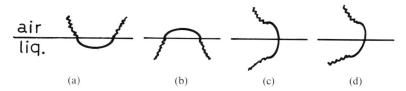

FIGURE 10.6. Four possible orientations of bowlic molecules with respect to the air–liquid interface: (a) up; (b) down; (c) vertical; and (d) tilted. Note that (a) and (b) are identical for a discotic molecule since there is no up–down asymmetry there.

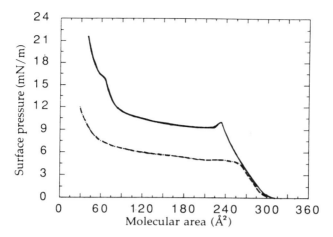

FIGURE 10.7. Surface pressure π versus molecular area A isotherms. The subphase is purified water (pH = 5.7); temperature is 21 °C. Solid curve, 8HATB; broken curve, 9HETB [22].

CO—), in that it takes more energy to remove the polar group and the whole molecule from the subphase for 8HATB than for 9HETB.

In the plateau region, anisotropic domains and an isotropic phase are found to coexist [23,40]. Within these domains, the molecules assume the vertical orientation similar to that found in some discotics [41]. Looking from the top of the film, the shape of the domains is needle-like and dot-like for 8HATB and 9HETB, respectively. From the light reflectivity study, both types of domains are found to be multilayered; the needle-like domains consist of molecules arranged in rectilinear rows; in the dot-like domains the molecules form concentric circles [40], with each circle like the donut mesophase first proposed by Lin [10]. This concentric-donut arrangement is partly deduced from the rate of domain growth [39]. In our opinion, both the concentric-donut and the (two-dimensional) onion phases are consistent with experimental data. It would be interesting to find out which one is the real arrangement. Or, maybe both phases are possible and can even coexist, i.e., there are actually two types of dot-like domains. Furthermore, if the concentric-donut arrangement is real then it may be possible to find the real donuts, or a concentric-donut with a hole in the middle. Similarly, the rectilinear rows in the needle-like domains may be understood as a set of parallel stringbeans [10], and we should look for single stringbeans in the Langmuir films. The possibility of finding single stringbeans or donuts should be higher at the high A region in which the concentration of the bowlic is lower.

Finally, it is worth pointing out that Langmuir films of simpler molecules have been shown to be fruitful ground in the study of pattern formation [42]. Similar and new phenomena can be expected when bowlic Langmuir films are used.

10.4   Bowlic Polymers

The bowlic monomers can be linked up to form main-chain, side-chain, or columnar bowlic polymers as shown in Fig. 10.8. These bowlic polymers have not been synthesized, but their counterparts using discotic monomers already exist [43]. Summarized below are theoretical studies on these very important materials, the bowlic polymers. Results in Sections 10.4.1 and 10.4.2 are equally applicable to the conventional polar polymers.

10.4.1   The Discrete Model for Dilute Solutions

In dilute polymer solutions of sufficiently low concentrations the polymers are separated from each other. We can then consider the conformations and behavior of a single main-chain bowlic (or polar) polymer in the presence of an electric field **E**. As proposed by Lam [24], a possible discrete model for this case is shown in Fig. 10.9. The description given below follows that of [24,26].

In the discrete model, the length of each monomer is b and the length of the chain is Lb. This model assumes a biased random walk of L steps, corresponding to L repeated monomers, such that the energy for a monomer parallel (antiparallel) to **E** causes energy $-\mu E$ (μE) where μ is the electric dipole moment of each monomer. A monomer perpendicular to **E** causes zero energy. Each bending from parallel to antiparallel, or vice versa, is called a hairpin which causes energy h. A pair of parallel (antiparallel) adjacent

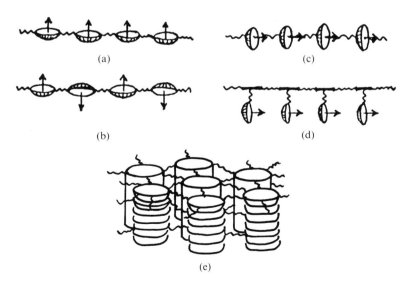

FIGURE 10.8. Three possible types of bowlic polymers: (a)–(c) main-chain; (d) side-chain; and (e) columnar. (a)–(d) are from [24].

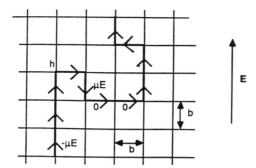

FIGURE 10.9. The discrete model of polar main-chain polymers in an external electric field **E**. The lattice may be one to three dimensions.

monomers have energy γ $(-\gamma')$. All μ, γ, γ', and h are positive. The lattice is three dimensional (3D) in general, but can be reduced to 2D or 1D for simplicity, or in special cases.

This model has not been solved analytically yet. However, it can be solved easily by computer simulation, e.g., by the Monte Carlo method. In the 1D continuum limit two special cases of this model have been solved exactly. First, for $E = 0$, the model reduces to that of Zwanzig and Lauritzen [44] which was solved exactly in the limit of $L \to \infty$. This solution shows a second-order phase transiton with the number of hairpins acting as the order parameter. Second, for $E \neq 0$ and $\gamma = \gamma' = 0$, the model reduces to that of Gunn and Warner [45] which was also solved exactly, but does not show any phase transition. As pointed out by Lauritzen and Zwanzig [46], the continuum version of a physically discrete model may contain spurious results.

The 2D Case

In order to solve the model analytically, the model is reduced to a simplified 2D (or 1D) version as shown in Fig. 10.10. This version allows the bending to

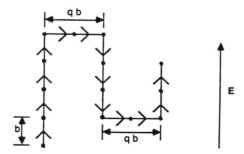

FIGURE 10.10. The simplified 2D (or 1D) model. q is a constant and $q = 0$ reduces the model to 1D. Here the horizontal bonds are in one direction only.

342   L. Lam

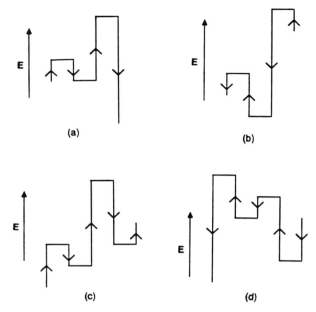

(a)

(b)

(c)

(d)

FIGURE 10.11. The four cases of the simplified 2D (or 1D) model of polar main-chain polymers. The diagram shown here corresponds to $N = 2$. (a) Case A: first monomer up, last monomer down. (b) Case B: first monomer down, last monomer up. (c) Case C: first monomer up, last monomer up. (d) Case D: first monomer down, last monomer down.

go in one direction only. To further simplify the model, the interaction between the monomers are assumed to be negligible, i.e., we assume $\gamma = \gamma' = 0$.

The partition function of the polar polymer is given by

$$Q(L, E) = Q_A(L, E) + Q_B(L, E) + Q_C(L, E) + Q_D(L, E), \quad (10.8)$$

where Q_A, Q_B, Q_C, and Q_D are calculated from the four cases shown in Fig. 10.11, given by

$$Q_A(L, E) = \sum_{N=1}^{\infty} \sum_{X_1=1}^{\infty} \cdots \sum_{X_N=1}^{\infty} \sum_{Y_1=1}^{\infty} \cdots \sum_{Y_N=1}^{\infty} \delta_{L, \sum X_j + \sum Y_j + (2N-1)q} \exp(-\beta U_A),$$

$$Q_B(L, E) = \sum_{N=1}^{\infty} \sum_{X_1=1}^{\infty} \cdots \sum_{X_N=1}^{\infty} \sum_{Y_1=1}^{\infty} \cdots \sum_{Y_N=1}^{\infty} \delta_{L, \sum X_j + \sum Y_j + (2N-1)q} \exp(-\beta U_B),$$

$$Q_C(L, E) = \sum_{N=1}^{\infty} \sum_{X_1=1}^{\infty} \cdots \sum_{X_{N+1}=1}^{\infty} \sum_{Y_1=1}^{\infty} \cdots \sum_{Y_N=1}^{\infty} \delta_{L, \sum X_j + \sum Y_j + 2Nq} \exp(-\beta U_C),$$

$$Q_D(L, E) = \sum_{N=1}^{\infty} \sum_{X_1=1}^{\infty} \cdots \sum_{X_{N+1}=1}^{\infty} \sum_{Y_1=1}^{\infty} \cdots \sum_{Y_N=1}^{\infty} \delta_{L, \sum X_j + \sum Y_j + 2Nq} \exp(-\beta U_D),$$

$$(10.9)$$

with

$$U_A = (2N - 1)h - \mu E(\sum X_j - \sum Y_j),$$
$$U_B = (2N - 1)h - \mu E(-\sum X_j + \sum Y_j),$$
$$U_C = 2Nh - \mu E(\sum X_j - \sum Y_j),$$
$$U_D = 2Nh - \mu E(-\sum X_j + \sum Y_j).$$

(10.10)

Here X_i is the ith segment where the dipoles are pointing up, and Y_i is the ith segment where the dipoles are pointing down. In cases A and B, there are $N + 1$ segments that are pointing up and N segments that are pointing down; therefore, there are $2N - 1$ hairpins. In cases C and D, there are N equal segments pointing up and down; therefore there are $2N$ hairpins.

From (10.9) and (10.10), it is clear that

$$Q_B(L, E) = Q_A(L, -E), \qquad Q_D(L, E) = Q_C(L, -E). \qquad (10.11)$$

Consequently, (10.8) becomes

$$Q(L, E) = Q_A(L, E) + Q_A(L, -E) + Q_C(L, E) + Q_C(L, -E). \quad (10.12)$$

Each term in (10.12) can be evaluated through the Z-transform (see [26] for details), resulting in

$$Q_A(L, E) = Q_A(L, -E) = \frac{f}{(L - q - 2)!} \lim_{z \to 0} \frac{d^{L-q-2}}{dz^{L-q-2}} \left[\frac{1}{F(z)} \right],$$

$$Q_C(L, E) = \frac{e^{u_0}}{(L - 1)!} \lim_{z \to 0} \frac{d^{L-1}}{dz^{L-1}} \left[\frac{1}{F(z)} \right] - \frac{1}{(L - 2)!} \lim_{z \to 0} \frac{d^{L-2}}{dz^{L-2}} \left[\frac{1}{F(z)} \right],$$

(10.13)

$$Q_C(L, -E) = \frac{e^{-u_0}}{(L - 1)!} \lim_{z \to 0} \frac{d^{L-1}}{dz^{L-1}} \left[\frac{1}{F(z)} \right] - \frac{1}{(L - 2)!} \lim_{z \to 0} \frac{d^{L-2}}{dz^{L-2}} \left[\frac{1}{F(z)} \right],$$

where

$$f \equiv e^{-\beta h}, \qquad u_0 \equiv \mu \beta E, \qquad (10.14)$$

and

$$F(z) \equiv z^2 - 2(\cosh u_0)z + 1 - f^2 z^{2(q+1)}. \qquad (10.15)$$

Therefore (10.12) can be written as

$$Q(L, E) = 2[H_L - (\cosh u_0)H_{L-1} + fH_{L-q-2}], \qquad (10.16)$$

where

$$H_m \equiv \lim_{z \to 0} \frac{d^m}{dz^m} \left[\frac{1}{F(z)} \right]. \qquad (10.17)$$

In general, $F(z) = 0$ is a $2(q + 1)$th-order algebraic equation which can be solved exactly if $2(q + 1) \leq 4$. Therefore, $F(z)$ can only be solved exactly, and $Q(L, E)$ can only be expressed analytically, for $q \leq 1$.

The 1D Case

If q is set to zero, the 2D model reduces to a 1D model. The partition function for the 1D model is given by

$$Q = \frac{2}{(1-f^2)(z_2-z_1)}\left[(\cosh u_0)\left(\frac{1}{z_1^L}-\frac{1}{z_2^L}\right)+(f-1)\left(\frac{1}{z_1^{L-1}}-\frac{1}{z_2^{L-1}}\right)\right],$$
(10.18)

where

$$z_{2,1} = \frac{(\cosh u_0)\pm\sqrt{(\cosh^2 u_0)-(1-f^2)}}{1-f^2}.$$
(10.19)

For $E = 0$, the Helmholtz free energy A and the average number of hairpins $\bar{n}$ are given by

$$A \equiv -\frac{1}{\beta}\ln Q = -\frac{1}{\beta}[\ln 2 + (L-1)\ln(1+f)],$$
(10.20)

$$\bar{n} = \frac{\partial A}{\partial h} = \frac{(L-1)f}{1+f}.$$
(10.21)

Consequently, $\bar{n}$ increases monotonically from 0 to $(L-1)/2$ as temperature T increases from zero to infinity. This means that the polymer will bend more often when the temperature increases.

The susceptibility χ is defined by $\partial P/\partial E$, where P is the polarization given by

$$P \equiv -\frac{\partial A}{\partial E} = L\mu(\sinh u_0)[\sinh^2 u_0 + \exp(-2\beta h)]^{-1/2}.$$
(10.22)

We then obtain

$$\chi = L\mu^2\beta(\cosh u_0)[\sinh^2 u_0 + \exp(-2\beta h)]^{-1/2}$$
$$- L\mu^2\beta(\cosh u_0)(\sinh^2 u_0)[\sinh^2 u_0 + \exp(-2\beta h)]^{-3/2}.$$
(10.23)

As $E \to 0$, we have $u_0 \to 0$, $\cosh u_0 \to 1$, and $\sinh u_0 \to 0$. This implies

$$\chi \to L\mu^2\beta\exp(\beta h),$$
(10.24)

which is identical to Eq. (5) of [45] if l there is identified as b in our model.

The 1D model result can also be obtained directly by mapping the model to the 1D Ising model with nearest neighbor interactions. This can be seen easily by stretching the chain in Fig. 10.10 into a straight line (with $q = 0$) and identifying the up (down) polarization in each monomer as an up (down) spin. Note that parallel adjacent spins have energy zero and antiparallel adjacent spins have energy h. Mathematically, the partition function $Q(L, E)$ is given by

$$Q(L, E) = \sum_{\sigma_1=-1}^{1}\sum_{\sigma_2=-1}^{1}\cdots\sum_{\sigma_L=-1}^{1}\exp(\beta\xi),$$
(10.25)

where

$$\xi = -\frac{h}{2}\sum_{j=1}^{L-1}(\sigma_j\sigma_{j+1} - 1) - \mu E \sum_{j=1}^{L}\sigma_j. \qquad (10.26)$$

Comparing (10.25) and (10.26) with the corresponding expressions in [47], the partition function here is seen to be equivalent to the free boundary partition function of the 1D Ising model with the following substitutions:

$$\exp\left(-\frac{\beta h}{2}(L-1)\right)Z_{1,N}^F \to Q(L, E), \qquad N \to L,$$

$$E_1 \to \frac{h}{2}, \qquad H \to \mu E \equiv \frac{u_0}{\beta}. \qquad (10.27)$$

Applying these substitutions and from Eq. (2.14) of [47], we obtain

$$Q(L, E) = \exp\left(-\frac{\beta h}{2}(L-1)\right)Z_{1,N}^F$$

$$= f^{(L-1)/2}[\lambda_+^{L-1}\{\cosh u_0 + (\sinh^2 u_0 + f)(\sinh^2 u_0 + f^2)^{-1/2}\}$$

$$+ \lambda_-^{L-1}\{\cosh u_0 - (\sinh^2 u_0 + f)(\sinh^2 u_0 + f^2)^{-1/2}\}], \quad (10.28)$$

where $f \equiv \exp(-\beta h)$, and

$$\lambda_\pm \equiv f^{-1/2}[\cosh u_0 \pm \sqrt{\sinh^2 u_0 + f^2}],$$

$$= f^{-1/2}[\cosh u_0 \pm \sqrt{\cosh^2 u_0 - (1 - f^2)}]. \qquad (10.29)$$

Substituting (10.29) into (10.28) results in the same expression of (10.18) for $Q(L, E)$.

Discussions

In principle, analytical solutions for the simplified discrete model given by (10.16) can also be evaluated explicitly for the 2D ($q = 1$) case—a task that remains to be done. Other physical quantities such as $\bar{n}$ and χ can then be calculated from the partition function as is done in the 1D case.

The assumption that $\gamma = \gamma' = 0$ means that the interactions among molecules are ignored and the result shows no phase transition. If necessary, the influence of the other molecules on a single molecule can be treated using the mean field approximation. For this purpose, the same formulation described above can be used; we simply replace E by $E + aP$, where a is a proportional constant and P is the polarization, the order parameter. The existence of phase transitions can indeed be obtained this way, or by an alternative approach which maps the problem to a quantum mechanical differential equation [27] (see Section 10.4.2).

10.4.2   The Worm-Like Model for Bowlic Nematic Polymers

In the worm-like model [27], the main-chain polar polymer molecule is assumed to be like a worm, influenced by other molecules through a self-consistent mean field. The functional integral technique is used to describe the orientational distribution of the tangent to the polymer, and the problem is mapped into a diffusion equation of a particle on the surface of a unit sphere in a dipolar mean field.

Specifically, the bending of the chain will cost an energy

$$\int_0^L \frac{\varepsilon}{2}\left[\frac{d\mathbf{u}(s)}{ds}\right]^2 ds, \tag{10.30}$$

where ε is the bending constant, L is the length of the polymer, and $\mathbf{u}$ is the tangent unit vector along the chain at arclength s. The interaction in the polar polymer is assumed to be dominated by the first rank $P_1(\cos\theta)$-type potential, instead of the second rank $P_2(\cos\theta)$-type as in conventional liquid crystal polymers [48,49]. Here θ is the angle between a polymer segment and the director in the nematic phase. Let us assume that the semiflexible chains in the polymer favor a long-range order of parallel alignment. After summing along the chain, the dipolar potential in the mean field approximation is given by

$$U_q = -\int_0^L b\bar{P}_1 P_1(\cos\theta)\, ds, \tag{10.31}$$

where b is the coupling constant and is positive in this case, P_1 is the first Legendre polynomial, and $\bar{P}_1$ is the polar nematic order parameter given by

$$\bar{P}_1 = \frac{1}{L}\left\langle \int_0^L ds\, P_1(z)\right\rangle. \tag{10.32}$$

As shown in [27], for a uniaxial system, $\bar{P}_1$ can be calculated from the Green function G such that

$$\bar{P}_1 = \int_0^L \frac{ds}{L} \frac{\iiint dz\, dz'\, dz''\, G(z', z; L, s)G(z, z''; s, 0)P_1(z)}{\iiint dz\, dz'\, dz''\, G(z', z; L, s)G(z, z''; s, 0)}, \tag{10.33}$$

where $z = \cos\theta$, and

$$G(\theta, \theta'; s, s') = \sum_{n=0}^{\infty} \psi_n(\theta)\psi_n(\theta') \exp(-\lambda_n D|s - s'|), \tag{10.34}$$

where the eigenfunction $\psi_n(\theta)$ is given by

$$\left[\lambda_n + \frac{1}{\sin\theta}\frac{d}{d\theta}\left(\sin\theta\frac{d}{d\theta}\right) + gP_1(\cos\theta)\right]\psi_n(\theta) = 0, \tag{10.35}$$

with $g \equiv \beta b\bar{P}_1/D = 2P_1/\tilde{T}^2$; the reduced temperature $\tilde{T} \equiv kT/(b\varepsilon)^{1/2}$ and $\beta \equiv 1/kT$. Here λ_n is the eigenvalue and D^{-1} is the persistence length.

A perturbation calculation near the polar nematic–isotropic phase transition gives

$$\bar{P}_1 = \alpha \tilde{T}^2 (1 - \tfrac{3}{2}\tilde{T}^2)^{1/2}, \tag{10.36}$$

where $\alpha = (\tfrac{45}{22})^{1/2}$, and the transition temperature for an infinite chain $\tilde{T}_c = (\tfrac{2}{3})^{1/2} = 0.816$. For a finite chain of length L

$$\tilde{T}_c(L) = \tilde{T}_c \left[1 - \frac{1}{2LD}(1 - \exp(-2LD)) \right]. \tag{10.37}$$

The term in square brackets is actually equal to $\langle R^2 \rangle / (LD^{-1})$, the ratio of the mean square dimension of the polymer to its value when L is infinite in the isotropic state.

In the strong nematic limit, either at low temperature or with a strong polar potential, an asymptotic calculation gives

$$\bar{P}_1 = \tfrac{4}{3} \cos^2 \left[\tfrac{1}{3} \cos^{-1} \left(-\frac{3\sqrt{3}}{4} \tilde{T} \right) \right]. \tag{10.38}$$

Equations (10.37) and (10.38) compare pretty well with the numerical results [27], as shown in Fig. 10.12.

For a 2D system such as in thin polymer films or in Langmuir films, the problem can be solved exactly [50]. In 2D, (10.35) becomes

$$\left(\frac{d^2}{d\theta^2} + g \cos \theta + \lambda_n \right) \psi_n(\theta) = 0. \tag{10.39}$$

Under the transformations $\Lambda_n = 4\lambda_n$, $q = 2g$, and $\xi = (\pi - \theta)/2$, (10.39) be-

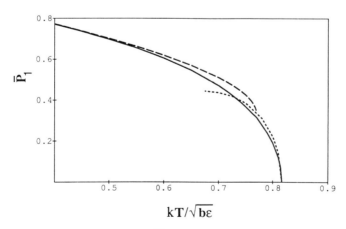

FIGURE 10.12. The order parameter $\bar{P}_1$ as a function of the reduced temperature $\tilde{T}$. The dotted and dashed lines represent the perturbation and asymptotic results, respectively. The solid line is the numerical result [27].

348   L. Lam

comes the Mathieu equation

$$\left\{\frac{d^2}{d\xi^2} + [\Lambda_n - 2q\cos(2\xi)]\right\}\psi_n(\xi) = 0. \tag{10.40}$$

The lowest eigenvalue Λ_0, in a power series of q, is approximately given by

$$\Lambda_0 = -\tfrac{1}{8}q^2 + \frac{7}{512}q^4, \tag{10.41}$$

which contributes to the free energy for the long chain

$$F/(kTLD) = \lambda_0 + \frac{\bar{P}^2}{\tilde{T}^2}, \tag{10.42}$$

where the second term on the right-hand side arises from the mean field theory. Consequently, we obtain a Landau form of the free energy

$$F/(kTLD) = \frac{1}{16}(\tilde{T}^2 - 2)q^2 + \frac{7}{512}q^4$$

$$= (\tilde{T}^2 - 2)\frac{\bar{P}_1^2}{\tilde{T}^4} + \frac{7}{2}\frac{\bar{P}_1^4}{\tilde{T}^8}, \tag{10.43}$$

where k is the Boltzmann constant. Note that there is no cubic term. Conse-

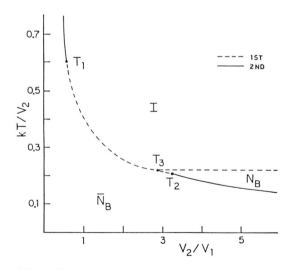

FIGURE 10.13. Phase diagram of bowlic monomers. I represents the isotropic liquid phase, N_B the nonpolar nematic phase, and $\bar{N}_B$ the polar nematic phase. T_1 and T_2 are tricritical points and T_3 is a triple point. The intermolecular interaction potential is given by $V_{ij} = -V_1 P_1(\cos\theta_{ij}) - V_2 P_2(\cos\theta_{ij})$, with $V_1 > 0$ and $V_2 > 0$. Columnar phases are not included [21]. A similar phase diagram is expected for bowlic polymers.

quently, the polar nematic–isotropic phase transition, in 2D here as in 3D above, is second order. The temperature dependence of the coefficient of the $\bar{P}_1$ terms gives the transition temperature $T_c = \sqrt{2}$. Minimizing F with respect to $\bar{P}_1$ gives

$$\bar{P}_1 = \sqrt{2/7}\,\tilde{T}^2(1 - \tilde{T}^2/2)^{1/2}. \tag{10.44}$$

Electric field effects are also discussed in [27]. Expressions for the susceptibility χ are obtained; the critical exponent γ equals 1, as expected for a mean field theory; and the order parameter is given numerically. For a polar polymer, giant dielectric response is possible. Finally, when both P_1 and P_2 interactions are allowed, we expect a phase diagram similar to that given by Leung and Lin [21] for bowlic monomers. The phase diagram contains two tricritical points and one triple point, and two types of bowlic nematics (polar and nonpolar) in addition to the isotropic liquid phase (Fig. 10.13). See [27] for further discussion.

10.4.3   Ultrahigh T_c Bowlic Superconductors

In the past few years there has been tremendous progress in raising the transition temperature T_c of superconductors [51] (see Fig. 10.14). The discovery of the high T_c superconducting copper oxides [52] was followed by that of the superconducting fullerenes [53].

As shown in Fig. 10.14, while the highest T_c of the molecular superconductors is still behind that of the cuprates, the rate of progress of the two categories are comparable. Historically, development of organic superconductors was prompted by the theoretical study of Little [54]. With the con-

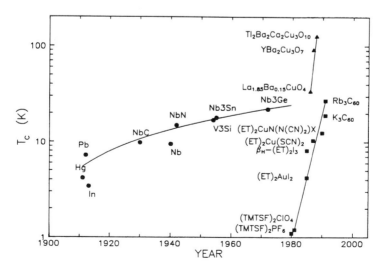

FIGURE 10.14. Progress of superconductivity over the century in metallic, cuprate, and molecular superconductors [51].

ventional phonon-mediated mechanism replaced by the excitonic mechanism, Little predicted that the superconducting T_c can be raised to greater than 1000 K (see [25]). The problem with this scheme is that the proposed molecules are very difficult, if not impossible, to synthesize; more importantly, even if they were synthesized there is no guarantee that they will stack up to form a 1D or a quasi-1D filament, as required by the theory.

To overcome these problems and with the excitonic mechanism in mind, ultrahigh T_c bowlic superconductors were proposed by Lam [24]. (The scheme should also work for the discotics, but the chance is higher in the bowlics [24,25]. See also [55].) The idea is that in the bowlic columnar phase, the columns are already there and they are empty tubes. We can then add metal (e.g., Cu) or transition metal atoms into the tubes by doping. Alternatively, we may actually attach such atoms to the centers of the bowlic molecules (as has been done successfully in the rodics and discotics). Either way, when the conditions are right we will have conducting or superconducting bowlic mesophases or crystals [25]. In the case of bowlic polymers shown in Fig. 10.8(e), the addition of metal or transition metal into the tubes will turn the polymer into a system with an infinite number of useful layers, which tends to raise further the superconducting T_c as demonstrated recently in the case of $(Sr_{1-x}Ca_x)_{1-y}CuO_2$ with a T_c of 110 K [56]. See [24,25] for more specific discussion.

10.5   Conclusions

Since the prediction of bowlics [5] three types of bowlic monomers have already been synthesized (see Table 10.1). The existence of the sign change of optical anisotropy with temperature [11] is a property unique to the bowlics but not to the discotics. On the other hand, when nematic and cholesteric discotics are found experimentally the same cannot be said for the bowlics. Maybe what is needed is new bowlic molecules with shallower bowls; this will lessen the tendency of the molecules to form columns, and encourage them to be separated from each other as required in the nematic or cholesteric phases. Note that lyotropic nematics bowlics seem to exist [15].

Ordinary polar molecules tend to form dimers due to the dipole–dipole interaction, but in the case of bowlics the steric effect will be able to overcome this. Consequently, bowlic (polar) nematics may be the *only* kind of ferroelectric liquid possible. And once existing, these bowlic nematics can be used for very fast electrooptical displays or switches. In view of these very important basic and applied interests, there is no doubt in our mind that further synthesizing effort is worthwhile.

Compared to the search for bowlic nematics, the task of synthesizing bowlic polymers is much more straightforward—the monomers are already there. Bowlic nematic polymers are in fact easier to form [27], and they could be stronger than the Kevlar.

It should be emphasized that, apart from the great interest in their liquid crystalline properties, the bowlics in bulk or as a Langmuir film can serve as a useful and novel physical system in the study of dynamical instabilities and pattern formation [10], such as those found in ferromagnetic liquids [57]. Hollow cage molecules, with or without foreign atoms inside the cages, can be formed with bowlic molecules as the building blocks, similar to the speleands and cryptophanes formed with one or two cyclotriveratrylenes [28].

Finally, unlike the rodics and discotics which were actually found or studied well before 1888 and 1977, the official dates of discoveries, respectively [58], we believe that bowlics are completely new and modern. In fact, they are so new that the measurement of their physical properties—be it viscosity, density, acoustical, optical, dielectric, or magnetic—has hardly started. Very exciting work in bowlics, including the search for ultrahigh T_c superconductors, awaits the experimentalists in physics, chemistry, and materials science. Furthermore, as happened in the case of bowlic monomers, theory is once again ahead of experiment in the study of bowlic polymers.

References

1. S. Chandrasekhar, B.K. Sadashiva, and K.A. Suresh, Pramana **9**, 471 (1977).

2. Lin Lei (L. Lam), Phys. Rev. Lett. **43**, 1604 (1979); in *Liquid Crystals*, edited by S. Chandrasekhar (Heyden, London, 1980).

3. For example, Lin Lei et al., Phys. Rev. Lett. **49**, 1335 (1982).

4. D.R. Nelson and B.I. Halperin, Phys. Rev. B **19**, 2457 (1979).

5. Lin Lei, Wuli (Beijing) **11**, 171 (1982).

6. Lin Lei, Mol. Cryst. Liq. Cryst. **91**, 77 (1983).

7. A.M. Levelut, J. Malthête, and A. Collet, J. Phys. (Paris) **47**, 351 (1986).

8. H. Zimmermann, R. Poupko, Z. Luz, and J. Billard, Z. Naturforsch. **40a**, 149 (1985).

9. J. Malthête and A. Collet, Nouv. J. Chemie **9**, 151 (1985).

10. Lin Lei, Mol. Cryst. Liq. Cryst. **146**, 41 (1987).

11. H. Zimmermann, R. Poupko, Z. Luz, and J. Billard, Z. Naturforsch. **41a**, 1137 (1986).

12. L.Y. Wang and X.F. Pei, J. Tsinghua Univ. **28** (S4), 80 (1988).

13. J. Malthête and A. Collet, J. Amer. Chem. Soc. **109**, 7544 (1987).

14. H. Zimmermann, R. Poupko, Z. Luz, and J. Billard, Liq. Cryst. **3**, 759 (1988).

15. H. Zimmermann, R. Poupko, Z. Luz, and J. Billard, Liq. Cryst. **6**, 151 (1989).

16. J.M. Buisine, H. Zimmermann, R. Poupko, Z. Luz, and J. Billard, Mol. Cryst. Liq. Cryst. **151**, 391 (1987); High Pressure Sci. Tech. (Kiev) **4**, 232 (1989).

17. R. Poupko, Z. Luz, N. Spielberg, and H. Zimmermann, J. Am. Chem. Soc. **111**, 6094 (1989).

18. L.Y. Wang, Z.M. Sun, X.F. Pei, and Y.P. Zu, Chem. Phys. **142**, 335 (1990); X.F. Pei, M.S. Thesis, Tsinghua University, Beijing, 1988.

19. M. Sarkar, N. Spielberg, K. Praefcke, and H. Zimmerman, Mol. Cryst. Liq. Cryst. **203**, 159 (1991).

20. X.J. Wang, K. Tao, J.A. Zhao, and L.Y. Wang, Liq. Cryst. **5**, 563 (1989).

21. K.M. Leung and Lin Lei, Mol. Cryst. Liq. Cryst. **146**, 71 (1987).

22. A. El Abed, A. Hochapfel, H. Hasmonay, J. Billard, H. Zimmermann, Z. Luz, and P. Peretti, Thin Solid Films **210/211**, 93 (1992).

23. A. El Abed, Ph.D. Thesis, Université Rene Descartes (Paris V), 1992.

24. L. Lam, Mol. Cryst. Liq. Cryst. **155**, 531 (1988).

25. L. Lam, in *3rd Asia Pacific Physics Conference*, edited by Y.W. Chan, A.F. Leung, C.N. Yang, and K. Young (World Scientific, Singapore, 1988).

26. Y.S. Yung, M.S. Thesis, San Jose State University, 1989.

27. X.J. Wang and L. Lam, Liq. Cryst. **11**, 411 (1992).

28. A Collet, Tetrahedron **43**, 5725 (1987).

29. E. Fontes, P.A. Heiney, and W.H. de Jeu, Phys. Rev. Lett. **61**, 1202 (1988).

30. A. Keller, in *Polymers, Liquid Crystals and Low-Dimensional Solids*, edited by N. March and M. Tosi (Plenum, New York, 1984); D. Friedman, in *Spiral Symmetry*, edited by I. Hargittai and C.A. Pickowver (World Scientific, River Edge, 1992).

31. L. Lam and H.S. Lakkaraju, unpublished (1989).

32. G.W. Gray and J.W. Goodby, *Smectics Liquid Crystals: Textures and Structures* (Leonard Hill, London, 1984), Plate 117; C. Baehr, M. Ebert, G. Frick, and J.H Wendorff, Liq. Cryst. **7**, 601 (1990).

33. J.S. Langer, Rev. Modern Phys. **52**, 1 (1980); D. Kessler, J. Koplik, and H. Levine, Adv. in Phys. **37**, 255 (1988); *Dynamics of Curved Fronts*, edited by P. Pelcé (Academic Press, San Diego, 1988).

34. Y.R. Shen, private communication.

35. L. Lam, in *Liquid Crystals—West '89*, edited by L. Lam (Society of Archimedes, San Jose, 1989).

36. See, for example, S. Chandrasekhar, *Liquid Crystals* (Cambridge University Press, Cambridge, 1977).

37. L. Lam, in *Solitons in Liquid Crystals*, edited by L. Lam and J. Prost (Springer-Verlag, New York, 1992).

38. L. Lam, in *Wave Phenomena*, edited by L. Lam and H.C. Morris (Springer-Verlag, New York, 1989); see also *Solitons in Liquid Crystals*, edited by L. Lam and J. Prost (Springer-Verlag, New York, 1992), Chapters 4 and 5.

39. A. El Abed, private communication (1992).

40. A. El Abed, P. Muller, P. Peretti, F. Gallet, and J. Billard, J. Phys. II (Paris) **3**, 51 (1993); A. El Abed, P. Peretti, and J. Billard, Liq. Cryst. **14**, 1607 (1993).

41. O. Albrecht, W. Cumming, W. Kreuder, A. Laschewsky, and H. Ringdorf, Colloid. Polymer. Sci. **264**, 659 (1986).

42. C.M. Knobler, Science **249**, 870 (1990).

43. W. Kreuder and H. Ringsdorf, Mackromol. Chem., Rapid Commun. **4**, 807 (1983); W. Kreuder, H. Ringsdorf, and P. Tschirner, *ibid.* **6**, 367 (1985); G. Wenz, *ibid.* **6**, 577 (1985); O. Herrmann-Schönherr, J.H. Wendorff, H. Ringsdorf, and P. Tschirner, *ibid.* **7**, 97 (1986).

44. R. Zwanzig and J.I. Lauritzen, J. Chem. Phys. **48**, 3351 (1968).

45. J.M.F. Gunn and M. Warner, Phys. Rev. Lett. **58**, 393 (1987).

46. J.I. Lauritzen and R. Zwanzig, J. Chem. Phys. **52**, 3740 (1970).

47. B.M. McCoy and T.T. Wu, *The Two-Dimensional Ising Model* (Harvard University Press, Cambridge, 1973).

48. M. Warner, J.M.F. Gunn, and A. Baumgärtner, J. Phys. A **18**, 3007 (1985).

49. X.J. Wang and M. Warner, J. Phys. A **19**, 2215 (1986).

50. X.J. Wang and L. Lam, unpublished.

51. D. Jerome, Condensed Matter News **1** (4), 11 (1992).

52. Phys. Today, June 1991.

53. A.F. Hebard, Phys. Today **45** (11), 26 (1992).

54. W.A. Little, Phys. Rev. A **134**, 1416 (1964).

55. Superconductor Week, October 19, 1987, p. 4.

56. M. Azuma et al., Nature **356**, 775 (1992). See also M. Laguës et al., Science **262**, 1850 (1993).

57. R.E. Rosensweig, *Ferrohydrodynamics* (Cambridge University Press, Cambridge, 1985).

58. V. Vill, Condensed Matter News **1** (5), 25 (1992).

Index